Progress in Mathematics
Volume 118

Izu Vaisman

Lectures on the Geometry of Poisson Manifolds

Springer Basel AG

Author:

Izu Vaisman
Department of Mathematics
and Computer Sciences
University of Haifa
Haifa 31999
Israel

A CIP catalogue record for this book is available from the Library of Congress, Washington, D.C., USA

Deutsche Bibliothek Cataloging-in-Publication Data

Vaisman, Izu:
Lectures on the geometry of poisson manifolds / Izu Vaisman.
– Basel ; Boston ; Berlin : Birkhäuser, 1994
(Progress in mathematics ; Vol. 118)
ISBN 978-3-0348-9649-8 ISBN 978-3-0348-8495-2 (eBook)
DOI 10.1007/978-3-0348-8495-2
NE: GT

Originally published by Birkhäuser Verlag in 1994
Softcover reprint of the hardcover 1st edition 1994
Printed on acid-free paper produced of chlorine-free pulp

ISBN 978-3-0348-9649-8

9 8 7 6 5 4 3 2 1

Table of Contents

Acknowledgement vii

0 Introduction 1

1 The Poisson bivector and the Schouten-Nijenhuis bracket 5

1.1 The Poisson bivector 5
1.2 The Schouten-Nijenhuis bracket 6
1.3 Coordinate expressions 9
1.4 The Koszul formula and applications 12
1.5 Miscellanea 16

2 The symplectic foliation of a Poisson manifold 19

2.1 General distributions and foliations 19
2.2 Involutivity and integrability 21
2.3 The case of Poisson manifolds 25

3 Examples of Poisson manifolds 31

3.1 Structures on $\mathbb{R}^n$. Lie-Poisson structures 31
3.2 Dirac brackets 36
3.3 Further examples 38

4 Poisson calculus 41

4.1 The bracket of 1-forms 41
4.2 The contravariant exterior differentiations 43
4.3 The regular case 48
4.4 Cofoliations 54
4.5 Contravariant derivatives on vector bundles 55
4.6 More brackets 57

5 Poisson cohomology 63

5.1 Definition and general properties 63
5.2 Straightforward and inductive computations 67
5.3 The spectral sequence of Poisson cohomology 72
5.4 Poisson homology 77

6 An introduction to quantization 83

6.1 Prequantization 83
6.2 Quantization 87
6.3 Prequantization representations 89
6.4 Deformation quantization 92

7 Poisson morphisms, coinduced structures, reduction 97

7.1 Properties of Poisson mappings 97
7.2 Reduction of Poisson structures 101
7.3 Group actions and momenta 107
7.4 Group actions and reduction 110

8 Symplectic realizations of Poisson manifolds 115

8.1 Local symplectic realizations 115
8.2 Dual pairs of Poisson manifolds 121
8.3 Isotropic realizations 123
8.4 Isotropic realizations and nets 128

9 Realizations of Poisson manifolds by symplectic groupoids 135

9.1 Realizations of Lie-Poisson structures 135
9.2 The Lie groupoid and symplectic structures of T^*G 137
9.3 General symplectic groupoids 143
9.4 Lie algebroids and the integrability of Poisson manifolds 147
9.5 Further integrability results 153

10 Poisson-Lie groups 161

10.1 Poisson-Lie and biinvariant structures on Lie groups 161
10.2 Characteristic properties of Poisson-Lie groups 164
10.3 The Lie algebra of a Poisson-Lie group 169
10.4 The Yang-Baxter equations 172
10.5 Manin triples 178
10.6 Actions and dressing transformations 182

References 189

Index 203

Acknowledgement

The author acknowledges with gratitude the support of his home university, the University of Haifa, Israel, where most of the work on this book was done, as well as the hospitality of the U. F. R. de Mathématiques, Université de Lille I, France, where the writing of the book was finished. The author is also grateful to Alan Weinstein, Yvette Kosmann-Schwarzbach and Grant Cairns for their helpful remarks.

0 Introduction

Everybody having even the slightest interest in analytical mechanics remembers having met there the Poisson bracket of two functions of $2n$ variables (p_i, q^i)

$$\{f,g\} = \sum_{i=1}^{n} \left(\frac{\partial f}{\partial p_i} \frac{\partial g}{\partial q^i} - \frac{\partial f}{\partial q^i} \frac{\partial g}{\partial p_i} \right), \tag{0.1}$$

and the fundamental role it plays in that field. In modern works, this bracket is derived from a symplectic structure, and it appears as one of the main ingredients of symplectic manifolds. In fact, it can even be taken as the defining element of the structure (e.g., [Tl1]). But, the study of some mechanical systems, particularly systems with symmetry groups or constraints, may lead to more general Poisson brackets.

Therefore, it was natural to define a mathematical structure where the notion of a Poisson bracket would be the primary notion of the theory, and, from this viewpoint, such a theory has been developed since the early 1970*s*, by A. Lichnerowicz, A. Weinstein, and many other authors (see the references at the end of the book). But, it has been remarked by Weinstein [We3] that, in fact, the theory can be traced back to S. Lie himself [Lie].

The basic properties of the bracket (0.1) are that:

(i) it yields the structure of a Lie algebra on the space of functions, i.e.,

$$\{f,g\} = -\{g,f\} \qquad \text{(skew-symmetry)} \tag{0.2}$$

$$\{f,\{g,h\}\} + \{g,\{h,f\}\} + \{h,\{f,g\}\} = 0 \tag{0.3}$$

(the Jacobi identity);

(ii) it has a natural compatibility with the usual associative product of functions, which is

$$\{h, fg\} = \{h,f\}g + f\{h,g\} \tag{0.4}$$

(i.e., the Leibniz rule of derivation).

These facts are of an algebraic nature, and it is natural to define an abstract *Poisson algebra* as an associative commutative algebra endowed with a Lie bracket that satisfies (0.2), (0.3) and (0.4). A systematic algebraic study of Poisson algebras has not yet been done. (See, however, Huebschmann's works [Hu1], [Hu2].)

Anyhow, in applications it is important that the elements of the algebra be functions, and, then, we are led to give

0.1. Definition. Let $\mathcal{C}$ be a category of *spaces* which have a distinguished associated linear space of functions. An object of $\mathcal{C}$ will be called a *Poisson space* if its distinguished space of functions is endowed with the structure of a Poisson algebra.

In particular, by taking $\mathcal{C}$ to be the category of smooth (C^∞), analytic or algebraic manifolds and maps, endowed with the space of smooth, analytic or algebraic functions, respectively, the corresponding Poisson spaces will be the *C^∞-Poisson manifolds, analytic Poisson manifolds, algebraic Poisson manifolds*, etc.

As a matter of fact, except for some isolated works (e.g., [Co1], [Lh5], etc.), only the C^∞-Poisson manifolds have been studied in a systematic way, in particular, in the foundational papers of A. Weinstein and his collaborators. This study includes two branches, geometry and mechanics. The second one will not be discussed in these lectures. More exactly, the aim of the present lectures is to give an exposition of the fundamental results of the differential geometry of C^∞-Poisson manifolds. Having said this, we shall omit hereafter the label C^∞.

As we already mentioned, the starting point of the theory of the Poisson manifolds is symplectic geometry. Not only do the symplectic manifolds offer the most basic Poisson bracket, but the geometry of these manifolds is the source of ideas on which the new theory, that of the Poisson manifolds, is based. Moreover, in a certain sense, via the so-called *symplectic realizations*, the Poisson structures always come from symplectic structures.

However, we shall not start with a preliminary chapter on symplectic geometry but, we shall assume that the reader, besides from being familiar with the general geometry of differentiable manifolds, has already acquired the necessary knowledge about this subject. The reason is that, on one hand, such a preliminary chapter would hardly be enough, and, on the other hand, there are already several books, some of them even textbooks, of symplectic geometry, such as [AM], [LM], [We1], [Va8], etc.

Of course, one could wonder whether the subject of the geometry of Poisson manifolds is already delimited well enough as to deserve a book addressed to those who want to study it. We believe that it is, and we hope that the present book will justify this belief. There is no doubt that the subject is important, which goes for both theory and applications. The latter are to classical and quantum physics. A large number of research papers about Poisson manifolds have already been published. (The reference list at the end of this book includes only papers quoted in the text and it is not at all exhaustive. Nevertheless, it contains more than one hundred such papers.) The basic results and the main lines of the theory are clear by now, and our work will give full details only as far as these basic results are concerned, while limiting ourselves to a sketchy exposition and bibliographical references only for further developments. But, until now, all the results, both basic and advanced are scattered in a lot of journals and conference proceedings, published in a variety of languages. We hope, therefore, that a unified exposition, like the present one, will prove both timely and useful.

And now, finally, a word about generalizations, that have already been studied by some authors. We do not intend to discuss them in any detail, just mention some of them.

It is a natural idea to generalize the notion of a Poisson manifold by relaxing condition (0.4), and asking instead that $\{f, g\}$ be just an operation of the local type, in the sense that

$$\text{support } \{f, g\} \subseteq (\text{support } f) \cap (\text{support } g). \tag{0.5}$$

Then, the set of the smooth functions on the manifold has the structure of a *local Lie algebra* [Kv], the operation $\{\ ,\ \}$ is called a *Jacobi bracket*, and a manifold endowed with such a bracket is a *Jacobi manifold*. The study of the Jacobi manifolds was made by A. Lichnerowicz and his collaborators [Lh3], [GL], [DLM], etc. The basic examples of Jacobi manifolds that are not Poisson manifolds are the *contact manifolds* (e.g., [LM], [Bl]) and the *locally conformal symplectic manifolds* (e.g., [Lf], [Va7]).

Finally, we mention the technically more complicated generalization which is the notion of a *Dirac manifold* studied by A. Weinstein and T. Courant [CW], [Cr].

1 The Poisson Bivector and the Schouten-Nijenhuis Bracket

1.1 The Poisson bivector

In this chapter we shall discuss various ways of representing a Poisson structure on a differentiable manifold. Let M^n be a Poisson manifold. The basic remark is the following obvious consequence of (0.4): $\{f, .\}$ is a derivation of $C^\infty(M)$. Hence $\forall f \in C^\infty(M)$ there exists a well defined vector field X_f such that

$$\{f, g\} = X_f g = -X_g f = dg(X_f) = -df(X_g). \tag{1.1}$$

X_f will be called the *Hamiltonian vector field* of f.

From (1.1) it follows that the bracket $\{f, g\}$ is determined by a skew-symmetric bilinear form on T^*M. In other words, there exists a C^∞ tensor field $w \in \Lambda^2 TM$ such that[1]

$$\{f, g\} = w(df, dg) = w^{ij} \frac{\partial f}{\partial x^i} \frac{\partial g}{\partial x^j} \tag{1.2}$$

where (x^i) are local coordinates on M, and we use the Einstein summation convention, a convention which we shall use overall in this book. w will be called the *Poisson bivector* of $(M, \{\ ,\ \})$.

In the usual way, w has an associated homomorphism

$$\# : T^*M \to TM, \tag{1.3}$$

where $\#\alpha \stackrel{\text{def}}{=} \alpha^\#$ is defined by

$$\beta(\alpha^\#) = w(\alpha, \beta) \quad (\alpha, \beta \in T^*M) \tag{1.4}$$

Notice, in particular, notice that $\forall f \in C^\infty(M)$ one has $(df)^\# = X_f$.

Now, the only condition which still has to be satisfied by w is the Jacobi identity (0.3), and an easy calculation yields

$$w^{hi}\partial_h w^{jk} + w^{hj}\partial_h w^{ki} + w^{hk}\partial_h w^{ij} = 0. \tag{1.5}$$

---

[1] Our conventions for the exterior calculus are those of [Mt] and not those of [KN], i.e., for instance, the value of a decomposable p-form on p arguments equals the corresponding determinant, without the factor $1/(p!)$, etc.

Consequently, a Poisson structure on M is equivalent to a bivector w that satisfies (1.5) and, in what follows we shall also speak of the Poisson manifold (M, w).

For instance, it is well known that if (M, ω) is a symplectic manifold, where ω is a closed nondegenerate 2-form, M has a Poisson structure given by

$$\{f, g\} = \omega(X_f, X_g)$$

where the Hamiltonian vector fields are defined by

$$i(X_f)\omega = -df$$

(e.g., [LM]). It follows easily that the corresponding bivector $w = \#_\omega \omega$, where $\#_\omega : T^*M \to TM$ is the natural isomorphism associated with the symplectic form ω extended to differential forms. (The operator $\#$ associated with w is then equal to $-\#_\omega$.)

1.2 The Schouten-Nijenhuis bracket

It is interesting to find an invariant meaning of (1.5) (besides (0.3), of course), and this is provided by the *Schouten-Nijenhuis bracket* [Nj] [BV2], to be considered here shortly. (The reader is referred to [BV2] for a detailed study.) In this section, M is an arbitrary differentiable manifold. Let $\mathcal{V}^i(M)$ ($\mathcal{V}^0(M) = C^\infty(M)$) denote the space of *i-vectors* (i.e., skew symmetric contravariant tensor fields of type $(i, 0)$), and $\mathcal{V}(M) = (\bigoplus_{i=0}^{n} \mathcal{V}^i(M), \wedge)$ be the *contravariant Grassmann algebra* of M. It is well known that $\forall X \in \mathcal{V}^1(M)$ there is a well defined *Lie derivative* operator L_X which, in particular, acts on $\mathcal{V}^q(M)$ by

$$(L_X Q)(x_0) = \frac{d}{dt}\bigg|_{t=0} \{\exp(-tX)_* Q(\exp tX(x_0))\} \tag{1.6}$$

($x_0 \in M$, $Q \in \mathcal{V}^q(M)$), and is such that $L_X Y = [X, Y]$ for $X, Y \in \mathcal{V}^1(M)$.

Accordingly, it is natural to define $\forall X_1, \ldots, X_p \in \mathcal{V}^1(M)$ the operation

$$[X_1 \wedge \ldots \wedge X_p, Q] = \sum_{i=1}^{p} (-1)^{i+1} X_1 \wedge \ldots \wedge \hat{X}_i \wedge \ldots \wedge X_p \wedge [X_i, Q], \tag{1.7}$$

where $[X_i, Q] \overset{\text{def}}{=} L_{X_i} Q$, and the hat $\wedge$ denotes the absence of the corresponding factor. Now, using (1.7) we can prove

1.1. Theorem. *There is a well defined unique* $\mathbb{R}$*-bilinear local type extension of the Lie derivative* L_X *to an operation*

$$[\ ,\ ] : \mathcal{V}^p(M) \times \mathcal{V}^q(M) \to \mathcal{V}^{p+q-1}(M), \tag{1.8}$$

such that (1.7) *holds good. This operation has the following properties*

$$[P, Q] = (-1)^{pq}[Q, P], \tag{1.9}$$

$$[P, Q \wedge R] = [P, Q] \wedge R + (-1)^{pq+q} Q \wedge [P, R], \tag{1.10}$$

$$(-1)^{p(r-1)}\,[P,[Q,R]] + (-1)^{q(p-1)}\,[Q,[R,P]] + \\ + (-1)^{r(q-1)}\,[R,[P,Q]] = 0. \tag{1.11}$$

Proof. The "local type" of (1.8) means that for $x_0 \in M$, $[P, Q](x_0)$ depends only on the restriction of the fields P, Q to a neighbourhood of x_0.

Suppose that Q of (1.7) is $Q = Y_1 \wedge \ldots \wedge Y_q$ ($Y_j \in \mathcal{V}^1(M)$). Then, since

$$L_X(Y_1 \wedge \ldots \wedge Y_q) = \sum_{j=1}^{q} Y_1 \wedge \ldots \wedge [X, Y_j] \wedge \ldots \wedge Y_q$$

formula (1.7) yields

$$[X_1 \wedge \ldots \wedge X_p, Y_1 \wedge \ldots \wedge Y_q] = \tag{1.12}$$

$$= (-1)^{p+1} \sum_{i=1}^{p} \sum_{j=1}^{q} (-1)^{i+j} [X_i, Y_j] \wedge \wedge X_1 \wedge \ldots \wedge \hat{X}_i \wedge \ldots \wedge X_p \wedge Y_1 \wedge \ldots \wedge \hat{Y}_j \wedge \ldots \wedge Y_q.$$

Furthermore, if we denote here $X_1 \wedge \ldots \wedge X_p = P$, it follows similarly that (1.12) can be interpreted as

$$[X_1 \wedge \ldots \wedge X_p, Y_1 \wedge \ldots \wedge Y_q] = \\ = (-1)^{pq} \sum_{j=1}^{q} (-1)^{j+1} Y_1 \wedge \ldots \wedge \hat{Y}_j \wedge \ldots \wedge Y_q \wedge [Y_j, P], \tag{1.13}$$

where $[Y, P] = L_Y$P.

Now, let us take some arbitrary elements $x_0 \in M$, $P \in \mathcal{V}^p(M)$, $Q \in \mathcal{V}^q(M)$. By the known technique of the partition of unity, if $p, q \geq 1$, there is an open neighbourhood U of x_0 where

$$P/_U = (X_1 \wedge \ldots \wedge X_p)/_U, Q/_U = (Y_1 \wedge \ldots \wedge Y_q)/_U \tag{1.14}$$

for some vector fields $X_1, \ldots, X_p, Y_1, \ldots Y_q$ on M. This obviously allows us to take (1.12) as the definition of the general bracket operation $[P, Q](x_0)$, and, because of (1.7) and (1.13), this operation depends only on P and Q, and not on the choice of the decompositions (1.14). It is also clear that this is the unique way to define a local type bracket where (1.7) holds. Moreover, if $p = q = 0$ we shall define $[P, Q] = 0$, and if $p \geq 1$, $q = 0$ we shall define $[P, Q] = [Q, P]$ by means of (1.7), which is easily seen to be equivalent to $[P, Q](\alpha_2, \ldots, \alpha_{p-1}) = P(dQ, \alpha_2, \ldots, \alpha_{p-1})$, for any 1-forms $\alpha_2, \ldots, \alpha_{p-1}$. This shows the invariance of $[P, Q]$ in the particular case under consideration.

Furthermore, the same formulas (1.7), (1.12) and (1.13) obviously imply (1.9). Then, (1.10) is a straightforward consequence of (1.7) and of the fact that

$$L_X(Q \wedge R) = (L_X Q) \wedge R + Q \wedge (L_X R),$$

if we just put $P = X_1 \wedge \ldots \wedge X_p$. Finally, (1.11) will be obtained by taking there

$$P = X_1 \wedge \cdots \wedge X_p, Q = Y_1 \wedge \cdots \wedge Y_q, R = Z_1 \wedge \cdots \wedge Z_r,$$

and by computing the necessary brackets via (1.12). A careful cancellation of terms, using in particular

$$[X_i, [Y_j, Z_k]] + [Y_j, [Z_k, X_i]] + [Z_k, [X_i, Y_j]] = 0$$

provides the result. If some of the numbers p, q, r vanish the results are even easier. We leave the details to the reader since writing them down is lengthy but technical only. The computation goes as follows:

$$[Q, R] = (-1)^{q+1} \sum_{j=1}^{q} \sum_{k=1}^{r} (-1)^{j+k} [Y_j, Z_k] \wedge Y_1 \wedge \ldots \wedge \hat{Y}_j \wedge \ldots \wedge Y_q \wedge$$

$$\wedge Z_1 \wedge \ldots \wedge \hat{Z}_k \wedge \ldots \wedge Z_r;$$

$$[P, [Q, R]] = (-1)^{p+q} \sum_{i=1}^{p} \sum_{j=1}^{q} \sum_{k=1}^{r} (-1)^{j+k+i+1} [X_i, [Y_j, Z_k]] \wedge$$

$$\wedge X_1 \wedge \ldots \wedge \hat{X}_i \wedge \ldots \wedge X_p \wedge Y_1 \wedge \ldots \wedge \hat{Y}_j \wedge \ldots \wedge Y_q \wedge$$

$$\wedge Z_1 \wedge \ldots \wedge \hat{Z}_k \wedge \ldots \wedge Z_r +$$

$$+ (-1)^q \sum_{i=1}^{p} \sum_{k=1}^{r} \sum_{s<j=1}^{q} (-1)^{j+k+i+s} ([X_i, Y_s] \wedge [Y_j, Z_k] -$$

$$- [X_i, Y_j] \wedge [Y_s \wedge Z_k]) \wedge X_1 \wedge \ldots \wedge \hat{X}_i \wedge \ldots \wedge X_p \wedge$$

$$\wedge Y_1 \wedge \ldots \wedge \hat{Y}_s \wedge \ldots \wedge \hat{Y}_j \wedge \ldots \wedge Y_q \wedge Z_1 \wedge \ldots \wedge \hat{Z}_k \wedge \ldots \wedge Z_p -$$

$$- \sum_{i=1}^{p} \sum_{j=1}^{q} \sum_{t<k=1}^{r} (-1)^{j+k+i+t} ([X_i, Z_t] \wedge [Y_j, Z_k] -$$

$$- [X_i, Z_k] \wedge [Y_j, Z_t]) \wedge X_1 \wedge \ldots \wedge \hat{X}_i \wedge \ldots \wedge X_p \wedge Y_1 \wedge \ldots \wedge$$

$$\wedge \hat{Y}_j \wedge \ldots \wedge Y_q \wedge Z_1 \wedge \ldots \wedge \hat{Z}_t \wedge \ldots \wedge \hat{Z}_k \wedge \ldots \wedge Z_r.$$

Then cyclic permutation, and an examination of signs will do. Q.e.d.

1.2. Definition. The operation $[\ ,\ ]$ of Theorem 1.1 is called the *Schouten-Nijenhuis bracket.*

1.3. Remarks. 1) Another definition of the Schouten-Nijenhuis bracket can be obtained as follows [Nj], [Lh2]. Remember the operation

$$(i(X_1 \wedge \ldots \wedge X_k)\omega)(\ldots) = \omega(X_1, \ldots, X_k, \ldots), \tag{1.15}$$

where ω is a differential form, and which also defines $i(P)\omega$, $\forall P \in \mathcal{V}^k(M)$. Then an algebraic computation based on (1.12) gives (see [BV2] for details)

$$\begin{aligned} i([P,Q])\omega = (-1)^{q(p+1)} i(P)d[i(Q)\omega] + \\ + (-1)^p i(Q)d[i(P)\omega] - i(P \wedge Q)d\omega, \end{aligned} \tag{1.16}$$

$\forall P \in \mathcal{V}^p(M)$, $\forall Q \in \mathcal{V}^q(M)$, $\forall \omega \in \Lambda^{p+q-1}(M)$. Clearly, (1.16) may also be used as the definition of the bracket. We prefer to prove (1.16) later on (Proposition 1.8) by a different computation.

2) If $\varphi : M \to N$ is a differentiable mapping, if P, P' are p-vector fields on M, N, respectively, that are φ-related (i.e., $\forall x \in M$, $\varphi_* P_x = P'_{\varphi(x)}$), and if the same holds for the q-vector fields Q, Q', then $[P,Q]$ and $[P',Q']$ are also φ-related. Indeed, by using local coordinates, it is easy to see that (P,P') $((Q,Q'))$ have local decompositions (1.14) such that the factors of $P(Q)$ are φ-related to the corresponding factors of P' (Q'). Then, our assertion follows from (1.12) since it is well known that this assertion is true for the Lie bracket of vector fields.

3) The Schouten-Nijenhuis bracket can be defined in a purely algebraic setting, and there is also a similar operation for symmetric contravariant tensors but we don't need it here. See [BV2] for a full development.

1.3 Coordinate expressions

Let (x^i) be local coordinates on M, and

$$\begin{aligned} P &= \frac{1}{p!} P^{i_1 \ldots i_p} \frac{\partial}{\partial x^{i_1}} \wedge \ldots \wedge \frac{\partial}{\partial x^{i_p}}, \\ Q &= \frac{1}{q!} Q^{j_1 \ldots j_q} \frac{\partial}{\partial x^{j_1}} \wedge \cdots \wedge \frac{\partial}{\partial x^{j_q}}. \end{aligned} \tag{1.17}$$

These expressions *decompose* the respective *multivector fields*, and we may apply (1.12) to these decompositions. The result is

$$
\begin{aligned}
[P,Q] = \frac{(-1)^p}{p!q!} \Bigg\{ & \sum_{s=1}^{q} (-1)^{s+1} Q^{j_1\ldots j_q} \frac{\partial P^{i_1\ldots i_p}}{\partial x^{j_s}} \frac{\partial}{\partial x^{i_1}} \wedge \ldots \wedge \frac{\partial}{\partial x^{i_p}} \wedge \\
& \wedge \frac{\partial}{\partial x^{j_1}} \wedge \cdots \wedge \frac{\hat{\partial}}{\partial x^{j_s}} \wedge \cdots \wedge \frac{\partial}{\partial x^{j_q}} - \\
& - (-1)^p \sum_{t=1}^{p} (-1)^t P^{i_1\ldots i_p} \frac{\partial Q^{j_1\ldots j_q}}{\partial x^{i_t}} \cdot \\
& \cdot \frac{\partial}{\partial x^{i_1}} \wedge \ldots \wedge \frac{\hat{\partial}}{\partial x^{i_t}} \wedge \ldots \wedge \frac{\partial}{\partial x^{i_p}} \wedge \frac{\partial}{\partial x^{j_1}} \wedge \ldots \wedge \frac{\partial}{\partial x^{j_q}} \Bigg\} = \\
= \frac{(-1)^p}{p!q!} \Bigg\{ & q Q^{u j_2\ldots j_q} \frac{\partial P^{i_1\ldots i_q}}{\partial x^u} \frac{\partial}{\partial x^{i_1}} \wedge \ldots \wedge \frac{\partial}{\partial x^{i_p}} \wedge \\
& \wedge \frac{\partial}{\partial x^{j_2}} \wedge \ldots \wedge \frac{\partial}{\partial x^{j_q}} + \\
& + (-1)^p p P^{u i_2\ldots i_p} \frac{\partial Q^{j_1,\ldots j_q}}{\partial x^u} \frac{\partial}{\partial x^{i_2}} \wedge \ldots \wedge \frac{\partial}{\partial x^{i_p}} \wedge \\
& \wedge \frac{\partial}{\partial x^{j_1}} \wedge \ldots \wedge \frac{\partial}{\partial x^{j_q}} \Bigg\}.
\end{aligned}
\tag{1.18}
$$

Hence, if we put

$$
[P,Q] = \frac{1}{(p+q-1)!} [P,Q]^{k_1,\ldots,k_{p+q-1}} \frac{\partial}{\partial x^{k_1}} \wedge \ldots \wedge \frac{\partial}{\partial x^{k_{p+q-1}}}, \tag{1.19}
$$

we have [Lh2]

$$
\begin{aligned}
[P,Q]^{k_1\ldots k_{p+q-1}} = & \frac{(-1)^p}{p!(q-1)!} \delta^{k_1\ldots\ldots k_{p+q-1}}_{i_1\ldots i_p j_2\ldots j_q} Q^{u j_2\ldots j_q} \frac{\partial P^{i_1\ldots i_p}}{\partial x^u} + \\
& + \frac{1}{(p-1)!q!} \delta^{k_1\ldots\ldots k_{p+q-1}}_{i_2\ldots i_p j_1\ldots j_q} P^{u i_2,\ldots i_p} \frac{\partial Q^{j_1\ldots j_q}}{\partial x^u},
\end{aligned}
\tag{1.20}
$$

where $\delta^{\cdots}_{\cdots}$ is the Kronecker multi-index i.e., 1 (-1) if the upper indices are an even (odd) permutation of the lower indices, and 0 otherwise [Lh1].

Now, if we apply these results to the calculation of $[w,w]$, and look at (1.5), we get

1.4. Proposition. *The bivector field $w \in \mathcal{V}^2(M)$ is a Poisson bivector on M iff $[w,w]=0$.*

We shall also develop a few more formulas concerning the Schouten-Nijenhuis bracket. Let us consider an arbitrary torsionless connection ∇ on the differentiable manifold M. Then, we may replace in (1.12)

$$
[X_i, X_j] = \nabla_{X_i} X_j - \nabla_{X_j} X_i.
$$

After this replacement, if we apply (1.12) to P, Q of (1.17), and use the skew symmetry of P, Q, the computation is practically the same as in (1.18), but it uses covariant derivatives instead of partial derivatives. Hence we get the formulas

$$\begin{aligned}[P,Q] = {} & \frac{1}{(p-1)!q!}P^{ui_2\ldots i_p}\nabla_u Q^{j_1\ldots j_q}\frac{\partial}{\partial x^{i_2}}\wedge\ldots\wedge\frac{\partial}{\partial x^{i_p}}\wedge \\ & \wedge\frac{\partial}{\partial x^{j_1}}\wedge\ldots\wedge\frac{\partial}{\partial x^{j_q}}+ \\ & +\frac{(-1)^p}{p!(q-1)!}Q^{uj_2\ldots j_q}\nabla_u P^{i_1\ldots i_p}\frac{\partial}{\partial x^{i_1}}\wedge\ldots\wedge\frac{\partial}{\partial x^{i_p}}\wedge \\ & \wedge\frac{\partial}{\partial x^{j_2}}\wedge\ldots\wedge\frac{\partial}{\partial x^{j_q}},\end{aligned} \tag{1.21}$$

$$\begin{aligned}[P,Q]^{k_1\ldots k_{p+q-1}} = {} & \frac{1}{(p-1)!q!}\delta^{k_1\ldots\ldots k_{p+q-1}}_{i_2\ldots i_p j_1\ldots j_q}P^{ui_2\ldots i_p}\nabla_u Q^{j_1\ldots j_q}+ \\ & +\frac{(-1)^p}{p!(q-1)!}\delta^{k_1\ldots\ldots k_{p+q-1}}_{i_1\ldots i_p j_2\ldots j_q}Q^{uj_2\ldots j_q}\nabla_u P^{i_1\ldots i_p}.\end{aligned} \tag{1.22}$$

If these formulas are applied to $[w,w]$ we get

1.5. Proposition. *Let (M,∇) be a manifold endowed with a torsionless linear connection. Then the bivector w defines a Poisson structure on M iff*

$$w^{hi}\nabla_h w^{jk} + w^{hj}\nabla_h w^{ki} + w^{hk}\nabla_h w^{ij} = 0 \tag{1.23}$$

1.6. Corollary. *Let (M,g) be a Riemannian manifold with the Levi-Civita connection ∇. Let w be a bivector on M. If there is a tensor field A^i_{jk} such that $A^i_{jk} = A^i_{kj}$, and*

$$\nabla_h w^{jk} + A^j_{uh}w^{uk} + A^k_{uh}w^{ju} = 0 \tag{1.24}$$

then w defines a Poisson structure on M.

Proof. The left hand side is the derivative of w with respect to the torsionless connection $\nabla + A$. Q.e.d.

The Poisson structure of Corollary 1.6 is parallel with respect to a certain torsionless linear connection. The converse question would be whether a given Poisson structure is parallel with respect to some torsionless connection. Such a connection will be called a *Poisson connection* [BF-], and we shall prove later on that any Poisson manifold (M,w) where rank w = const. has Poisson connections.

1.4 The Koszul formula and applications

Now, we establish another important formula, due to Koszul [Kz]. Let us again consider an arbitrary torsionless connection ∇ on M, and define the *generalized divergence* $D_\nabla Q$ of $Q \in \mathcal{V}^q(M)$ by

$$(D_\nabla Q)^{j_2 \ldots j_q} = \nabla_k Q^{k j_2 \ldots j_q}. \tag{1.25}$$

Then we have

1.7. Proposition. *If $P \in \mathcal{V}^p(M)$ and $Q \in \mathcal{V}^q(M)$, the following formula holds*

$$[P, Q] = D_\nabla(P \wedge Q) - (D_\nabla P) \wedge Q - (-1)^p P \wedge (D_\nabla Q). \tag{1.26}$$

Proof. If P and Q are represented by (1.17), we have

$$(P \wedge Q)^{k_1 \ldots k_{p+q}} = \frac{1}{p!q!} \delta^{k_1 \ldots\ldots k_{p+q}}_{i_1 \ldots i_p j_1 \ldots j_q} P^{i_1 \ldots i_p} Q^{j_1 \ldots j_q}. \tag{1.27}$$

Therefore, since $\delta^{\cdots}_{\cdots}$ is parallel for all ∇, we get

$$\begin{aligned} D_\nabla(P \wedge Q)^{k_2 \ldots k_{p+q}} &= \nabla_s (P \wedge Q)^{s k_2 \ldots k_{p+q}} = \\ &= \frac{1}{p!q!} \delta^{s k_2 \ldots\ldots k_{p+q}}_{i_1 \ldots i_p j_1 \ldots j_q} \{(\nabla_s P^{i_1 \ldots i_p}) Q^{j_1 \ldots j_q} + \\ &+ P^{i_1 \ldots i_p} (\nabla_s Q^{j_1 \ldots j_q})\}. \end{aligned} \tag{1.28}$$

Since $\delta^{\cdots}_{\cdots}$ are determinants, one gets easily (e.g. [Go])

$$\delta^{s k_2 \ldots k_v}_{h_1 \ldots . h_v} = \sum_{\alpha=1}^{v} (-1)^{\alpha+1} \delta^s_{h_\alpha} \delta^{k_2 \ldots k_v}_{h_1 \ldots \hat{h}_\alpha \ldots h_v}, \tag{1.29}$$

and using this in the case of (1.28), and because of the skew symmetry of the components of P and Q, it follows that

$$\begin{aligned} D_\nabla(P \wedge Q)^{k_2 \ldots k_{p+q}} &= \frac{1}{p!q!} \Big\{ p \delta^{k_2 \ldots\ldots k_{p+q}}_{i_2 \ldots i_p j_1 \ldots j_q} \cdot \\ &\cdot \left[(\nabla_s P^{s i_2 \ldots i_p}) Q^{j_1 \ldots j_q} + P^{s i_2 \ldots i_p} (\nabla_s Q^{j_1 \ldots j_q}) \right] + \\ &+ (-1)^p q \delta^{k_2 \ldots\ldots k_{p+q}}_{i_1 \ldots i_p j_2 \ldots j_q} \cdot \\ &\cdot \left[(\nabla_s P^{i_1 \ldots i_p}) Q^{s j_2 \ldots j_q} + P^{i_1 \ldots i_p} (\nabla_s Q^{s j_2 \ldots j_q}) \right] \Big\}. \end{aligned} \tag{1.30}$$

If (1.30) is compared to (1.22) and (1.25), it turns out to be exactly (1.26). Q.e.d.

1.8. Proposition. *The Schouten-Nijenhuis bracket and the exterior differential are related by means of the formula* (1.16).

Proof. In what follows we compute with P, Q of (1.17), and with the k-form

$$\omega = \frac{1}{k!}\omega_{h_1\ldots h_k} dx^{h_1} \wedge \ldots \wedge dx^{h_k}.$$

For $k = p-1$, we shall define the vector field $A(P,\omega)$ such that

$$A^k = \frac{1}{(p-1)!} P^{ki_2\ldots i_p}\omega_{i_2\ldots i_p},$$

and, of course, similar vector fields will be considered in all possible cases. Now, if ∇ is the covariant derivative used in (1.25) and (1.26), a straightforward derivation, and the use of the skew-symmetry of our tensor fields give

$$D_\nabla A(P,\omega) = i(D_\nabla P)\omega + i(P)d\omega,$$

where the *interior product* is that defined by (1.15). On the other hand, for $k = p$, (1.15) gives

$$i(P)\omega = \frac{1}{k!} P^{i_1\ldots i_p}\omega_{i_1\ldots i_p},$$

whence, for $k = p+q$, we get using (1.27) that

$$i(P \wedge Q)\omega = i(Q)i(P)\omega.$$

Then, if we take $k = p+q-1$, and compute $i([P,Q])\omega$ using (1.26), and the results established above, we get

$$i([P,Q])\omega = (-1)^{q(p-1)}i(P)d[i(Q)\omega] + (-1)^p i(Q)[di(P)\omega] - i(P\wedge Q)d\omega +$$
$$+ D_\nabla A(P\wedge Q,\omega) - (-1)^{q(p-1)}D_\nabla A(P,i(Q)\omega) - (-1)^p D_\nabla A(Q,i(P)\omega).$$

This is exactly (1.16), if the last three terms cancel. And, we see that they cancel, indeed, by computing via (1.27)

$$A^k(P\wedge Q,\omega) = \frac{1}{(p+q-1)!}\left(\frac{1}{p!q!}\delta^{kh_1\ldots\ldots h_{p+q}}_{i_1\ldots i_p j_1\ldots j_q} P^{i_1\ldots i_p} Q^{j_1\ldots j_q}\omega_{h_2\ldots h_{p+q}}\right),$$

and, then, by developing here the $\delta^{\cdots}_{\cdots}$ with (1.29). Q.e.d.

Formula (1.26) is particularly interesting if ∇ is assumed to be the Levi-Civita connection of a Riemannian metric g. Indeed, in this case we have the musical isomorphism $\#_g : T^*M \to TM$, and if we remember that the codifferential $\delta_g\alpha$ of a differential p-form α is given by (e.g. [Lh1])

$$(\delta_g\alpha)_{i_1\ldots i_{p-1}} = -g^{st}\nabla_t\alpha_{si_1\ldots i_{p-1}}, \tag{1.31}$$

and compare this with (1.25), we get

$$D_\nabla = -\#_g \delta_g \#_g^{-1}. \tag{1.32}$$

With this remark, and if we denote

$$\pi = \#_g^{-1} P, \kappa = \#_g^{-1} Q, \tag{1.33}$$

formula (1.26) becomes

$$[P, Q] = \#_g\{(\delta_g \pi) \wedge \kappa + (-1)^p(\pi \wedge \delta_g \kappa) - \delta_g(\pi \wedge \kappa)\}. \tag{1.34}$$

Conversely, this result is also interesting because it yields an explicit formula for the codifferential of an exterior product of forms namely,

$$\delta_g(\pi \wedge \kappa) = (\delta_g \pi) \wedge \kappa + (-1)^p \pi \wedge (\delta_g \kappa) - \#_g^{-1}[\#_g \pi, \#_g \kappa]. \tag{1.35}$$

Particularly, we get [Va9]

1.9. Proposition. *Let (M, g) be a Riemannian manifold, and ω a 2-form on M. Then $w = \#_g \omega$ defines a Poisson structure on M iff*

$$\delta_g(\omega \wedge \omega) = 2\omega \wedge (\delta_g \omega). \tag{1.36}$$

This result is useful if we prefer to represent the Poisson structure of a Poisson manifold by a 2-form instead of a bivector. Of course, this requires the use of an auxiliary Riemannian metric g. For instance, we have

1.10. Corollary. *Let (M, g) be a compact Riemannian symmetric space. Then, any harmonic 2-form of (M, g) provides M with a Poisson structure.*

Proof. It is well known that on a compact Riemannian symmetric space the exterior product of harmonic forms is harmonic (e.g., [Hl], [Ln]). Hence (1.36) holds. Q.e.d.

1.11. Corollary. *Let ω be a closed self-dual or anti-self-dual 2-form on a connected 4-dimensional Riemannian manifold (M, g). Then $\#_g \omega$ is a Poisson bivector on M iff ω is a form of constant length.*

Proof. By definition, the hypotheses mean that we have

$$d\omega = 0, \qquad *\omega = \pm\omega.$$

Since in our case $\delta_g = - * d*$ [Lh1], [Wl], we also have $\delta_g \omega = 0$, and we shall get a Poisson bivector $\#_g \omega$ iff $\delta_g(\omega \wedge \omega) = 0$. If we use the operators e and i, the exterior and interior product on forms, which, when used for a k-form λ, are related by [Lh1]

$$i(\lambda)\alpha = (-1)^{k(\deg \alpha - 1)} *^{-1} e(\lambda) * \alpha,$$

we have

$$\omega \wedge \omega = e(\omega)\omega = \pm e(\omega) * \omega = \pm * i(\omega)\omega = \pm * (\| \omega \|^2).$$

Hence the Poisson condition becomes $d(\| \omega \|^2) = 0$. Q.e.d.

1.12. Definition. Two Poisson structures of a manifold M are said to be *compatible* (e.g., [MM], [KS1]) if the corresponding bivectors w_1, w_2 satisfy the condition $[w_1, w_2] = 0$.

This notion is useful in the study of the integrability of mechanical systems. We quoted it here because formula (1.35) provides us with a compatibility condition expressed in terms of an auxiliary metric, namely,

$$\begin{aligned}\delta_g \#_g^{-1}(w_1 \wedge w_2) = (\delta_g \#_g^{-1} w_1) \wedge (\#_g^{-1} w_2) + \\ + (\#_g^{-1} w_1) \wedge (\delta_g \#_g^{-1} w_2).\end{aligned} \tag{1.37}$$

As in the case of Corollary 1.10 this condition yields

1.13. Corollary. *Any two harmonic 2-forms of a compact Riemannian symmetric space (M, g) define a pair of compatible Poisson structures of M.*

A main problem in this direction is to look for Poisson structures w which are compatible with the Poisson structure associated to a given symplectic form σ of M. For instance, we have

1.14. Proposition. *Let w be a Poisson bivector on a symplectic manifold (M^{2m}, σ). Then w is compatible with σ iff the 2-form $\lambda = \#_\sigma^{-1} w$ is closed.*

Proof. Let $(x^\alpha) = (x^i, x^{i^*})$ $(\alpha = 1, \ldots, 2m;\ i = 1, \ldots, m;\ i^* = i + m)$ be a *Darboux chart* for (M, σ) [LM] i.e., one for which

$$\sigma = \epsilon_{ij^*} dx^i \wedge dx^{j^*} \quad (\epsilon_{ij^*} = \delta_{ij}). \tag{1.38}$$

Then the local components of the form λ are given by

$$\lambda_{ij} = w^{i^*j^*}, \quad \lambda_{ij^*} = -w^{i^*j}, \quad \lambda_{i^*j} = -w^{ij^*}, \quad \lambda_{i^*j^*} = w^{ij}. \tag{1.39}$$

Using (1.20), and since the components ϵ^{i^*j}, etc. of $\#_\sigma \sigma$ are constant, the compatibility condition $[\#_\sigma \sigma, w] = 0$ becomes

$$\epsilon^{\theta\tau} \frac{\partial w^{\mu\nu}}{\partial x^\theta} + \epsilon^{\theta\mu} \frac{\partial w^{\nu\tau}}{\partial x^\theta} + \epsilon^{\theta\nu} \frac{\partial w^{\tau\mu}}{\partial x^\theta} = 0.$$

If we write down this condition for the indices i, i^* etc., and if we take into consideration (1.39), we see that we have exactly the relations that express $d\lambda = 0$. Q.e.d.

In the particular case where σ is the Kähler form of a Kähler metric g on M one has $\#_\sigma^{-1} w = C \#_g^{-1} w$, where C is the operator that transforms the arguments of a form by the complex structure tensor J (e.g., [Wl], [Lh1]). This remark leads to

1.15. Corollary. *(i) Let γ be a 2-form of a Kähler manifold (M, g, J). Then $\#_g\gamma$ is a Poisson bivector compatible with the Kähler form σ of $(M, g,)$ iff the following relations hold*

$$\delta_g(\gamma \wedge \gamma) = 2\gamma \wedge (\delta_g\gamma), \quad d(C\gamma) = 0. \tag{1.40}$$

(ii) If a harmonic 2-form γ is such that $\#_g\gamma$ is a Poisson structure of the compact Kähler manifold (M, g, J), then this structure is necessarily compatible with that defined by the Kähler form of (M, g, J).

The first part was already proven, and the second follows from the fact that, under the hypotheses, $C\gamma$ also must be harmonic (since, the Laplacian Δ commutes with $C[Wl]$), and closed. Q.e.d.

1.16. Remark. On a symplectic manifold (M, σ) it is possible to use a torsionless symplectic connection ∇ (i.e., $\nabla\sigma = 0$) in (1.22) and (1.26), and use it to compute Schouten-Nijenhuis brackets. On the other hand, one also has a $\tilde{*}$-operator, analogous to the Riemannian $*$, and a codifferential $\tilde{\delta} = -\tilde{*}d\tilde{*}$ (e.g. [LM]). A technical computation, which is the same as the one known in Riemannian geometry (e.g., [Lh1]), yields for this $\tilde{\delta}$ the formula (1.31), with g replaced by σ, and with the symplectic connection ∇. Therefore, we shall have also formulas like (1.32), (1.34), (1.36) and (1.37) with $\tilde{\delta}$ instead of δ_g. Again, this might be useful in the study of Poisson structures that are compatible with a given symplectic structure.

1.5 Miscellanea

Before ending this chapter we shall add a couple of further definitions that will be useful later on.

1.17. Definition. A mapping $\varphi : (M_1, w_1) \to (M_2, w_2)$ between two Poisson manifolds is called a *Poisson mapping* or a *Poisson morphism* if $\forall f, g \in C^\infty(M_2)$ one has

$$\{f \circ \varphi, g \circ \varphi\}_1 = \{f, g\}_2 \circ \varphi, \tag{1.41}$$

or, equivalently, the tensor fields w_1 and w_2 are φ-related. Furthermore, if φ is also a diffeomorphism it will be called a *Poisson automorphism* or *equivalence*.

We should notice that since the Poisson structures are of the contravariant type there are no pull-back or push-forward operations of such structures by a mapping φ, unless φ is, essentially, a diffeomorphism. We can however speak of a *Poisson submanifold* $(\tilde{M}, \tilde{w})$ of (M, w), namely, we shall use this name for a submanifold $\tilde{M}$ of M, if the immersion $\tilde{M} \to M$ is a Poisson mapping. Obviously, if the structure $\tilde{w}$ exists it is unique.

1.18. Definition. If (M_1, w_1) and (M_2, w_2) are Poisson manifolds, then $(M_1 \times M_2, w)$, where $w = w_1 \oplus w_2$, is obviously a new Poisson manifold. We call it the *Poisson product* of (M_1, w_1) and (M_2, w_2).

Clearly, the natural projections of a Poisson product onto its factors are Poisson mappings.

Finally, we should like to mention that the Schouten-Nijenhuis bracket is also useful for the characterization of the Jacobi manifolds. Indeed, if (0.4) is replaced by the weaker condition (0.5), it is still possible to get a coordinate expression of the bracket $\{f, g\}$ since, by a classical theorem of Peetre (e.g., [Na]), (0.5) implies that this bracket has to be given by linear differential operators with respect to both f and g. Furthermore, if the Jacobi identity is written down for such operators, one sees that they must be of order ≤ 1 [Kv], [GL]. Together with the invariance of the bracket, this means that the latter must be given by an expression of the form

$$\{f, g\} = A(df, dg) + f(Eg) - gE(f), \tag{1.42}$$

where A is a bivector field, and E is a vector field on the corresponding Jacobi manifold M.

Furthermore, a local coordinate computation yields (e.g., [GL])

$$\sum_{\mathrm{Cycl}(f,g,h)} \{\{f, g\}, h\} = \frac{1}{2}([A, A] - 2E \wedge A)(df, dg, dh) - \sum_{\mathrm{Cycl}(f,g,h)} h[E, A](df, dg), \tag{1.43}$$

so that a Jacobi manifold is a differentiable manifold M, endowed with a Lie bracket (1.42), such that

$$[A, A] = 2E \wedge A, \quad [E, A] = L_E A = 0, \tag{1.44}$$

where $[\ ,\ ]$ is the Schouten-Nijenhuis bracket.

2 The Symplectic Foliation of a Poisson Manifold

2.1 General distributions and foliations

Let (M^n, w) be a Poisson manifold. According to the definition, as given at the beginning of Chapter 1, we expect the set $\chi_w(M) \overset{\text{def}}{=} \{X_f \mid f \in C^\infty(M)\}$ of the Hamiltonian vector fields of M to play an important role. In order to discover it, let us define

$$\mathcal{S}_{x_0}(M) = \{v \in T_{x_0}M / \exists f \in C^\infty(M), X_f(x_0) = v\} \quad (x_0 \in M). \tag{2.1}$$

Notice that, since $X_f = \#(df)$, where $\#$ is the homomorphism (1.3), we have $\mathcal{S}_{x_0}(M) = \text{im}\#_{x_0}$. In order to speak of the set $\mathcal{S}(M)$ of all these planes, we shall adopt the following terminology, which is slightly different from that of the textbooks on differentiable manifolds.

2.1. Definition. A set of linear subspaces $\mathcal{S}(M) = \{\mathcal{S}_{x_0}(M)\}$ of the tangent spaces $T_{x_0}M$ is called a *(general) distribution*. Such a distribution is said to be *differentiable* (C^∞) if $\forall x_0 \in M$ there is a finite number of differentiable vector fields $X_1, \ldots, X_s \in \mathcal{S}(M)$, such that $\mathcal{S}_{x_0}(M) = \text{span}\ \{X_1(x_0), \ldots, X_s(x_0)\}$.

It is obvious that $\mathcal{S}(M)$ given by formula (2.1) is a differentiable distribution, and it is called the *characteristic distribution* of the Poisson structure w.

Now, coming back to the general case of Definition 2.1, put $\rho(x) = \dim \mathcal{S}_x$ $(x \in M)$. Sometimes, one also refers to $\rho(x)$ as the *rank* of $\mathcal{S}(M)$ at x. Clearly, if $\mathcal{S}(M)$ is differentiable $\rho(x)$ is a *lower semicontinuous* function, since $\rho(x)$ cannot decrease in a neighbourhood of x. In case $\rho(x) = \text{const.}$, $\mathcal{S}(M)$ is a distribution in the usual sense, and we call it a *regular distribution*. In the general case, $x \in M$ will be a *regular point* if x is a local maximum of $\rho(x)$ or, equivalently, $\rho(x) = \text{const.}$ on an open neighbourhood of x. All the other points of M will be called *singular points of* $\mathcal{S}(M)$. The set $\mathcal{D}$ of the regular points of $\mathcal{S}(M)$ is obviously open and, also, it is dense in M since, if $x_0 \in M \backslash \mathcal{D}$, and U is any neighbourhood of x_0, U contains regular points ($\rho/_U$ must have a maximum because it is integer valued and bounded) i.e., $x_0 \in \overline{\mathcal{D}}$. But $\mathcal{D}$ may not be connected as shown by

2.2. Example. $M = \mathbb{R}^2$, $\mathcal{S}_{(x,y)}(M) = \text{span}\{\frac{\partial}{\partial x}, \varphi(y)\frac{\partial}{\partial y}\}$, where $\varphi(y)$ is a C^∞-function which is 0 for $y \leq 0$, and positive for $y > 0$ (e.g., $\varphi(y) = 0$ for $y \leq 0$, and $\varphi(y) = e^{-(1/y^2)}$ for $y > 0$). Then the singular points are those of the x-axis, and the connected components of $\mathcal{D}$ are the half-planes $y > 0$ (where $\rho = 2$), and $y < 0$ (where $\rho = 1$).

2.3 Remark. No complementary distribution of $\mathcal{S}(M)$ with respect to TM is differentiable if $\mathcal{S}(M)$ has singular points. Indeed, the dimension of the complementary planes will not be a lower semicontinuous function.

It is clear that the notions of an *integral submanifold*, an integral *submanifold of maximal dimension* (along it $\mathcal{S}$ must have a constant rank!), and a *completely integrable distribution* can be defined just as in the regular case. For instance, $\mathcal{S}$ of Example 2.2 is completely integrable. It is also possible to define as in the regular case the *leaves* of the completely integrable distribution $\mathcal{S}$, and to prove that there is a leaf through every point, which leads us to call such a distribution a *(general) foliation.* Of course, now, not all the leaves have the same dimension, as it was for *regular foliations.* For instance, in Example 2.2, the leaf through the point $(x_0, y_0 \leq 0)$ is 1-dimensional (a line parallel to the x-axis), while that through $(x_0, y_0 > 0)$ is 2-dimensional (the upper half-plane). (A reader who feels uncomfortable with these notions can find them, for instance, in [Mt] and, for the general case, in [LM].)

In the regular case, the classical Frobenius theorem yields a necessary and sufficient condition of complete integrability. But, as we see from the following example due to Sussmann [Su], this theorem may not hold in the general case.

2.4. Example. $M = \mathbb{R}^2$, $\mathcal{S}_{(x,y)}(M) = \operatorname{span}\{\frac{\partial}{\partial x}, \varphi(x)\frac{\partial}{\partial y}\}$, with the same function φ as in Example 2.2. Here, the Frobenius involutivity condition holds since we have

$$\left[\frac{\partial}{\partial x}, \varphi(x)\frac{\partial}{\partial y}\right] = \begin{cases} 0 & \text{for } x \leq 0, \\ \frac{\partial \ln \varphi}{\partial x}\varphi(x)\frac{\partial}{\partial y} & \text{for } x > 0. \end{cases} \tag{2.2}$$

But, there are no leaves through the points of the y-axis.

Sussmann [Su] and Stefan [St] have found another simple condition which ensures complete integrability in all cases.

2.5. Definition. A differentiable general distribution $\mathcal{S}(M)$ is said to be χ_0-*invariant* (or just *invariant* if no confusion is possible) if there is a set $\chi_0\mathcal{S}(M)$ of vector fields on M such that $\forall x \in M$, the values of the fields of χ_0 at x span $\mathcal{S}_x(M)$, and $\forall X \in \chi_0$, $\forall t \in \mathbb{R}$, and $\forall x \in M$ such that $(\exp tX)(x)$ is defined, one has

$$(\exp tX)_*[\mathcal{S}_x(M)] = \mathcal{S}_{(\exp tX)(x)}(M). \tag{2.3}$$

For instance, this condition is easily seen to hold in Example 2.2 because the orbits of the defining vector fields of $\mathcal{S}(M)$ stay in the leaves. (The orbits of $\varphi(y)(\partial/\partial y)$ through $(x_0, y_0 \leq 0)$ are points, and those through $(x_0, y_0 > 0)$ are vertical half-lines.) But, in Example 2.4, condition (2.3) is obviously violated for translations parallel to the x-axis and which cross the y-axis.

Now, the basic result in this context is

2.6. Theorem. (Sussmann-Stefan-Frobenius). *A general differentiable distribution $\mathcal{S}(M)$ is completely integrable iff it is an invariant distribution.*

Proof. If $\mathcal{S}(M)$ is completely integrable, we may take χ_0 = the space of all the C^∞ vector fields of $\mathcal{S}(M)$. Condition (2.3) follows because the orbits of $X \in \chi_0$ are contained in the leaves.

Now, let us prove the converse. Take $x_0 \in M$, where $\rho(x_0) = r$, and take r vector fields $X_1, \ldots, X_r \in \chi_0$ such that their values at x_0 yield a basis of $\mathcal{S}_{x_0}(M)$. Then, it is clear that there exists a small enough open neighbourhood W of $0 \in \mathbb{R}^r$ such that the mapping

$$\exp(t_r X_r) \circ \ldots \circ \exp(t_1 X_1)(x_0) : W \to M \tag{2.4}$$

is well defined $\forall (t_1, \ldots, t_r) \in W$. This mapping is obviously differentiable, and sends the tangent basis $\{\frac{\partial}{\partial t_i}|_0\}$ of $T_0\mathbb{R}^r$ to $(X_1(x_0), \ldots, X_r(x_0))$, therefore, it is an immersion on some $W \subseteq \mathbb{R}^r$, and yields an r-dimensional submanifold of M through x_0. The tangent spaces of this submanifold are given by the actions of the flows of (2.4) on $\mathcal{S}_{x_0}(M)$, and (2.3) ensures that these tangent spaces must be $\mathcal{S}_x(M)$. Q.e.d.

2.7. Remark. A non-differentiable distribution may be *completely integrable* without being invariant in the sense of Definition 2.5, as shown by the following example due to Dazord [Dz1]: $M = \mathbb{R}^2$ with the following leaves: the individual points not on the x-axis, and the x-axis as a whole. The tangent distribution of these leaves has no nonzero differentiable vector field.

2.2 Involutivity and integrability

At this point, let us mention that the classical notion of *involutivity* can be extended to general distributions as follows

2.8. Definition. A (general) distribution $\mathcal{S}(M)$ is said to be *involutive* if there exists a Lie subalgebra $\chi_0[\mathcal{S}(M)]$ of $\chi(M) = \mathcal{V}^1(M)$ with the Lie bracket, such that the span of the values at x of the fields of $\chi_0[\mathcal{S}(M)]$ equals $\mathcal{S}_x M$, $\forall x \in M$.

Since the bracket of vector fields tangent to a submanifold N is also tangent to N, we see that a completely integrable distribution is involutive by means of χ_0= the space of all the C^∞ vector fields of $\mathcal{S}(M)$. But, of course, the converse can be proven only in the regular case (Frobenius). On the other hand, if $\chi_0[\mathcal{S}(M)]$ is a finite dimensional Lie algebra, $\mathcal{S}(M)$ is completely integrable, even if non-regular, since the leaves will be the orbits of a corresponding Lie group action on M [Pl].

The following Viflyantsev-Frobenius theorem [Vf] gives an interesting condition for the complete integrability of a general differentiable distribution.

2.9. Theorem. (Viflyantsev-Frobenius). *The differentiable distribution $\mathcal{S}(M)$ is completely integrable iff it satisfies the following two conditions:*

a) *$\mathcal{S}(M)$ is involutive;*

b) *$\forall x_0 \in M, \forall v \in \mathcal{S}_{x_0}(M)$, there is a path $x(t), x(0) = x_0$, which is tangent to v and such that $\rho(x(t))$ = const.*

Proof. Complete integrability obviously implies a) and b). Indeed, involutivity was already justified, and, then, the vector v of b) is tangent to the leaf through x_0, which has a constant dimension.

In order to obtain the converse result, we shall prove first that hypothesis b) ensures the fact that whenever X is a differentiable vector field in $\mathcal{S}(M)$ the integral curve $x(t)$ of X through $x_0 \in M$ also satisfies the condition $\rho(x(t)) =$ const.

This obviously happens if $X(x_0) = 0$, so we take $X(x_0) \neq 0$, and we also take an open neighbourhood U of x_0 which has local coordinates $(x^1, \ldots, x^n)$ with $x^i(x_0) = 0$, $\frac{\partial}{\partial x^1} = X$, and $\forall y \in U$, $\rho(y) \geq \rho(x_0)$. Then we may think of U as of a neighbourhood of 0 in $\mathbb{R}^n$.

Now, let us take an arbitrary point $p(x_0^1, 0, \ldots 0)$, and an open cubical neighbourhood $V_p \subseteq U$, centered at p and of width 2ϵ ($\epsilon > 0$). It is enough to assume that $x_0^1 > 0$. The idea is to prove that $\exists z \in V_p$ with $\rho(z) = \rho(x_0)$, since, then, p will be a limit of such points, hence $\rho(p) \leq \rho(x_0)$ i.e., $\rho(p) = \rho(x_0)$ because of the properties of U.

Denote $\delta = \epsilon/(2x_0^1)$. Take $\tilde{x}(t)$ to be a path which satisfies condition b) for $v = X(x_0)$. Then, $\exists z_1 = \tilde{x}(t_1)$ such that $x^1(z_1) < \epsilon$, and the Euclidean distance

$$d(\tilde{x}(\tau), \mathrm{pr}_{x^1}\tilde{x}(\tau)) < \delta x^1(\tilde{x}(\tau)), \tag{$*$}$$

where $0 \leq \tau \leq t_1$, and pr_{x^1} denotes the projection on the x^1-axis. (This is true because a curve and its tangent line have a *contact of order* 1.) Now, we can repeat this procedure for a curve $\tilde{\tilde{x}}$ which satisfies *b*) for $X(z_1)$, and get $z_2 = \tilde{\tilde{x}}(t_2)$ such that one has $x^1(z_2) \geq x^1(z_1)$, $x^1(z_2) - x^1(z_1) < \epsilon$, and $\forall \tau$, $0 \leq \tau \leq t_2$,

$$d\left(\tilde{\tilde{x}}(\tau),\ \mathrm{pr}_{\text{onto } || \text{ to } x^1 \text{ through } z_1}\tilde{\tilde{x}}(\tau)\right) \leq \delta\left(x^1(\tilde{\tilde{x}}(\tau)) - x^1(z_1)\right). \tag{$**$}$$

And so on, we build a sequence of points $z_1, z_2, \ldots$, and a continuous piecewise differentiable path $u(\tau)$ which consists of $\tilde{x}(\tau)$ between x_0 and z_1, $\tilde{\tilde{x}}(\tau)$ between z_1, z_2, etc. such that $\rho(u(\tau)) = \rho(x_0)$, and the steps are small enough not to get too far away of the x^1-axis, in the sense that we have for all τ

$$d\left(u(\tau), \mathrm{pr}_{x^1}u(\tau)\right) \leq \delta x^1(u(\tau)). \tag{$+$}$$

Indeed, along $\tilde{x}(\tau)$ (+) is just ($*$), along $\tilde{\tilde{x}}(\tau)$ we obtain (+) by

$$d\left(\tilde{\tilde{x}}(\tau),\ \mathrm{pr}_{x^1}\tilde{\tilde{x}}(\tau)\right) \leq d\left(\tilde{\tilde{x}}(\tau), \mathrm{pr}_{(\text{on } || \ x^1 \text{ through } z_1)}\tilde{\tilde{x}}(\tau)\right) +$$

$$+ d(z_1, \mathrm{pr}_{x^1}z_1) \leq \delta\left(x^1(\tilde{\tilde{x}}(\tau)) - x^1(z_1)\right) + \delta x^1(z_1) = \delta x^1(\tilde{\tilde{x}}(\tau)),$$

etc.

Now, $x^1(z_k)$ is a non-decreasing sequence of positive numbers, and if all these numbers were $< x_0^1$, $\alpha = \lim_{k\to\infty} x^1(z_k) = \sup\{x^1(z_k)\} < x_0^1$ would exist, and with it a limit point z of the sequence z_k on $u(\tau)$ (because, obviously, x^1 may serve as a parameter of $u(\tau)$ since its tangents at z_k are parallel to the x^1-axis). But then, we could go one step further in the previous construction starting from z. This would further extend $u(\tau)$, and provide us with points whose existence contradicts the definition of α.

Hence $u(\tau)$ gets to a point z_k such that $x^1(z_k) \geq x_0^1$, while $x^1(z_{k-1}) < x^1$. Accordingly,

$$d(z_{k-1}, pr_{x^1} z_{k-1}) \leq \delta x^1(z_{k-1}) < \delta x_0^1 = \frac{\epsilon}{2x_0^1} x_0^1 = \frac{\epsilon}{2},$$

and, on the other hand

$$x_0^1 - x^1(z_{k-1}) < x^1(z_k) - x^1(z_{k-1}) < \epsilon.$$

These inequalities yield $z_{k-1} \in V_p$ and we are done with the proof of the fact that ρ is constant along any integral path of a vector field in $\mathcal{S}(M)$.

Now, we can finish the proof by using the same induction procedure as in the proof of the classical Frobenius theorem given in [Sg] as one does in [Vf], or we can prove that $\mathcal{S}(M)$ is invariant, and apply Theorem 2.6.

We shall take the second course, and prove the following

2.9′. Lemma. *Let $\mathcal{S}(M)$ be a differentiable general distribution of M^n, and let X be a vector field such that i) $\forall x_0 \in M$, $\dim \mathcal{S}(M)/_{\exp tX(x_0)} =$ const.; ii) there exists a $C^\infty(M)$-module $\chi_0(\mathcal{S}(M))$ of vector fields of $\mathcal{S}(M)$ such that, $\forall x_0 \in M$, $\mathcal{S}_{x_0}(M)$ is the span of the values of the fields of $\chi_0(\mathcal{S}(M))$ at x_0, and $\forall Y \in \chi_0(\mathcal{S}(M))$, $[X, Y] \in \mathcal{S}(M)$. Then, the flow of X leaves $\mathcal{S}(M)$ invariant, i.e., $\forall v \in \mathcal{S}_{x_0}(M)$, $(\exp tX)_*(v) \in \mathcal{S}(M)$.*

Proof. Clearly, it suffices to prove that $\forall Y \in \chi_0(\mathcal{S}(M))$, the local fields $Y_t = (\exp tX)_* Y$, defined for t small enough, and on some neighborhood U of $x_0 \in M$, belong to $\mathcal{S}(M)$.

Let us assume that $\dim \mathcal{S}_{x_0}(M) = \rho_0$, and $\dim \mathcal{S}(M)/_U \geq \rho_0$, and let X_a $(a = 1, \ldots, \rho_0)$ be vector fields of $\chi_0(\mathcal{S}(M))$ that are independent over U. Particularly, because of *i)*, $\{X_a(\exp tX(x_0))\}$ is a basis of the planes of $\mathcal{S}(M)$ along the path $\exp tX(x_0)$. Furthermore, let ω_u $(u = 1, \ldots, n - \rho_0)$ be independent 1-forms on U which define a basis of the annihilator of span $\{X_a\}$ hence, $\omega_u = 0$ are independent linear equations of $\mathcal{S}(M)$ along $\exp tX(x_0)$.

Now, consider the functions

$$\varphi_u(t) = [\omega_u(Y_t)](\exp tX(x_0)) = [(\exp tX)^* \omega_{u,\exp tX(x_0)}](Y_{x_0}). \tag{2.5}$$

In view of the usual definition of the Lie derivative of a differential form (e.g., [Mt]), we get

$$\begin{aligned}&\frac{d\varphi_u(t)}{dt} = \\ &= \lim_{s\to 0}\frac{(\exp(t+s)X)^*\omega_{u,\exp(t+s)X(x_0)}(Y_{x_0}) - (\exp tX)^*\omega_{u,\exp tX(x_0)}(Y_{x_0})}{s} = \\ &= [(\exp tX)^*(L_X\omega_u)](Y_{x_0}) = [(L_X\omega_u)(Y_t)](\exp tX(x_0)).\end{aligned} \tag{2.6}$$

But, the general formula

$$(L_X\omega_u)(Z) = X(\omega_u(Z)) - \omega_u([X,Z]), \tag{2.7}$$

where Z is any vector field of M, allows us to conclude that hypothesis ii) implies

$$(L_X\omega_u)_{\exp tX(x_0)} = 0 \qquad (\text{modulo } \omega_u = 0). \tag{2.8}$$

Indeed, if $\omega_u(Z) = 0$, $Z = \zeta^a X_a$ on U, and, since $\chi_0(\mathcal{S}(M))$ is a $C^\infty(M)$-module and taking a smaller U, we may see $Z/_U$ as the restriction of a vector field of $\chi_0(\mathcal{S}(M))$ to U. Then, by ii), $[X,Y] \in \mathcal{S}(M)$, and we shall have $\omega_u([X,Y]) = 0$ along $\exp tX(x_0)$, thereby justifying (2.8).

Accordingly, we must have

$$(L_X\omega_u)_{\exp tX(x_0)} = \alpha_u^v(t)\omega_{v,\exp tX(x_0)} \tag{2.9}$$

for some coefficients α_u^v $(u,v = 1,\dots,n-\rho_0)$, and (2.6) becomes

$$\frac{d\varphi_u(t)}{dt} = \alpha_u^v(t)\varphi_v(t). \tag{2.10}$$

Therefore, $\varphi_u(t)$ satisfy a system of linear homogeneous ordinary differential equations and, since it is obvious that $\varphi_u(0) = 0$, we must have $\varphi_u(t) \equiv 0$. By (2.5), this means $Y_t \in \mathcal{S}(M)/_{\exp tX(x_0)}$. Q.e.d.

We should notice that the result of Lemma 2.9′ may not be true if hypothesis i) is dropped. Indeed, in the previous Example 2.4, a translation along the x-axis from $x_1 > 0$ to $x_2 < 0$ does not preserve $\mathcal{S}(M)$ in spite of the fact that hypothesis ii) is satisfied for $X = \partial/\partial x$.

Now, if we come back to the proof of Theorem 2.9, and if we use the previous lemma for the $\chi_0(\mathcal{S}(M))$ of the definition of the involutivity of $\mathcal{S}(M)$ (Definition 2.8), and with $X \in \chi_0(\mathcal{S}(M))$, it becomes clear that $\mathcal{S}(M)$ is an invariant distribution, and Theorem 2.9 is proven. Q.e.d.

In practice, what is more useful is the following simpler result (e.g., [KSM]) whose proof is a combination of Theorem 2.6, and Lemma 2.9′ only:

2.9″. Theorem. (The generalized Frobenius theorem). *A general differentiable distribution $\mathcal{S}(M)$ is completely integrable iff there exists a Lie subalgebra χ_0 of the Lie algebra of all the C^∞ vector fields of M such that i) $\forall x_0 \in M$, $\mathcal{S}_{x_0}(M)$* $=$ span $\{X(x_0)/X \in \chi_0\}$; *ii) $\forall x_0 \in M$, $\forall X \in \chi_0$,* $\dim \mathcal{S}(M)/_{\mathrm{expt}X(x_0)} =$ const.

A further clarification of the local structure of a general foliation is contained in

2.10. Theorem. (Dazord [Dz1]). *Let $\mathcal{S}(M)$ be a completely integrable distribution of M, and $x_0 \in M$ with $\rho(x_0) =$* r. *Then, x_0 has an open neighbourhood U with local coordinates v^i such that $\forall x \in U$, $\frac{\partial}{\partial v^1}|_x, \ldots, \frac{\partial}{\partial v^r}|_x \in \mathcal{S}_x(M)$. More exactly, there is a chart $\psi : U \to W \times V$, where the factors are open neighborhoods of* 0 *in* $\mathbb{R}^r$ and $\mathbb{R}^{n-r}$, *respectively, such that $\psi_*(\mathcal{S}(M)) = \mathbb{R}^r \times \Delta_V$, where Δ_V is a general completely integrable distribution of V with $\Delta_V(0) = 0$.*

Proof. If S is a local r-dimensional integral submanifold of $\mathcal{S}(M)$ through x_0, we have a neighborhood U of x_0 endowed with local coordinates u^i ($i = 1, \ldots, n = \dim M$) such that $u^i(x_0) = 0$ and S has the local equations $u^{r+1} = 0, \ldots, u^n = 0$. Let B be the submanifold of U given by $u^1 = 0, \ldots, u^r = 0$, which is transversal to $\mathcal{S}(M)$ at x_0. Then, for a neighborhood W of $0 \in \mathbb{R}^r$, we define $\varphi : W \times B \to U$ by

$$\varphi(v^1, \ldots, v^r, b) = \exp(v^1 X_1) \circ \ldots \circ \exp(v^r X_r)(b), \tag{2.11}$$

where $(v^1, \ldots, v^r) \in W$, and $b \in$ B. Clearly $\varphi_*(0, x_0)$ is non-singular. Hence, after restricting the neighborhoods if necessary, φ is a diffeomorphism that yields the local chart $\psi = \varphi^{-1}$, if B is identified with a neighborhood V as needed in the theorem. Correspondingly, $\psi_*(\mathcal{S}(M))$ will be a completely integrable distribution on $W \times V$.

The vector $\frac{\partial}{\partial v^a}$ is the ψ_*-image of the tangent vector of the line $v^a =$ variable, the other coordinates in (2.11) remaining constant. Since $\mathcal{S}(M)$ is invariant in the sense of Definition 2.5 (because it is completely integrable), (2.11) shows that these tangent vectors belong to $\mathcal{S}(M)$. Therefore $\frac{\partial}{\partial v^a}$ belong to $\psi_*(\mathcal{S}(M))$, and this latter distribution must be invariant under v^a-translations. Accordingly, $\psi_*(\mathcal{S}(M))$, that decomposes as indicated by the theorem at the points of $\{0\} \times V$ in a natural way, will be obtained by such translations, and it will have the required form everywhere. Q.e.d.

2.11. Remark. The germ of the manifold (V, Δ_V) is called the *transversal germ* of $\mathcal{S}(M)$ at x_0. One can prove that it is well defined up to a diffeomorphism [Dz1]. See, later on, in Theorem 2.16 a proof of a similar but more complicated property in the case of Poisson manifolds. In the case of a regular foliation $\Delta_V = 0$.

2.3 The case of Poisson manifolds

Now, let us come back again to the Poisson manifold (M, w) and, first, notice the

2.12. Theorem. *The characteristic distribution $\mathcal{S}(M)$ of the Poisson manifold (M, w) is completely integrable, and the Poisson structure induces symplectic structures on the leaves of $\mathcal{S}(M)$.*

Proof. The Jacobi identity, together with formula (1.1) imply

$$X_{\{f,g\}} = [X_f, X_g], \tag{2.12}$$

hence, $\chi_w(M)$, defined at the beginning of this chapter is a Lie algebra and $\mathcal{S}(M)$ is an involutive distribution. On the other hand, and for the same reasons (1.1) and (0.4), we have

$$(L_{X_h} w)(df, dg) = X_h(\{f, g\}) - \{X_h f, g\} - \{f, X_h g\} = 0, \tag{2.13}$$

which shows that a Hamiltonian field is an *infinitesimal Poisson automorphism*. Hence, by a classical result (e.g., [Mt], p. 148 transposed to multivectors) $\exp(tX_h)$ are Poisson automorphisms, and the complete integrability of $\mathcal{S}(M)$ follows either from Theorem 2.6 or from Theorem 2.9.

In this case it is usual to refer to $\dim \mathcal{S}_x(M)$ as the *rank* of the Poisson structure at x, since, obviously, this is exactly the rank of $\#_w$. Clearly, this rank must be even, and the leaves $S(x)$ of $\mathcal{S}(M)$ $(x \in M)$ are even dimensional submanifolds of M. If $\tilde{f}, \tilde{g} \in C^\infty(S(x))$, we can extend them (locally) to functions $f, g \in C^\infty(M)$, and define

$$\{\tilde{f}, \tilde{g}\}(x) = \{f, g\}(x) = X_f(x)g.$$

The result depends on $\tilde{g}$ only since it computes along the integral curve of X_f through x, and this curve belongs to $S(x)$. For a similar reason, the result will depend on $\tilde{f}$, and not on f. Clearly, this yields a Poisson structure on $S(x)$ which makes $S(x)$ into a Poisson submanifold of M. Clearly, we have $\mathcal{S}_x(S(x)) = T_x S(x)$. Hence, the Poisson structure of $\mathcal{S}(x)$ is *nondegenerate*, and it is the contravariant expression of a symplectic structure. Q.e.d.

The previous theorem suggests

2.13. Definition. The leaves of $\mathcal{S}(M)$ are called the *symplectic leaves* of the Poisson manifold M, and $\mathcal{S}(M)$ is also said to be the *symplectic foliation* of M.

2.14. Theorem. *Let M^n be a differentiable manifold, and $\mathcal{S}(M)$ a general foliation such that a) every leaf S of $\mathcal{S}(M)$ is endowed with a symplectic structure ω_S, b) if $f \in C^\infty(M)$, the vector field X_f defined by $X_f(x) =$ the Hamiltonian vector field of $f/_{S(x)}$ on $(S(x), \omega_{S(x)})$ at x is a differentiable vector field on M. Then M has a unique Poisson structure whose symplectic foliation is $\mathcal{S}(M)$.*

Proof. Use the vector fields X_f in order to define $\{f, g\} = X_f g$. Q.e.d.

In other words, a Poisson structure can be defined by its symplectic foliation instead of the Poisson bivector.

2.15. Definition. If the symplectic foliation of (M, w) is regular i.e., if rank $w =$ const., the Poisson manifold M is called *regular*.

Of course, the symplectic foliation of (M, w) has adapted coordinates in the sense of Theorem 2.10, but a more careful analysis shows that such coordinates can be chosen in a way that "canonizes" also the Poisson bivector. More exactly, the local structure of a Poisson manifold M at $x_0 \in M$ is described by the following *splitting theorem* due to A. Weinstein [We3]

2.16. Theorem. *Let (M^n, w) be a Poisson manifold, $x_0 \in M$, and $\rho(x_0) = 2h$. Then x_0 has an open neighbourhood U in M such that $(U, w/_U)$ is Poisson equivalent by a mapping φ to a product $S \times N$, where S is a $2h$-dimensional symplectic manifold, and N is a Poisson manifold of rank 0 at $\varphi(x_0)$. Moreover, the factors S and N are unique up to local equivalences.*

Proof. Essentially, the existence follows like Darboux' theorem. It is clear for $h = 0$. If $h \neq 0$, there is a local function p_1 with $X_{p_1} \neq 0$ at x_0 and on U, and we can represent this field as $X_{p_1} = \partial/\partial q^1$. In other words, we find two functions p_1, q^1 on U such that $\{p_1, q^1\} = 1$.

Implicitly, X_{p_1}, X_{q^1} are linearly independent and $[X_{p_1}, X_{q^1}] = X_{\{p_1,q^1\}} = 0$. Hence, we can make a new coordinate transformation, to $x^1, \ldots, x^n$, such that $X_{p_1} = \partial/\partial x^1$, $X_{q^1} = \partial/\partial x^2$, and, therefore,

$$\{p_1, x^\lambda\} = \{q^1, x^\lambda\} = 0, \quad \lambda = 3, \ldots, n. \tag{2.14}$$

But, if we look at the Jacobian of the functions $(p_1(x^i), q^1(x^i), x^3, \ldots, x^n)$ we see that it is nonzero. Hence, we have local coordinates $(p_1, q^1, x^3, \ldots, x^n)$ on U, which satisfy (2.14), and also $\{p_1, q^1\} = 1$. These relations allow us to compute the local coordinates of X_{p_1}, X_{q^1}, and we get $X_{p_1} = \frac{\partial}{\partial q^1}, X_{q^1} = -\frac{\partial}{\partial p_1}$. Therefore, we also have

$$\frac{\partial}{\partial p_1}\{x_i, x_j\} = -\{q_1, \{x_i, x_j\}\} = -\{q_1, x_i\}x_j - \{q_1, x_j\}x_i = 0$$

for all $i, j = 3, \ldots, n$, and similarly $\frac{\partial}{\partial q^1}\{x_i, x_j\} = 0$. These facts show that there is a well defined induced Poisson structure in the variables $x^3, \ldots, x^n$, and we have a decomposition as the one asked by the Theorem but, with a 2-dimensional S, and a Poisson structure of rank $2h - 2$ on N. If this Poisson structure is treated in a similar way, then, after a total of h steps, we are done.

The S-factor of the decomposition is clearly the local leaf of $\mathcal{S}(M)$ through x_0, and its germ is unique. (Remember also that any two germs of $2h$-dimensional manifolds are symplectomorphic.)

Now, assume that there are two factors N and N' for the splitting theorem at x_0, and let (p_a, q^a, x^σ) $(a = 1, \ldots, h;\ \sigma = 2h+1, \ldots n)$ be the coordinates on $S \times N$ obtained in the existence proof. Since we are in a small neighbourhood of x_0, N and N' can be connected by a 1-parameter family of manifolds N_t, $0 \le t \le 1$, transversal to S at x_0, and defined by equations of the form

$$p_a - P_a(x^\sigma, t) = 0, \quad q^a - Q^a(x^\sigma, t) = 0, \tag{2.15}$$

which will give N for $t = 0$, and N' for $t = 1$.

In order to establish the equivalence of N and N', we shall look for a "time-dependent" Hamiltonian vector field X_{H_t} of a function $H_t(p_a, q^a, x^\sigma)$ with respect to the Poisson structure w of M such that $\exp(tX_{H_t})$ maps N_0 to N_t $(0 \le t \le 1)$. In view of the values of the Poisson brackets of p_a, q^a, x^σ as established in the existence part, we shall have

$$X_{H_t} = \frac{\partial H_t}{\partial p_a}\frac{\partial}{\partial q^a} - \frac{\partial H_t}{\partial q^a}\frac{\partial}{\partial p_a} + w^{\sigma\tau}(x)\frac{\partial H_t}{\partial x^\sigma}\frac{\partial}{\partial x^\tau},$$

and we want the vector field $X_{H_t} + \frac{\partial}{\partial t}$ to be tangent to (2.15). This happens if

$$\begin{aligned} \frac{\partial H_t}{\partial q_a} &= -\frac{\partial P_a}{\partial t} + w^{\sigma\tau}(x)\frac{\partial P_a}{\partial x^\sigma}\frac{\partial H_t}{\partial x^\tau}, \\ \frac{\partial H_t}{\partial p_a} &= \frac{\partial Q_a}{\partial t} - w^{\sigma\tau}(x)\frac{\partial Q^a}{\partial x^\sigma}\frac{\partial H_t}{\partial x^\tau}, \end{aligned} \tag{2.16}$$

modulo (2.15) i.e., along N_t.

Moreover, let us also ask that $H_t = 0$ along N_t. As shown by (2.15), the tangent space of N_t is spanned by the vector fields

$$Y_\sigma = \frac{\partial}{\partial x^\sigma} + \frac{\partial P_a}{\partial x^\sigma}\frac{\partial}{\partial p_a} + \frac{\partial Q^a}{\partial x^\sigma}\frac{\partial}{\partial q^a}.$$

Hence, $H_t = 0$ on N_t implies $Y_\sigma H_t = 0$ on N_t, and we can use this condition to compute the derivatives $\partial H_t/\partial x^\sigma$ along N_t, then insert the corresponding result in (2.16). This gives us an equivalent system of conditions for H_t namely, that along N_t one must have

$$\left(\delta_a^b + w^{\sigma\tau}(x)\frac{\partial P_a}{\partial x^\sigma}\frac{\partial Q^b}{\partial x^\tau}\right)\frac{\partial H_t}{\partial q^b} + w^{\sigma\tau}(x)\frac{\partial P_a}{\partial x^\sigma}\frac{\partial P_b}{\partial x^\tau}\frac{\partial H_t}{\partial p_b} = -\frac{\partial P_a}{\partial t},$$

$$w^{\tau\sigma}\frac{\partial Q^a}{\partial x^\sigma}\frac{\partial Q^b}{\partial x^\tau}\frac{\partial H_t}{\partial q^b} + \left(\delta_a^b - w^{\sigma\tau}(x)\frac{\partial Q^a}{\partial x^\sigma}\frac{\partial P_b}{\partial x^\tau}\right)\frac{\partial H_t}{\partial p_b} = \frac{\partial Q_a}{\partial t},$$

which is a linear system of equations with respect to $\partial H_t/\partial q^b$ and $\partial H_t/\partial p_b$. The matrix of this system is nondegenerate at x_0 because of $w^{\sigma\tau}(x_0) = 0$. Therefore, this matrix is also nondegenerate on U (restricted if necessary), and we obtain some well defined prescribed values

$$\frac{\partial H_t}{\partial q^a} = A_a(x^\sigma, t), \quad \frac{\partial H_t}{\partial p_a} = B_a(x^\sigma, t)$$

along N_t.

Accordingly, the function

$$H_t = (q^a - Q^a(x,t))A_a(x,t) + (p_a - P_a(x,t))B_a(x,t),$$

for instance, satisfies the required conditions.

Finally, since X_{H_1} is a Hamiltonian field, its flow defines Poisson automorphisms, and we discovered one such automorphism mapping N to N'. Of course, this automorphism preserves the leaves of (M, w) and their symplectic structures, hence it will map the intersection of these leaves with N into those with N'. But these intersections are exactly the symplectic leaves of the Poisson structures given by the splitting theorem on N and N', respectively. Hence these two Poisson structures are equivalent. Q.e.d.

2.17. Definition. The germ of the Poisson manifold N, defined up to equivalence, is called the *germ of the transverse Poisson structure* of the symplectic leaf S, or of (M, w), at x_0.

2.18. Definition. The adapted coordinates (p_a, q^a, x^σ) above are called *canonical* for the Poisson manifold M at x_0, and satisfy the following relations

$$\begin{aligned} &\{p_a, p_b\} = 0; \quad \{p_a, q^b\} = \delta^b_a; \quad \{q^a, q^b\} = 0, \\ &\{p_a, x^\sigma\} = 0; \quad \{q_a, x^\sigma\} = 0. \end{aligned} \tag{2.17}$$

2.19. Remark. If x_0 is a regular point of $(M, \mathcal{S}(M))$, the Poisson structure of the transversal germ is trivial (the distribution $\mathcal{S}(N)$ is zero), hence, the canonical coordinates also satisfy the condition

$$\{x^\sigma, x^\tau\} = 0. \tag{2.18}$$

Notice again that the existence of canonical coordinates was proven in Theorem 2.16.

Now, we can fulfill a promise made in Chapter 1 and prove

2.20. Theorem. *A Poisson manifold (M, w) has Poisson connections iff it is regular.*

Proof. That the existence of a Poisson connection (i.e., a torsionless connection ∇ with $\nabla w = 0$) implies regularity is obvious. Conversely, it is classical (e.g., [Va6]) that a symplectic manifold has a family of symplectic connections (i.e., torsionless connections that make the symplectic structure parallel), and these are also Poisson connections. For a regular Poisson manifold (M, w) we take a neighbourhood U_x, $x \in M$, where one has the canonical coordinates of Remark 2.19. Then, $U_x \approx S \times N$ where S is a symplectic manifold, and a pair (Γ_1, Γ_2), where Γ_1 is a symplectic connection of S, and Γ_2 is a torsionless connection of N, yields a Poisson connection of U_x. Now, if M is covered by such neighborhoods U_x, and if we define a connection on M by gluing up the Poisson connections of the U_x by means of a subordinated partition of unity, we get a global Poisson connection on M. Q.e.d.

Finally, let us mention that the existence of the symplectic foliation suggests another interesting notion, namely, the objects such as functions, tensor fields, etc., which are constant along the leaves of the symplectic foliation $\mathcal{S}(M)$ will be said to be *projectable.* This name comes from the fact that they can be projected on the space of leaves. In particular, a projectable function $f \in C^\infty(M)$ is one such that $\forall g \in C^\infty(M)$ one has $X_g f = 0$ i.e., f belongs to the center of the Poisson algebra $C^\infty(M)$. Such functions are also called *Casimir functions* [We3], because of an example that involves a Lie algebra (see Chapter 3, later on).

Many of the results of the present chapter can be extended, correspondingly, to Jacobi manifolds (M, A, E) (see the end of Chapter 1). For such a manifold, Hamiltonian vector fields X_f can be defined by

$$X_f = \#_A(df) + fE, \tag{2.19}$$

and one has (e.g., [GL]) $[X_f, X_g] = X_{\{f,g\}}$. Hence, if the characteristic distribution $\mathcal{S}(M)$ is defined like in the Poisson case by $\mathcal{S}_x(M) = \text{span}\ \{X_f\}_x$ $(x \in M)$, then Theorem 2.9 can be used, and it tells us that the (general) distribution $\mathcal{S}(M)$ is integrable. But, of course, we do not necessarily get a symplectic structure on its leaves, even if these are even-dimensional. Namely, the even-dimensional leaves of $\mathcal{S}(M)$ are either symplectic or locally conformal symplectic manifolds, and the odd-dimensional leaves are contact manifolds [GL], [DLM].

Furthermore, a splitting theorem like Theorem 2.16, and the notion of a transversal structure can also be obtained in the Jacobi case, where, of course, one must distinguish between two different classes of points $x \in M$, those that belong to an even-dimensional leaf of $\mathcal{S}(M)$, and those that belong to an odd-dimensional leaf. These results were developed by Dazord, Lichnerowicz and Marle [DLM].

3 Examples of Poisson Manifolds

3.1 Structures on $\mathbb{R}^n$. Lie-Poisson structures

So far, we have seen that the symplectic manifolds are Poisson manifolds, and that harmonic 2-forms define Poisson structures on compact Riemannian symmetric spaces. In this chapter, we shall discuss in a more systematic way a number of geometrically interesting Poisson structures.

a) A simple way to obtain Poisson structures on spaces $\mathbb{R}^n$ is by inserting coefficients in the canonical symplectic structure. For instance, on $\mathbb{R}^2$, the bivector

$$w = h(x,y)\frac{\partial}{\partial x} \wedge \frac{\partial}{\partial y} \tag{3.1}$$

obviously defines a Poisson structure. It has rank 0 at the points where $h = 0$, and rank 2 elsewhere, therefore, the symplectic leaves are the points where $h = 0$, and the connected components of the open subset $h \neq 0$ of $\mathbb{R}^2$. This structure generalizes to $\mathbb{R}^{2n+p} = \{(x^k, y^k, z^h)\}$ $(k = 1, \ldots, n;\ h = 1, \ldots, p)$ by putting

$$w = \sum_{k=1}^{n} h_k(x^k, y^k)\frac{\partial}{\partial x^k} \wedge \frac{\partial}{\partial y^k}. \tag{3.2}$$

Particularly, the bivector

$$w = \frac{\partial}{\partial x^1} \wedge \frac{\partial}{\partial x^2} + \varphi(x^4)\frac{\partial}{\partial x^3} \wedge \frac{\partial}{\partial x^4}, \tag{3.3}$$

where $\varphi(x^4) = 0$ for $x^4 \leq 0$, and $\varphi(x^4) > 0$ for $x^4 > 0$, defines a Poisson structure on $\mathbb{R}^4$, where the symplectic leaf through a point with $x^4 > 0$ is the upper half-space $x^4 > 0$, while the leaves through points with $x^4 < 0$ are the planes $x^3 = ct$, $x^4 = ct$. Non-trivial transversal germs appear only at points where $x^4 = 0$, and the Poisson structure of such a germ is $\varphi(x^4)\frac{\partial}{\partial x^3} \wedge \frac{\partial}{\partial x^4}$.

b) The most general Poisson bivector on $\mathbb{R}^n = \{(x_i)\}$ $(i = 1, \ldots, n)$ is, of course, of the form

$$w = \frac{1}{2}w_{ij}\frac{\partial}{\partial x_i} \wedge \frac{\partial}{\partial x_j}, \qquad [w, w] = 0. \tag{3.4}$$

Particularly, any constant coefficients w_{ij} will do, and we call these *constant Poisson structures* [Bh].

Hence, the following natural question is that of finding the *linear Poisson structures*, where, by definition

$$w_{ij} = c_{ij}^k x_k \qquad (c_{ij}^k = -c_{ji}^k). \tag{3.5}$$

In this case, (1.5) becomes

$$c_{jk}^h c_{hi}^p + c_{ki}^h c_{hj}^p + c_{ij}^h c_{hk}^p = 0. \tag{3.6}$$

In other words c_{ij}^k must be the structure constants of an n-dimensional Lie algebra.

Of course, the same result holds on any n-dimensional vector space V instead of $\mathbb{R}^n$, if (x_i) are seen as linear coordinates, and c_{ij}^k change in accordance with the change of the basis of V. The corresponding Lie algebra can be realized as the dual space V^* with the bracket

$$[f, g] = \{f, g\} \qquad (f, g \in V^*), \tag{3.7}$$

where $\{f, g\} \in V^*$ because of (3.5).

Therefore, conversely, if $\mathcal{G}$ is any finite dimensional Lie algebra, its dual space $\mathcal{G}^*$ has a natural Poisson structure, given by (3.5) or, equivalently, by

$$\{\varphi, \psi\}(\gamma) = \gamma([d_\gamma\varphi, d_\gamma\psi]), \tag{3.8}$$

where $\gamma \in \mathcal{G}^*$, and $\varphi, \psi \in C^\infty(\mathcal{G}^*)$, therefore, $d_\gamma\varphi$, $d_\gamma\psi \in \mathcal{G}$, since we can regard them as linear mappings from $T_\gamma\mathcal{G}^* = \mathcal{G}^*$ to $\mathbb{R}$. This is called the *Lie-Poisson structure* of $\mathcal{G}^*$ [We3], and one has

3.1. Proposition [We3]. *The symplectic leaves of the Lie-Poisson structure of $\mathcal{G}^*$ are the orbits of the coadjoint representation of any connected Lie group G whose Lie algebra is $\mathcal{G}$.*

Proof. We shall regard $\mathcal{G}^*$ as V, and $\mathcal{G}$ as V^*, and use the local coordinates x_k as in (3.5). Then, $\forall\varphi \in C^\infty(\mathcal{G}^*)$ we get the Hamiltonian field

$$X_\varphi = \sum_i \frac{\partial\varphi}{\partial x_i} X_{x_i},$$

and this shows that the characteristic distribution $\mathcal{S}$ of the Lie-Poisson structure is generated by the set χ_l of the Hamiltonian vector fields of the linear functions on $\mathcal{G}^*$, while formula (3.7) shows that χ_l is a Lie algebra, and $f \to X_f$ $(\forall f \in \mathcal{G} = V^*)$ is an epimorphism $\pi : \mathcal{G} \to \chi_l$ with the kernel $\mathcal{C}(\mathcal{G})$ = the center of $\mathcal{G}$.

Now, let us look at the group G of the proposition, and remember what the coadjoint representation is. We shall denote the elements of G by italics. $\forall g, h \in G$ one defines Adg $(h) = ghg^{-1}$, and the differential of the mapping Ad$g : G \to G$ at 0 is the *adjoint representation* Ad$g : \mathcal{G} \to \mathcal{G}$. The utilization of the same notation for the mapping and its differential is harmless in this case.

Furthermore the *coadjoint representation* is provided by the linear mappings $\mathrm{Coad}g : \mathcal{G}^* \to \mathcal{G}^*$, where

$$\mathrm{Coad}g(\gamma)(f) = \gamma(\mathrm{Ad}g^{-1}(f)) \qquad (\gamma \in \mathcal{G}^*, f \in \mathcal{G}), \tag{3.9}$$

or, equivalently, if we regard f as an element of V^*,

$$f \circ \mathrm{Coad}g = (\mathrm{Ad}g^{-1})(f) \in V^*. \tag{3.10}$$

Since it is classical that $\mathrm{Ad}g^{-1}$ is an isomorphism of $\mathcal{G}$, (3.10) and (3.7) show that $\mathrm{Coad}g$ is compatible (i.e., it satisfies (1.41)) with the Poisson bracket of linear functions on $\mathcal{G}^*$. But, $\forall\gamma \in \mathcal{G}^*$, $\{\varphi, \psi\}(\gamma) = \{d_\gamma\varphi, d_\gamma\psi\}(\gamma)$, where $\varphi, \psi \in C^\infty(\mathcal{G}^*)$, and the differentials are seen as linear mappings by the identification $T_\gamma\mathcal{G}^* = \mathcal{G}^*$. Hence a mapping $\mathcal{G}^* \to \mathcal{G}^*$ which is compatible with the brackets of the linear functions is just a Poisson mapping. Therefore, the coadjoint representation of G on $\mathcal{G}^*$ is by linear Poisson mappings, and symplectic leaves of $\mathcal{G}^*$, i.e., orbits of the Lie algebra χ_l are sent by $\mathrm{Coad}g$ into symplectic leaves.

The action of G on $\mathcal{G}^*$ by Coad, is not effective, but the corresponding action of $G/_{\mathrm{Center}(G)}$ is, and its *infinitesimal action*, which sends the elements of $\mathcal{G}/_{\mathrm{Center}(\mathcal{G})}$ into *fundamental vector fields* on $\mathcal{G}^*$, *injects* $\mathcal{G}/_{\mathrm{Center}(G)}$ into the Lie algebra of all the vector fields of $\mathcal{G}^*$. Because of the previous invariance property of the symplectic leaves, $\mathcal{G}/_{\mathrm{Center}(\mathcal{G})}$ is, in fact, injected in χ_l, and since their dimensions are the same, the infinitesimal action of $G/_{\mathrm{Center}(G)}$ is χ_l. Then, it is clear that the orbits of Coad G are exactly the symplectic leaves of $\mathcal{G}^*$.

As a matter of fact, the result is even more straightforward since we can see that $\forall f \in \mathcal{G} = V^*$ *the fundamental infinitesimal transformation (vector field)* associated to f by the coadjoint representation is just X_f. Indeed, if Y_f is the requested fundamental field, its value at $\gamma \in \mathcal{G}^*$ is

$$\begin{aligned} Y_f(\gamma) &= -\frac{d}{dt}\Big|_{t=0} (\mathrm{Coad}(\exp tf)(\gamma)) = \\ &= \frac{d}{dt}\Big|_{t=0} (\gamma \circ \mathrm{Ad}(\exp tf)) = \gamma([f, .]) \overset{(3.8)}{=} \{f, .\}_\gamma = X_f(\gamma), \end{aligned} \tag{3.11}$$

where $[\ ,\ ]$ is the Lie bracket in $\mathcal{G}$, and $\{\ ,\ \}$ is the Poisson bracket in $\mathcal{G}^*$.

The minus sign has been inserted in (3.11) as a matter of definition. Namely, following [LM], for instance, for an action of a Lie group G on M, we define the associated infinitesimal action of $X \in \mathcal{G}$ by the tangent vectors of the curves given by $\exp(-tX)$ acting on M. This is done in order to ensure that $[X, Y] \in \mathcal{G}$ has the usual Lie bracket as its image.

Hence, the *infinitesimal action* on $\mathcal{G}^*$ which is associated with Coad is exactly that of the Lie algebra χ_l, and the orbits of Coad are exactly those of χ_l. Q.e.d.

3.2. Remarks. 1) It is a Corollary of Proposition 3.1 that the orbits of any coadjoint action of a connected Lie group are homogeneous symplectic manifolds, a fact which is known as the Kirillov-Kostant-Souriau theorem [AM].

2) The name *Casimir functions* (see the end of Chapter 2) was first used in the case of the Lie-Poisson structures.

3) The study of the transverse Poisson structure at $\gamma \in \mathcal{G}^*$ is interesting and complicated enough, and we shall not discuss it here [We4].

The notion of a Lie-Poisson structure leads to an important problem which we shall only indicate here. Let (M, w) be a Poisson manifold, and x_0 a rank zero point of M. Then, if (x^i) $(x^i(x_0) = 0)$ are local coordinates around x_0, there is a Taylor expansion

$$w^{ij} = c_k^{ij} x^k + o(2), \tag{3.12}$$

and if we insert (3.12) into the Poisson condition (1.5) we see that c_k^{ij} are the structure constants of a certain Lie algebra $\mathcal{G}_{x_0}(w)$. Moreover, using (3.12) it follows immediately that if $\mathcal{F}^1_{x_0} \overset{\text{def}}{=}$ the set of functions of $C^\infty(M)$ that vanish at x_0, and $\mathcal{F}^2_{x_0} \overset{\text{def}}{=}$ the set of functions of $\mathcal{F}^1_{x_0}$ whose first derivatives vanish at x_0, then $\mathcal{F}^1_{x_0}$ is a Lie subalgebra of $(C^\infty(M), \{\ ,\ \})$, and $\mathcal{F}^2_{x_0}$ is an ideal of the previous algebra, and that $\mathcal{G}_{x_0}(w)$ is exactly the Lie algebra structure induced on the quotient algebra $\mathcal{F}^1_{x_0}/\mathcal{F}^2_{x_0}$, a quotient which can be identified with the cotangent space $T^*_{x_0}M$, as it is well known. This shows that $\mathcal{G}_{x_0}(w)$ does not depend on the choice of the local coordinates (x^i), and the Lie-Poisson structure of $\mathcal{G}_{x_0}(w)$ is called the *linear approximation* of w at x_0 [We3].

3.3. Definition. A Poisson structure of rank 0 at x_0 is said to be *linearizable* at x_0 if it is Poisson equivalent to its linear approximation.

Furthermore, if $\text{rank}_{x_0} w \neq 0$, the notions of linear approximation and linearization will be applied to the transversal Poisson structure of w at x_0.

Now, the problem is to decide whether a given Poisson structure is linearizable or not. This linearization problem implies some fine analysis which is not yet fully understood [We3], [We4], [Co1], [Co2], [Dz2], [Du].

c) Next, we may ask for the Poisson structures on $\mathbb{R}^n$ such that

$$w_{ij} = a_{ij} + c_{ij}^k x_k, \qquad (a_{ij} = -a_{ji}, c_{ij}^k = -c_{ji}^k). \tag{3.13}$$

These will be called *affine Poisson structures (modified Lie-Poisson structures* in [LM]). Now, the Poisson condition (1.5) still yields that c_{ij}^k must be the structure constants of a Lie algebra $\mathcal{G}$, hence our problem is in fact that of looking for affine Poisson structures on the dual $\mathcal{G}^*$ of the Lie algebra $\mathcal{G}$. Furthermore, we must also have

$$a_{hi} c_{jk}^h + a_{hj} c_{ki}^h + a_{hk} c_{ij}^h = 0, \tag{3.14}$$

a condition which has the following invariant interpretation: a_{ij} yields a 2-form α on $\mathcal{G}$, which is invariant because w is a tensor, and (3.14) is

$$\alpha([f,g],h)+\alpha([g,h],f)+\alpha([h,f],g)=0, \tag{3.14'}$$

where $f,g,h \in \mathcal{G}$, and the bracket is that of $\mathcal{G}$. In terms of Lie algebra cohomology [CE], this condition means that α is a *Chevalley-Eilenberg* 2-*cocycle* of $\mathcal{G}$, and it will be called the *constant part* of w. With this notation, if $\gamma \in \mathcal{G}^*$ and $\varphi, \psi \in C^\infty(\mathcal{G}^*)$, we have

$$\{\varphi,\psi\}_w(\gamma)=\gamma([d_\gamma\varphi, d_\gamma\psi])+\alpha(d_\gamma\varphi, d_\gamma\psi). \tag{3.15}$$

3.4. Proposition [Bh]. *The affine Poisson structures of $\mathcal{G}^*$ are in one-to-one correspondence with the central extensions $\tilde{\mathcal{G}}$ of $\mathcal{G}$ by $\mathbb{R}$.*

Proof. By definition the expression "$\tilde{\mathcal{G}}$ is a central extension of $\mathcal{G}$ by $\mathbb{R}$" means that one has a well defined exact sequence of Lie algebras

$$0 \to \mathbb{R} \xrightarrow{\iota} \tilde{\mathcal{G}} \xrightarrow{\pi} \mathcal{G} \to 0, \tag{3.16}$$

where $\iota(\mathbb{R})$ belongs to the center of $\tilde{\mathcal{G}}$. This shows that $\tilde{\mathcal{G}}=\mathcal{G}\oplus\mathbb{R}$ as a vector space, and that we must have

$$[f\oplus t, g\oplus\tau]^\sim = [f\oplus 0, g\oplus 0]^\sim = [f,g]\oplus\alpha(f,g), \tag{3.17}$$

where α is a 2-form on $\mathcal{G}$, which satisfies (3.14′). Hence, there is a one-to-one correspondence between affine Poisson structures on $\mathcal{G}$ and central extensions $\tilde{\mathcal{G}}$.

Moreover, from (3.15) and (3.17) it is also clear that the mapping π of (3.16) is a Poisson mapping between the Lie-Poisson structure of $\tilde{\mathcal{G}}$, and the affine Poisson structure of $\mathcal{G}$. Q.e.d.

Just as in the case of the Lie-Poisson structures, it follows now that the characteristic distribution $\mathcal{S}(M)$ is generated by the subalgebra χ_a of the Hamiltonian vector fields of the *affine functions* on $\mathcal{G}^*$ (i.e., of the form $a^i x_i + b$). The computation (3.11) can also be repeated, but it shows that $\forall f \in \mathcal{G}$ the fundamental field Y_f of the coadjoint action, at $\gamma \in \mathcal{G}^*$ is

$$Y_f(\gamma)=X_f(\gamma)-\alpha_f, \tag{3.18}$$

where $\alpha_f(g)\overset{\text{def}}{=}\alpha(f,g)$. Since for an affine function $\varphi = f + \text{const.}$, $X_\varphi = X_f$, (3.18) shows that the characteristic distribution of w is obtained from the tangent distribution of the orbits of the coadjoint action of a connected Lie group G with Lie algebra $\mathcal{G}$ by translations with α_f. In Section IV.4 of [LM] (see also [Me]), one discusses in detail a case when the symplectic distribution of w is given by the orbits of an affine representation of G on $\mathcal{G}^*$ whose linear part is the coadjoint representation.

Finally, we mention that some research on *quadratic Poisson structures* on $\mathbb{R}^n$ i.e., where the components of w are homogeneous quadratic polynomials in x_i, has also been done [BR], [DuH], [LX].

3.2 Dirac brackets

Now, let us go over from $\mathbb{R}^n$, to more general manifolds, and discuss the example of the so-called *Dirac brackets*, usually encountered in the study of constrained mechanical systems [Sn1]. For a general theory of Dirac structures see [CW] and [Cr]. Here, we give a generalized version of the Dirac brackets following essentially [Lh2]. Thereby, we get a new example of Poisson manifolds:

d) Let M^{2n} be a differentiable manifold which is endowed with a non-degenerate differential 2-form ω i.e., (M^{2n}, ω) is an *almost-symplectic manifold.* Of course, it will be symplectic if ω is closed. Furthermore, assume that M is also endowed with a regular $2k$-dimensional foliation $\mathcal{S}$ such that the pullback of ω to every leaf is a closed symplectic form. In this case, S will be called a *leafwise symplectic foliation* of M.

It is not difficult to give examples where M is a symplectic manifold e.g., $M = \mathbb{R}^{2n}$, $\omega = \sum_{i=1}^{n} dx^i \wedge dy^i$, and $\mathcal{S}$ is defined by $x^{k+a} =$ const., $y^{k+a} =$ const. $(a = 1, \ldots, n-k)$. Following is a non-symplectic example.

3.5. Example. Let N^{2n+1} be a *contact manifold* with the *contact form* θ [LM] i.e., $\theta \wedge d\theta^n \neq 0$ everywhere. Then $M = N \times S^1$, $\omega = d\theta + \theta \wedge \lambda$, where λ is the length element of S^1, is obviously an almost symplectic manifold. N has a well defined vector field E which is determined by the conditions

$$\theta(E) = 1, \qquad i(E)d\theta = 0,$$

and *lifts* to M. Then, one also has the vector field A tangent to S^1, that lifts to M. Obviously $[E, A] = 0$, hence span $\{E, A\}$ is an integrable distribution which consists of symplectic planes, since $\omega(E, A) \neq 0$. Moreover, ω pulls back to a closed form on each leaf because the leaves are 2-dimensional. Therefore, span $\{E, A\}$ yields a leafwise symplectic foliation of M.

Now, we come back to the general case, and prove

3.6. Proposition. *The leafwise Poisson bracket of the leafwise symplectic foliation $\mathcal{S}$ of the almost symplectic manifold (M, ω) yields a Poisson structure of M.*

Proof. The only thing to be proven is that, $\forall f \in C^\infty(M)$, the Hamiltonian vector field along the leaves X_f is differentiable. Since ω is nondegenerate, we have the differentiable field $\tilde{X}_f$ defined by

$$i(\tilde{X}_f)\omega = -df.$$

On the other hand, since the leaves of $\mathcal{S}$ are symplectic, the ω-orthogonal distribution $T'\mathcal{S}$ of $\mathcal{S}$ is complementary to $T\mathcal{S}$ in TM, and it is differentiable, because the foliation $\mathcal{S}$ is regular. Hence $TM = T\mathcal{S} \oplus T'\mathcal{S}$, and it follows easily that X_f is the $T\mathcal{S}$-component of $\tilde{X}_f$. Q.e.d.

3.7. Definition. The Poisson bracket of the structure in Proposition 3.6 is called the *Dirac bracket* of $(M, \omega, \mathcal{S})$.

Clearly, with this bracket M becomes a regular Poisson manifold, and $\mathcal{S}$ is its symplectic foliation.

The Dirac brackets are not only important for applications; they also induce all the regular Poisson structures. Namely, we have

3.8. Theorem [Va10]. *Let (M, w) be a regular Poisson manifold. Then M can be embedded as a Poisson submanifold of an almost symplectic manifold (N, ω) endowed with a Dirac bracket.*

Proof. Since w is regular, its symplectic foliation $\mathcal{S}$ is regular, and we may choose a differentiable decomposition

$$TM = T\mathcal{S} \oplus \nu\mathcal{S}. \tag{3.19}$$

The total space N of the dual vector bundle $\nu^*\mathcal{S}$ will be the manifold that we are looking for.

In order to see this, let (x^a, y^u) ($a = 1, \ldots, \dim M$-rank w; $u = 1, \ldots$,rank w) be local coordinates on M such that $T\mathcal{S}$=span$\{\frac{\partial}{\partial y^u}\}$ (the Frobenius theorem), and accordingly

$$\nu\mathcal{S} = \text{span}\left\{X_a = \frac{\partial}{\partial x^a} - t_a^u \frac{\partial}{\partial y^u}\right\}, \tag{3.20}$$

for some local functions t_a^u (e.g., [Va2]). By taking the dual cobasis, we get

$$T^*\mathcal{S} = \text{ span } \{\theta^u = dy^u + t_a^u dx^a\}; \quad \nu^*\mathcal{S} = \text{ span } \{dx^a\}. \tag{3.21}$$

Hence $\xi \in \nu^*\mathcal{S}$ has an expression $\xi = \xi_a dx^a$, and (x^a, y^u, ξ_a) are local coordinates on N. Moreover, it is clear that $\xi = \xi_a dx^a$ is a well defined 1-form on N, which has the following geometric significance: it is obtained by gluing up the *Liouville forms* [AM], [LM] of the cotangent bundles of the local transversal submanifolds of $\mathcal{S}$.

In the sequel, we shall denote by Ann the annihilator of a plane, i.e., the space of 1-forms that vanish on that plane.

Clearly, $\nu^*\mathcal{S} = \text{Ann } T\mathcal{S} = \ker \#$, because w is skew-symmetric, and we get $w(\alpha, \beta) = 0$ if one of the arguments is in $\nu^*\mathcal{S}$. Thereby, we see that the Poisson bivector must be of the form

$$w = \frac{1}{2} w^{uv} \frac{\partial}{\partial y^u} \wedge \frac{\partial}{\partial y^v} \qquad (w^{vu} = -w^{uv}). \tag{3.22}$$

Accordingly, if for $u, v \in TM$ we put

$$\omega_0(u, v) = w(\kappa, \nu), \tag{3.23}$$

where $\#\kappa = pr_{TS}u$, $\#\nu = pr_{TS}v, \omega_0$ will be a well defined 2-form on M, and its local expression will be

$$\omega_0 = \frac{1}{2}\omega_{uv}\theta^u \wedge \theta^v, \tag{3.24}$$

where $\omega_{uv}w^{us} = \delta^s_v$. It is clear that the pullbacks of ω_0 to the leaves of $\mathcal{S}$ are exactly the symplectic structures induced on the leaves by w.

Finally, let $\pi : N \to M$ be the natural vector bundle projection. Then

$$\omega = \pi^*\omega_0 + d\xi \tag{3.25}$$

is a nondegenerate 2-form on N, which defines an almost symplectic structure. Furthermore, $\pi^{-1}(\mathcal{S})$ is the foliation of N defined by $x^a =$ const., $\xi_a =$ const., and it is clear that ω induces symplectic structures on the leaves of $\pi^{-1}(\mathcal{S})$. Therefore, N has a well defined Dirac bracket associated with ω and $\pi^{-1}(\mathcal{S})$, and we shall denote by $\tilde{w}$ the corresponding Poisson bivector. Now, M is embedded in N as the zero section of $\nu^*\mathcal{S}$, and it has the local equations $\xi_a = 0$. Hence, M consists of a union of leaves of $\pi^{-1}(\mathcal{S})$ namely, the leaves of $\mathcal{S}$ itself. This obviously shows that w and $\tilde{w}$ are related by this embedding, and, therefore, that M is a Poisson submanifold of $(N, \tilde{w})$ as defined in Section 1.5. Q.e.d.

3.3 Further examples

e) Another general example is offered by certain submanifolds N of a given, not necessarily regular, Poisson manifold (M, w) which are such that w *induces* a Poisson structure on them, without making them, however, Poisson submanifolds always. Namely, we have

3.9. Proposition. *Let (M, w) be a Poisson manifold, and let N be a submanifold of M such that for every symplectic leaf S of M the intersection $N \cap S$ is a symplectic submanifold of S. Then, the symplectic structures of these submanifolds yield a Poisson structure induced by w in N.*

Proof. We shall apply Theorem 2.14. If $\varphi \in C^\infty(N)$, let $X^N_\varphi(y)$ $(y \in N)$ be the Hamiltonian field associated with the symplectic structure of $N \cap S(y)$. The only thing which we must justify is the differentiability of the field X^N_φ. Clearly, $N \cap S$ can be seen simultaneously as a submanifold of S, N, and M, and we are allowed to extend φ to $\tilde{\varphi} \in C^\infty(M)$ (it suffices to do this locally). The tangent spaces $T(N \cap S)$ define an integrable general distribution on N, and the connected components of $N \cap S$ are its leaves. On the other hand, in TS there is a symplectically orthogonal distribution νS of $T(N \cap S)$, along $N \cap S$, and, like in the proof of Proposition 3.6, $X^N_\varphi(y)$, is the projection of $X^M_{\tilde{\varphi}}(y)$ (which is in T_yS) onto $T(N \cap S)$. This projection preserves differentiability. Q.e.d.

A particular case of this results is

3.10. Proposition [We3]. *Let N be a submanifold of the Poisson manifold (M, w), such that: i) N is transversal to the symplectic leaves of w; ii) $\forall x \in N$, $T_x N \cap \mathcal{S}_x(M)$ is a symplectic subspace of $\mathcal{S}_x(M)$. Then, there is a naturally induced Poisson structure in N.*

Proof. The hypotheses of Proposition 3.10 obviously imply those of Proposition 3.9. Q.e.d.

3.11. Remark. In [We3] the hypotheses of Proposition 3.10 are expressed by the following equivalent conditions: *i)* $(\ker \#_x) \cap \operatorname{Ann}(T_x N) = \{0\}$; *ii)* $[\#(\operatorname{Ann} T_x N)] \cap T_x N = \{0\}$ $(x \in N)$. Indeed, since w is skew-symmetric, $\ker \#_x = \operatorname{Ann} \mathcal{S}_x(M)$, and *i)* shows that no nonzero 1-form vanishes on $T_x N + \mathcal{S}_x(M)$ i.e., this latter union is n-dimensional $(n = \dim M)$, and the two sub-spaces are transversal. Then, *ii)* means that no nonzero vector $v \in T_x N \cap \mathcal{S}_x M$ can be symplectically orthogonal to $T_x N \cap \mathcal{S}_x M$ in $\mathcal{S}_x M$, since this orthogonality relation would mean exactly $v \in [\#(\operatorname{Ann} T_x N)] \cap T_x N$. This fact means that the intersection is symplectic in $\mathcal{S}_x M$.

f) In accordance with Corollary 1.6, another interesting situation is that of a Riemannian manifold (M^m, g) which has a parallel bivector w i.e., $\nabla w = 0$ for the Levi-Civita connection ∇ of (M, g). Then, of course, w defines a Poisson structure on M. On the other hand, $\#_g^{-1} w$ is a g-parallel 2-form, and one is in a well-known geometric situation ([Lh1], §115), which we may describe as follows.

Since w is parallel, it must have a constant rank and $\operatorname{im}\#_w$ is a field of parallel planes on M i.e., the symplectic foliation $\mathcal{S}(M)$ is g-parallel. Hence, the Poisson structure w is regular, the g-orthogonal plane field $\nu\mathcal{S}$ of $\mathcal{S}$ is also parallel and integrable, M is, locally, a product manifold with the coordinates y^u along $\mathcal{S}(M)$, and x^a along $\nu\mathcal{S}$, and g is reducible as (e.g., [KN] vol. I, §IV 5, 6)

$$g = g' + g'' \quad (g' = g_{ab}(x)dx^a dx^b,\ g'' = g_{uv}(y)dy^u dy^v). \tag{3.26}$$

Moreover, $\forall f \in C^\infty(M)$, $X_f x^a = 0$, and, by taking $f = x^b, y^u$, we deduce $w^{ab} = 0$; $w^{bu} = -w^{ub} = 0$. Then, the induced symplectic form of the leaves of $\mathcal{S}$ has the local components, say, ω_{uv} where $\omega_{uv} w^{sv} = \delta^s_u$, and these components, as well as w^{uv} depend only on the y^u since we have (because of (3.26))

$$\frac{\partial w^{uv}}{\partial x^a} = \nabla_a w^{uv} = 0.$$

Now, let us look at the symplectic vector bundle $T\mathcal{S}$ over M, with the symplectic structure defined by ω. It is classical (e.g. [Lh1, §96], [Va8]) that it has the ω-compatible complex structure $J = A(-A^2)^{-1/2}$, where A is given by $g''(AX, Y) = \omega(X, Y)$. Since $(-A^2)^{1/2}$ can be expressed as a polynomial in A (e.g., [Gt]), J, extended by 0 to $\nu\mathcal{S}$, is a g-parallel tensor field on M, and it depends on (y^u) alone hence, J is g''-parallel along the leaves of $\mathcal{S}(M)$. Then, the same facts hold for the Hermitian metric $h(X, Y) = \omega(X, JY)$, and we see that there is a Kähler metric on the leaves of $\mathcal{S}(M)$.

On M, the metric $\tilde{g} = g' + h$ is therefore a *partially Kähler metric* [Va5], which is parallel with respect to g, and, according to these results, we have

3.12. Proposition. *The parallel Poisson structures of a Riemannian manifold* (M, g) *are exactly those defined by the Kähler foliation of the* g*-parallel partially Kähler metrics of* M *(if any).*

To close this chapter let us mention that it is easy to construct nontrivial Poisson structures on any manifold. Indeed, for $m = 2n + p$ the Poisson bivector (3.2) will have a compact support in $\mathbb{R}^m$ if the coefficients h_k have a compact support in $\mathbb{R}^m$, and we may transfer such a compactly supported Poisson structure to any manifold M^m by means of a chart. The obtained Poisson structure can then be extended by zero to the whole manifold M. But the existence problem becomes nontrivial if some stronger conditions are demanded, in particular, if one is looking for regular Poisson structures.

For instance, S^4 has no regular nontrivial Poisson structures. Indeed there is no such structure of rank 4 since there is no symplectic structure, and there is no structure of rank 2 since, otherwise, the orientations of the symplectic foliation $\mathcal{S}$, and of its transversal distribution $\mathcal{S}'$ would join to give an almost complex structure on S^4, and it is known that such a structure does not exist. Of course, S^4 also has no structures of the type mentioned in Corollary 1.10 since it has no nonzero harmonic 2-forms (the second Betti number $b_2(S^4) = 0$).

4 Poisson Calculus

4.1 The bracket of 1-forms

An interesting feature of the Poisson manifolds is the existence of a contravariant version of the differential calculus of forms, and we develop this calculus here. It is based on the possibility to extend the Poisson bracket to 1-forms, as it was discovered by several authors independently [GD], [MM], etc. (See more references in [KSM2].) We shall denote by $\Lambda^k M$ the space of differential forms of degree k on a differentiable manifold M.

4.1. Theorem. *Let (M, w) be a Poisson manifold. Then, there exists a unique $\mathbb{R}$-bilinear, skew symmetric operation $\{\ ,\ \} : \Lambda^1 M \times \Lambda^1 M \to \Lambda^1 M$ such that*

$$\{df, dg\} = d\{f, g\} \qquad (f, g \in C^\infty(M)) \tag{4.1}$$

$$\{\alpha, f\beta\} = f\{\alpha, \beta\} + (\alpha^\# f)\beta \qquad (f \in C^\infty(M), \alpha, \beta \in \Lambda^1 M, \alpha^\# = \#\alpha). \tag{4.2}$$

This operation is given by the general formula

$$\{\alpha, \beta\} = L_{\alpha^\#}\beta - L_{\beta^\#}\alpha - d(w(\alpha, \beta)). \tag{4.3}$$

Furthermore, it provides $\Lambda^1 M$ with a Lie algebra structure such that $\# : \Lambda^1 M \to \mathcal{V}^1(M)$ is a Lie algebra homomorphism i.e.,

$$\{\alpha, \beta\}^\# = [\alpha^\#, \beta^\#]. \tag{4.4}$$

Proof. If the operation exists, it must be of the local type i.e., $\{\alpha, \beta\}(x_0)$ will depend on β (therefore, on α too) around x_0 only, $\forall x_0 \in M$. Indeed, if $\beta_1/_U = \beta_2/_U$ for an open neighbourhood U of x_0, and if f is a C^∞-function that vanishes outside U, and equals 1 on a compact neighbourhood $V_{x_0} \subset U$, then $\gamma = f\beta_1 = f\beta_2$ is well defined on M, and we have by (4.2)

$$\begin{aligned}\{\alpha, \gamma\}(x_0) = \{\alpha, f\beta_1\}(x_0) = f(x_0)\{\alpha, \beta_1\}(x_0) + (\alpha^\# f)(x_0)\beta_1(x_0) = \\ = \{\alpha, \beta_1\}(x_0).\end{aligned}$$

Similarly, $\{\alpha, \gamma\}(x_0) = \{\alpha, \beta_2\}(x_0)$ i.e., $\{\alpha, \beta_1\}(x_0) = \{\alpha, \beta_2\}(x_0)$, as claimed.

The next remark is that, because of the skew-symmetry, we have something corresponding to (4.2) for the first factor, and, generally, we will have $\forall h, f \in C^\infty(M)$

$$\{h\alpha, f\beta\} = hf\{\alpha.\beta\} + h(\alpha^\# f)\beta - f(\beta^\# h)\alpha. \tag{4.5}$$

Hence, if the operation exists, we may compute it by means of the local coordinates x^i as follows. Take $\alpha = \alpha_i dx^i$, $\beta = \beta_j dx^j$. Then

$$\begin{aligned}\{\alpha,\beta\} &= \{\alpha_i dx^i, \beta_j dx^j\} = \alpha_i\beta_j d\{x^i, x^j\}+\\ &+ \alpha_i(X_{x^i}\beta_j)dx^j - \beta_j(X_{x^j}\alpha_i)dx^i =\\ &= \alpha_i\beta_j(dw^{ij}) + \alpha_i w^{ik}\frac{\partial\beta_j}{\partial x^k}dx^j - \beta_j w^{jk}\frac{\partial\alpha_i}{\partial x^k}dx^i =\\ &= d(w^{ij}\alpha_i\beta_j) - w^{ij}\beta_j d\alpha_i - w^{ij}\alpha_i d\beta_j +\\ &+ \alpha_i w^{ik}\frac{\partial\beta_j}{\partial x^k}dx^j - \beta_j w^{jk}\frac{\partial\alpha_i}{\partial x^k}dx^i.\end{aligned}$$

It is easy to check that this result means

$$\begin{aligned}\{\alpha,\beta\} &= d(w(\alpha,\beta)) + i(\alpha^{\#})d\beta - i(\beta^{\#})d\alpha =\\ &= L_{\alpha^{\#}}\beta - L_{\beta^{\#}}\alpha - dw(\alpha,\beta).\end{aligned} \tag{4.6}$$

Hence, if $\{\ ,\ \}$ exists, it must be given by (4.3) and (4.6), and the result is unique. On the other hand, it is trivial to check that the operation (4.3) satisfies (4.1) and (4.2), and this justifies the existence part of the theorem.

Now, in order to justify (4.4), we first use (4.1) and (2.12) to get

$$\{df, dg\}^{\#} = (d\{f,g\})^{\#} = X_{\{f,g\}} = [X_f, X_g], \tag{4.7}$$

which is a particular case of (4.4). Then, we notice that the same proof as for the uniqueness of $\{\ ,\ \}$ shows that $\{\ ,\ \}^{\#}$ is the unique $\mathbb{R}$-bilinear, skew-symmetric operation $\Lambda^1 M \times \Lambda^1 M \to \mathcal{V}^1(M)$ which satisfies (4.7) as well as the rule

$$\{\alpha, f\beta\}^{\#} = f\{\alpha,\beta\}^{\#} + (\alpha^{\#}f)\beta^{\#}$$

deduced from (4.2). But $[\alpha^{\#}, \beta^{\#}]$ is also an operation that satisfies the same conditions. Hence, uniqueness implies (4.4).

Finally, for the mentioned Lie algebra structure, it suffices of course to check the Jacobi identity for 1-forms of the type adf, bdg, cdh. Using (4.2), repeatedly, we get

$$\begin{aligned}\{\{adf, bdg\}, cdh\} &= abcd\{\{fg\},h\} + ac(X_f b)d\{g,h\} -\\ &- bc(X_g a)d\{f,h\} - c(X_h(ab))d\{f,g\} + cX_h(bX_g a)df -\\ &- cX_h(a(X_f b))dg + [abX_{\{f,g\}}c + a(X_f b)(X_g c) -\\ &- b(X_g a)(X_f c)]dh.\end{aligned}$$

Now, since the Poisson bracket of functions satisfies the Jacobi identity, and using (2.12) it is easy to check that

$$\{\{adf, bdg\}, cdh\} + \{\{bdg, cdh\}, adf\} + \{\{cdh, adf\}, bdg\} = 0.$$

Q.e.d.

4.2. Remark. The results of Theorem 4.1 mean that the triple $(T^*M, \{\ ,\ \}, \#)$ is a *Lie algebroid* in the sense of Pradines [Pr1]. This is a notion which we shall need more seriously in a later chapter and we shall defer its definition until then.

4.2 The contravariant exterior differentiation

The existence of the bracket of 1-forms defined above allows us to mimic the usual definitions of the operators d, $i(X)$, L_X, ∇_X [BV1,2], [KSM2]. First, we define the contravariant exterior differential $\sigma : \mathcal{V}^k(M) \to \mathcal{V}^{k+1}(M)$ for $Q \in \mathcal{V}^k(M)$ by

$$\begin{aligned}(\sigma Q)(\alpha_0, \dots, \alpha_k) = \sum_{i=0}^{k} (-1)^i \alpha_i^\# (Q(\alpha_0, \dots, \hat{\alpha}_i, \dots, \alpha_k)) + \\ + \sum_{i<j=0}^{k} (-1)^{i+j} Q(\{\alpha_i, \alpha_j\}, \alpha_0, \dots, \hat{\alpha}_i, \dots, \hat{\alpha}_j, \dots, \alpha_k),\end{aligned} \tag{4.8}$$

where $\alpha_i \in \Lambda^1 M$, and the hat denotes missing arguments.

Formally, (4.8) is exactly the same as the expression of the exterior differential of forms d (e.g. [Mt]) hence its algebraic consequences will be the same and we have

$$\sigma(Q_1 \wedge Q_2) = (\sigma Q_1) \wedge Q_2 + (-1)^{\deg Q_1} Q_1 \wedge (\sigma Q_2), \tag{4.9}$$

$$\sigma^2 = 0. \tag{4.10}$$

For a similar reason, and with (4.4), if we define $\# : \Lambda^k M \to \mathcal{V}^k M$ by

$$\lambda^\# (\alpha_1, \dots, \alpha_k) = (-1)^k \lambda(\alpha_1^\#, \dots, \alpha_k^\#) \quad (\lambda \in \Lambda^k M) \tag{4.11}$$

(the sign was chosen such that the formula is correct for $k = 1$), we shall obtain

$$\sigma(\lambda^\#) = -(d\lambda)^\#. \tag{4.12}$$

Furthermore, if we put [BV1,2]

$$(i_\alpha Q)(\alpha_1, \dots, \alpha_{k-1}) = Q(\alpha, \alpha_1, \dots, \alpha_{k-1}) \tag{4.13}$$

$$(L_\alpha Q)(\alpha_1, \dots, \alpha_k) = \alpha^\# (Q(\alpha_1, \dots, \alpha_k)) - \\ - \sum_{i=1}^{k} Q(\alpha_1, \dots, \{\alpha, \alpha_i\}, \dots, \alpha_k), \tag{4.14}$$

we shall have the usual relations

$$\begin{aligned} i_{\{\alpha,\beta\}} &= L_\alpha i_\beta - i_\beta L_\alpha, & L_{\{\alpha,\beta\}} &= L_\alpha L_\beta - L_\beta L_\alpha, \\ L_\alpha &= i_\alpha \sigma + \sigma i_\alpha, & L_\alpha \sigma &= \sigma L_\alpha. \\ i_\alpha \lambda^\# &= (-1)^k (i_{\alpha^\#}\lambda)^\#, & L_\alpha \lambda^\# &= (-1)^k (L_{\alpha^\#}\lambda)^\#. \end{aligned} \tag{4.15}$$

Let us also notice the immediate formula

$$(L_\alpha Q - L_{\alpha^\#}Q)(\alpha_1,\ldots,\alpha_k) = \sum_{i=1}^{k} Q(\alpha_1,\ldots,L_{\alpha^\#}\alpha_i - \{\alpha,\alpha_i\},\ldots,\alpha_k), \tag{4.16}$$

which has as immediate consequence (use (4.3)) that $\forall f \in C^\infty(M)$

$$L_{df}Q = L_{X_f}Q \tag{4.17}$$

i.e., we get a true Lie derivative in this case. This fact is not true for an arbitrary 1-form α.

4.3. Proposition [BV1,2], [KSM2]. *For every $Q \in \mathcal{V}^k(M)$ one has*

$$\sigma Q = -[w, Q], \tag{4.18}$$

where the bracket is that of Schouten-Nijenhuis (Chapter 1).

Proof. For $k = 0$ (4.18) holds from the corresponding definitions. Then,

$$\begin{aligned} [w,X](df,dg) &= (L_X w)(df,dg) = X(w(df,dg)) - w(L_X df, dg) - \\ &- w(df, L_X dg) = X\{f,g\} + w(dg, d(Xf)) - w(df, d(Xg)) = \\ &= X\{f,g\} + \{g, Xf\} - \{f, Xg\} = X\{f,g\} + X_g X f - X_f X g = \\ &= -(\sigma X)(df,dg) \end{aligned}$$

(the last equality uses (4.8)) i.e., (4.18) holds for $Q \in \mathcal{V}^1(M)$.

For the general result, it suffices to take $Q = X_1 \wedge \ldots \wedge X_k$ ($X_i \in \mathcal{V}^1(M)$) and to use (4.9) and (1.7), (1.9). Q.e.d.

The operation (4.18) was first defined by Lichnerowicz [Lh2].

Using Proposition 4.3, and formula (1.20) we have the coordinate expression of σ:

$$\begin{aligned} (\sigma Q)^{i_0\ldots i_k} &= -\frac{1}{k!}\delta^{i_0\ldots i_k}_{hj_1\ldots j_k} w^{uh}\partial_u Q^{j_1\ldots j_k} - \\ &- \frac{1}{2(k-1)!}\delta^{i_0\ldots\ldots i_k}_{h_1h_2j_2\ldots j_k} Q^{uj_2\ldots j_s}\partial_u w^{h_1h_2}. \end{aligned} \tag{4.19}$$

As in (1.22) we can replace here the partial derivatives by covariant derivatives with respect to torsionless linear connections on M. We may also use (1.26), and write

$$\sigma Q = w \wedge D_\nabla Q + (D_\nabla w) \wedge Q - D_\nabla(w \wedge Q), \tag{4.20}$$

and, if we choose a Riemannian metric g on M, D_∇ will be given by (1.32), etc.

Now, we shall indicate that σ is related with an interesting differential operator on forms defined by Koszul [Kz] and Brylinski [Br] $\delta = \delta_w : \Lambda^k M \to \Lambda^{k-1} M$, which satisfies

$$\begin{aligned}
&\delta_w(f_0 df_1 \wedge \ldots \wedge df_k) = \\
&= \sum_{i=1}^{k} (-1)^{i+1} \{f_0, f_i\} df_1 \wedge \ldots \wedge d\hat{f_i} \wedge \ldots \wedge df_k + \\
&+ \sum_{i<j=1}^{k} (-1)^{i+j} f_0 d\{f_i, f_j\} \wedge df_1 \wedge \ldots \wedge d\hat{f_i} \wedge \ldots \wedge d\hat{f_j} \wedge \ldots \wedge df_k.
\end{aligned} \tag{4.21}$$

We shall say that δ is the *Poisson codifferential* of (M, w). The fact that (4.21) yields a well defined operator will be seen later on, by establishing the invariant coordinate formula (4.23′). A careful cancellation of terms in the corresponding expression of $\delta^2(f_0 df_1 \wedge \ldots \wedge df_k)$, using the Jacobi identity for $\{\ ,\ \}$, shows that $\delta^2 = 0$. Indeed, (4.21) can be written under the form

$$\delta(f_0 df_1 \wedge \ldots \wedge df_k) = i(X_{f_0})(df_1 \wedge \ldots \wedge df_k) + f_0 \delta(df_1 \wedge \ldots \wedge df_k),$$

whence we get

$$\begin{aligned}
\delta^2(f_0 df_1 \wedge \ldots \wedge df_k) &= \delta(i(X_{f_0})(df_1 \wedge \ldots \wedge df_k)) + \\
&+ i(X_{f_0})\delta(df_1 \wedge \ldots \wedge df_k) + f_0 \delta^2(df_1 \wedge \ldots \wedge df_k).
\end{aligned}$$

By looking at the expression of $\delta^2(df_1 \wedge \ldots \wedge df_k)$ given by (4.21), we get

$$\begin{aligned}
&\delta^2(df_1 \wedge \ldots \wedge df_k) = \\
&= \sum_{h<i<j=1}^{k} (-1)^{i+j+k} d(\text{Cycl. } \{\{f_i, f_j\}, f_h\}) \wedge \\
&\wedge df_1 \wedge \ldots \wedge d\hat{f_h} \wedge \ldots \wedge d\hat{f_i} \wedge \ldots \wedge d\hat{f_j} \wedge \ldots \wedge df_k = 0.
\end{aligned}$$

Furthermore, we get

$$\begin{aligned}
&i(X f_0)\delta(df_1 \wedge \ldots \wedge df_k) + \delta(i(X_{f_0})(df_1 \wedge \ldots \wedge df_k) = \\
&= \sum_{i<j=1}^{k} (-1)^{i+j} (\text{Cycl.}\{f_0, \{f_i, f_j\}\}) df_1 \wedge \ldots \wedge d\hat{f_i} \wedge \ldots \wedge d\hat{f_j} \wedge \ldots \wedge df_k + \\
&\quad + \text{ terms that cancel } = 0.
\end{aligned}$$

Now, if we remember the operator $i(Q)$ defined by formula (1.15), we can formulate

4.4. Proposition [BV2]. *$\forall\lambda \in \Lambda^k M$, and $\forall Q \in \mathcal{V}^{k-1}(M)$ one has*

$$i(\sigma Q)\lambda = i(Q)(\delta\lambda) + (-1)^k \delta(i(Q)\lambda). \tag{4.22}$$

Proof. We proceed by a componentwise computation with respect to local coordinates (x^i). Take

$$\lambda = \frac{1}{k!}\lambda_{i_1\ldots i_k} dx^{i_1} \wedge \ldots \wedge dx^{i_k}.$$

Then (4.21) yields

$$k!\delta\lambda = \sum_{j=1}^{k} (-1)^{j+1}\{\lambda_{i_1\ldots i_k}, x^{i_j}\} dx^{i_1} \wedge \ldots \wedge d\hat{x}^{i_j} \wedge \ldots \wedge dx^{i_k} +$$

$$+ \sum_{j<p=1}^{k} (-1)^{j+p} \lambda_{i_1\ldots i_k} d\{x^{i_j}, x^{i_p}\} \wedge dx^{i_1} \wedge \ldots \wedge d\hat{x}^{i_j} \wedge \ldots \wedge d\hat{x}^{i_p} \wedge \ldots \wedge dx^{i_k}.$$

Here, we have to replace $\{x^{i_j}, x^{i_p}\} = w^{i_j i_p}$, and $\{\lambda_{i_1\ldots i_k}, x^{i_j}\} = w^{hi_j}\partial_h \lambda_{i_1\ldots i_k}$, then notice that the terms of the first sum are all equal, and so are the terms of the second sum, because of the skew-symmetry of λ and w. The result will be that

$$\delta\lambda = \frac{1}{(k-1)!}(\delta\lambda)_{i_2\ldots i_k} dx^{i_2} \wedge \ldots \wedge dx^{i_k}$$

where

$$\begin{aligned}(\delta\lambda)_{i_2\ldots i_k} &= w^{hp}\partial_h \lambda_{pi_2\ldots i_k} - \frac{1}{2(k-2)!}\delta^{hj_3\ldots j_k}_{i_2\ldots i_k} \lambda_{uvj_3\ldots j_k} \partial_h w^{uv} = \\ &= w^{hp}\partial_h \lambda_{pi_2\ldots i_k} - \frac{1}{2}\sum_{s=2}^{k} (-1)^s \lambda_{uvi_2\ldots \hat{i}_s \ldots i_k} \partial_{i_s} w^{uv},\end{aligned} \tag{4.23}$$

which gives us the coordinate expression of the Brylinski differential. For the last equality in (4.23), use (1.29) i.e.

$$\delta^{hj_3\ldots j_k}_{i_2\ldots i_k} = \sum_{s=2}^{k} (-1)^s \delta^h_{i_s} \delta^{j_3\ldots j_k}_{i_2\ldots \hat{i}_s \ldots i_k}.$$

We can convince ourselves that (4.23) are indeed the components of a form by taking a torsionless connection ∇ on M, and by expressing the partial derivatives that enter in the last expression (4.23) by means of covariant derivatives. Then, the terms that contain the connection coefficients cancel, and we get the tensorial expression

$$(\delta\lambda)_{i_2\ldots i_k} = w^{hp}\nabla_h\lambda_{pi_2\ldots i_k} - \frac{1}{2}\sum_{s=2}^{k}(-1)^s\lambda_{uvi_2\ldots\hat{i}_s\ldots i_k}\nabla_{i_s}w^{uv} = \tag{4.23'}$$
$$= w^{hp}\nabla_h\lambda_{pi_2\ldots i_k} - \frac{1}{2(k-2)!}\delta^{hj_3\ldots j_k}_{i_2\ldots i_k}\lambda_{uvj_3\ldots j_k}\nabla_h w^{uv}.$$

Notice that this result proves that δ is well defined, which is not clear from (4.21). (See a different proof in [Br].)

Furthermore, take

$$Q = \frac{1}{(k-1)!}Q^{i_1\ldots i_{k-1}}\frac{\partial}{\partial x^{i_1}}\wedge\ldots\wedge\frac{\partial}{\partial x^{i_{k-1}}},$$

and use the definition of $i(Q)$ to get

$$(i(Q)\lambda)_h = \frac{1}{(k-1)!}Q^{i_1\ldots i_{k-1}}\lambda_{i_1\ldots i_{k-1}h}. \tag{4.24}$$

Finally, write down the terms of the formula (4.22) by using (4.24) and (4.23) to compute $\delta(i(Q)\lambda)$, (4.19) and (4.24) to compute $i(\sigma Q)\lambda$, and again (4.24), (4.23) for $i(Q)(\delta\lambda)$. Then the formula (4.22) checks straightforwardly. Q.e.d.

Formula (4.22) gives us a new definition of σ if δ is known, and conversely since, by (4.23), the last term of (4.22) is equal to $(-1)^k i(w)d(i(Q)\lambda))$, and we have

$$i(Q)\delta\lambda = i(\sigma Q)\lambda - (-1)^k i(w)d(i(Q)\lambda). \tag{4.22'}$$

Before going on, we also mention the following interesting facts about the operator δ:

a) In Remark 1.16 it has been mentioned that in the case of a symplectic manifold one has a symplectic codifferential $\tilde{\delta}$ which is expressible by (1.31) where g is to be replaced by the Poisson bivector associated with the symplectic structure, and ∇ by a symplectic torsionless connection of M. Now, if the same connection is used in (4.23′), we see that, in the case of a symplectic manifold, the Poisson codifferential is equal to the symplectic codifferential, a fact proven differently in [Br].

b) Another important remark [Kz], [Br] is that one has $d\delta + \delta d = 0$. This follows from (4.21) by a straightforward calculation, and it leads to a spectral sequence of differential forms of a symplectic manifold [Br].

c) The expression (4.23) of δ, and the classical coordinate expression of $d\lambda$ lead easily to the formula, $\delta = i(w)d - di(w)$ which was the definition given to δ in [Kz], where, also, the letter Δ was used instead of δ.

4.3 The regular case

The operator σ has an interesting structure in the case of a regular Poisson manifold (M, w), where $\dim M = m = 2n+s$, and rank $w = n$. In this case, the symplectic foliation $\mathcal{S}$ of M is regular, and we may choose a C^∞ transversal distribution $\nu\mathcal{S}$ as we did in the proof of Theorem 3.8. This gives us the decomposition (3.19) which, in turn, induces a decomposition of the multivector fields, and of the differential forms into *bihomogeneous components of type* (p, q). Namely, $Q \in \mathcal{V}^k(M)$ will have the type (p, q) if it belongs to $\wedge^q(T\mathcal{S}) \otimes \wedge^p(\nu\mathcal{S})$, and $\lambda \in \Lambda^k M$ will be of type (p, q) if it belongs to the dual of the previous space. With the notation of the formulas (3.20) and (3.21), the local expressions of such Q and λ will be of the form

$$Q = \frac{1}{p!q!} Q^{a_1 \ldots a_p u_1 \ldots u_q} X_{a_1} \wedge \ldots \wedge X_{a_p} \wedge \frac{\partial}{\partial y^{u_1}} \wedge \ldots \wedge \frac{\partial}{\partial y^{u_q}}, \tag{4.25}$$

$$\lambda = \frac{1}{p!q!} \lambda_{a_1 \ldots a_p u_1 \ldots u_q} dx^{a_1} \wedge \ldots \wedge dx^{a_p} \wedge \theta^{u_1} \wedge \ldots \wedge \theta^{u_q}. \tag{4.26}$$

Correspondingly, an operator will be *homogeneous of type* (a, b) if it sends elements of type (p, q) to elements of type $(p + a, q + b)$.

4.5. Proposition [Va9]. *On a regular Poisson manifold* (M, w), *and with respect to a given choice of* $\nu\mathcal{S}$, σ *has a well defined decomposition*

$$\sigma = \sigma' + \sigma'', \tag{4.27}$$

where type $\sigma' = (-1, 2)$, *and type* $\sigma'' = (0, 1)$.

Proof. Take $Q \in \mathcal{V}^{pq}(M)$ (an obvious notation!), and look at σQ as defined by (4.8), where the arguments α_i are homogeneous i.e., they have either type (1,0) or (0,1). The result is obviously 0 if either the number of the arguments of type (1,0) is larger than $p + 2$ or the number of the arguments of type (0,1) is larger than $q + 2$. Hence

$$\sigma = \sigma_{(2,-1)} + \sigma_{(1,0)} + \sigma_{(0,1)} + \sigma_{(-1,2)},$$

where the indices denote the type of the components.

Now, it is clear that

$$\Lambda^{1,0}(M) = \text{Ann}\, T\mathcal{S} = \ker \#_w \tag{4.28}$$

(see Remark 3.11 for the last equality), and the property

$$\{\alpha, \beta\}^\# = [\alpha^\#, \beta^\#]$$

shows that if at least one of the 1-forms α, β is of the type (1,0) so is $\{\alpha, \beta\}$ too. Moreover, if both α and β are of the type (1,0), formula (4.3) shows that $\{\alpha, \beta\} = 0$. These remarks allow us to see that σQ also vanishes if either $p + 2$ or $p + 1$ arguments are of the type (1,0) i.e., $\sigma_{(2,-1)} = 0$ and $\sigma_{(1,0)} = 0$). Then, we put $\sigma' = \sigma_{(-1,2)}$, $\sigma'' = \sigma_{(0,1)}$, and we are done. Q.e.d.

4.6. Remarks. 1) The above mentioned properties of $\ker \#$ mean that this space is an abelian ideal of the Lie algebra $(\Lambda^1 M, \{\ ,\ \})$.

2) By applying (4.8) to the corresponding number of arguments we get

$$(\sigma' Q)(\alpha_0, \ldots, \alpha_{p-2}, \beta_0, \ldots, \beta_{q+1}) = \tag{4.29}$$

$$= \sum_{i<j=0}^{q+1} (-1)^{i+j} Q(\{\beta_i, \beta_j\}, (\alpha_0, \ldots, \alpha_{p-2}, \beta_0, \ldots, \hat{\beta}_i, \ldots, \hat{\beta}_j, \ldots, \beta_{q+1})$$

$$(\sigma'' Q)(\alpha_0, \ldots, \alpha_{p-1}, \beta_0, \ldots, \beta_q) = \tag{4.30}$$

$$= \sum_{i=0}^{q} (-1)^{p+i} \beta_i^{\#} (Q(\alpha_0, \ldots, \alpha_{p-1}, \beta_0, \ldots, \hat{\beta}_i, \ldots, \beta_q) +$$

$$+ \sum_{i=0}^{p-1} \sum_{j=0}^{q} (-1)^{p+i+j} Q(\{\alpha_i, \beta_j\}, \alpha_0, \ldots, \hat{\alpha}_i, \ldots, \alpha_{p-1}, \beta_0, \ldots, \hat{\beta}_j, \ldots, \beta_q) +$$

$$+ \sum_{i<j=0}^{q} (-1)^{i+j} Q(\alpha_0, \ldots, \alpha_{p-1}, \{\beta_i, \beta_j\}, \beta_0, \ldots, \hat{\beta}_i, \ldots, \hat{\beta}_j, \ldots, \beta_q).$$

In these formulas, α have type (1,0) and β have type $(0,1)$. $\{\beta_i, \beta_j\}$ is not homogeneous and it has a decomposition

$$\{\beta_i, \beta_j\} = \{\beta_i, \beta_j\}' + \{\beta_i, \beta_j\}'' \tag{4.31}$$

into components of type (1,0) and (0,1), respectively. In (4.29), only the first component $\{\ ,\ \}'$ is relevant, and in (4.30) only the second one $\{\ ,\ \}''$. Notice also:

4.7. Exercise. If β_1, β_2 are 1-forms of type (0,1), and X is a vector field of type (1,0), one has

$$\{\beta_1, \beta_2\}(X) = (L_X w)(\beta_1, \beta_2). \tag{4.32}$$

Indeed, it will be enough to play a little bit with the formula (4.3), and the expression of the Lie derivative.

Now, let us also indicate an interesting particular class of manifolds.

4.8. Definition [Va9]. A regular Poisson manifold (M, w) will be called *transversally constant* if its symplectic foliation $\mathcal{S}$ has a transversal distribution $\nu\mathcal{S}$ such that every local vector field X which satisfies the properties: i) $X \in \nu\mathcal{S}$, ii) X is an infinitesimal automorphism of the foliation $\mathcal{S}$, is an infinitesimal automorphism of the Poisson structure w.

For instance, if $M = S \times N$, and w is defined by the Hamiltonian vector fields with respect to a fixed symplectic structure of S, then $\nu\mathcal{S} = TN$ has, obviously, the requested property, and w is transversally constant. In particular, the use of the canonical coordinates of Definition 2.18 shows that every regular Poisson manifold is locally transversally constant.

A more interesting example is given by the Dirac bracket of a triple $(M,\omega,\mathcal{S})$ where ω is a symplectic form. Then, we take $\nu\mathcal{S}$ to be the ω-orthogonal distribution of $\mathcal{S}$, and for $X \in \nu\mathcal{S}$, $Y_1, Y_2 \in T\mathcal{S}$, $d\omega(X,Y_1,Y_2) = 0$ is the same relation as $(L_X\omega)(Y_1,Y_2) = 0$. If $Y_1 = X_f$, $Y_2 = X_g$ (Hamiltonian vector fields with respect to the Dirac bracket) for two functions $f, g \in C^\infty(M)$, the previous relation becomes

$$X(\omega(X_f,X_g)) = \omega(X_f,[X,X_g]) - \omega(X_g,[X,X_f]).$$

Now, let us add the condition that X preserves the foliation $\mathcal{S}$ i.e., $[X,X_g]$, $[X,X_f] \in T\mathcal{S}$. Then, in the second member of the previous equality, X_f, X_g may be replaced by $\tilde{X}_f, \tilde{X}_g$ = the Hamiltonian vector fields of f, g with respect to the symplectic structure ω (see Proposition 3.6), and we get

$$\begin{aligned} X\{f,g\} &= dg([X,X_f]) - df([X,X_g]) = [X,X_f](g) - [X,X_g](f) = \\ &= XX_fg - X_fXg - XX_gf + X_gXf = 2X\{f,g\} - \{f,Xg\} + \{g,Xf\}, \end{aligned}$$

whence

$$X\{f,g\} = \{Xf,g\} + \{f,Xg\}.$$

Hence, X is a Poisson infinitesimal automorphism of the Dirac bracket, as requested.

Finally, by looking at the explanations given for the example f) of Chapter 3, we see immediately that any parallel Poisson structure of a Riemannian manifold is transversally constant. The corresponding transversal distribution is just the orthogonal distribution.

Now, what we want to prove is

4.9. Proposition [Va9]. *For a transversally constant regular Poisson manifold one has*

$$\sigma' = 0, \qquad \sigma = \sigma''. \tag{4.33}$$

Proof. It should be understood that σ', σ'' of (4.33) are with respect to the $\nu\mathcal{S}$ of Definition 4.8. Let β_1, β_2 be 1-forms of the type $(0,1)$, $x_0 \in M$, and $X_{x_0} \in \nu_{x_0}\mathcal{S}$. In order to evaluate $\{\beta_1,\beta_2\}_{x_0}(X_{x_0}) = \{\beta_1,\beta_2\}'_{x_0}(X_{x_0})$, we shall extend X_{x_0} to a local field X of type (1,0), and which preserves the foliation $\mathcal{S}$ of the Poisson structure. Then, formula (4.32), and Definition 4.8 yield $\{\beta_1,\beta_2\}'(X) = 0$. Therefore, the general relation $\{\beta_1,\beta_2\}' = 0$ holds good, and (4.33) follows from (4.29) and (4.27). Q.e.d.

We know already that one can also speak of bihomogeneous forms, and a procedure similar to the one used in Proposition 4.5 yields a type decomposition of the exterior differential d as follows (e.g., [Va2])

$$d = d' + d'' + d_{(2,-1)} \tag{4.34}$$

where type $d' = (1,0)$, type $d'' = (0,1)$, and d'' is just the exterior differential along the leaves of $\mathcal{S}$. The relation with the type of the multivector fields is given by

4.10. Proposition. i) $\#(\Lambda^{pq}M) = 0$ *if* $p > 0$, *and* $\#(\Lambda^{0q}M) = \mathcal{V}^{0q}(M)$; ii) $\sigma\lambda^{\#} = \sigma''\lambda^{\#} = -(d''\lambda)^{\#}$ *for all* $\lambda \in \Lambda^{*}M$.

Proof. The results are immediate consequences of (4.11) and (4.12). For the last part of i), if $Q \in \mathcal{V}^{0q}(M)$, there is a well defined form $\lambda \in \Lambda^{0q}M$ such that $Q = \lambda^{\#}$, where $\lambda(Y_1, \ldots, Y_q) = Q(\#^{-1}Y_1, \ldots, \#^{-1}Y_q)$ where $\#^{-1}Y_i$ exists (and it is defined up to $\ker \#$) since type $Y_i = (0,1)$. For ii), $\sigma'\lambda^{\#} = 0$ because it has a negative first degree. Q.e.d.

In order to obtain a more general relation, let us also take a Euclidean metric G on $\nu^{*}\mathcal{S}$, and the isomorphism

$$\tilde{\#} \stackrel{\text{def}}{=} \#_G \oplus \#_w : \nu^{*}\mathcal{S} \oplus T^{*}\mathcal{S} \to \nu\mathcal{S} \oplus T\mathcal{S}, \tag{4.35}$$

which extends to $\lambda \in \Lambda^{pq}M$ by

$$\lambda^{\tilde{\#}}(\alpha_1, \ldots, \alpha_p, \beta_1, \ldots, \beta_q) = (-1)^q \lambda(\alpha_1^{\#_G}, \ldots, \alpha_p^{\#_G}, \beta_1^{\#_w}, \ldots, \beta_q^{\#_w}). \tag{4.36}$$

In (4.36), again, type $\alpha_i = (1,0)$, type $\beta_i = (0,1)$. This extension is, of course, an isomorphism $\tilde{\#}$: $\Lambda^{pq}M \to \mathcal{V}^{pq}(M)$. For the next results, we also agree that, always, type $X = (1,0)$, type $Y = (0,1)$, and $\tilde{\#}\alpha = X$, $\tilde{\#}\beta = Y$.

4.11. Proposition. *For* $\lambda \in \Lambda^{pq}M$ *one has*

$$(\tilde{\#}^{-1}\sigma''\tilde{\#}\lambda = -d''\lambda + \mathcal{L}_G\lambda, \tag{4.37}$$

where $\mathcal{L}_G\lambda$ *is a form of type* $(p, q+1)$ *defined by*

$$\begin{gathered}(\mathcal{L}_G\lambda)(X_0, \ldots, X_{p-1}, Y_0, \ldots, Y_q) = \sum_{i=0}^{p-1}\sum_{j=0}^{q}(-1)^{p+i+j}. \\ \cdot \lambda([(L_{Y_j}G^{*})(X_i, .)]^{\#_G}, X_0, \ldots, \hat{X}_i, \ldots, X_{p-1}, Y_0, \ldots, \hat{Y}_j, \ldots, Y_q),\end{gathered} \tag{4.38}$$

where G^{*} *is the dual metric of* G *on* $\nu\mathcal{S}$.

Proof. Using (4.30), and our notational conventions, we get

$$(\sigma''\lambda^{\#})(\alpha_0,\dots,\alpha_{p-1},\beta_0,\dots,\beta_q) = (-1)^q\left\{\sum_{i=0}^{q}(-1)^{p+i}\cdot\right. \tag{4.39}$$

$$\cdot Y_i(\lambda(X_0,\dots,X_{p-1},Y_0,\dots,\hat{Y}_i,\dots,Y_q) + \sum_{i=0}^{p-1}\sum_{j=0}^{q}(-1)^{p+i+j}\cdot$$

$$\cdot\lambda(\{\alpha_i,\beta_j\}^{\#_G},X_0,\dots,\hat{X}_i,\dots,X_{p-1},Y_0,\dots,\hat{Y}_j,\dots,Y_q)+$$

$$\left.+\sum_{i<j}^{q}(-1)^{i+j}\lambda(X_0,\dots,X_{p-1},\{\beta_i,\beta_j\}^{\#_w},Y_0,\dots,\hat{Y}_i,\dots,\hat{Y}_j,\dots,Y_q)\right\}.$$

But, since β are of type $(0,1)$, $\{\beta_i,\beta_j\}^{\#_w} = [Y_i,Y_j]$. On the other hand, $\forall X$ of type (1,0) we have

$$\{\alpha_i,\beta_j\}(X) = -(L_{Y_j}\alpha_i)(X) = -Y_j(\alpha_i(X)) + \alpha_i([Y_j,X]), \tag{4.40}$$

and since $\alpha_i(X) = G^*(X_i,X)$, where G^* is induced by G on νS (this is equivalent to $\alpha_i^{\#_G} = X_i$), the previous equality becomes easily

$$\{\alpha_i,\beta_j\}(X) = -(L_{Y_j}G^*)(X_i,X) + \#_G^{-1}([X_i,Y_j])(X). \tag{4.41}$$

If these results are replaced in (4.39), we get

$$(\tilde{\#}^{-1}\sigma''\tilde{\#}\lambda)(X_0,\dots,X_{p-1},Y_0,\dots,Y_q) = -\left\{\sum_{i=0}^{q}(-1)^{p+i}\cdot\right.$$

$$\cdot Y_i(\lambda(X_0,\dots,X_{p-1},Y_0,\dots,\hat{Y}_i,\dots,Y_q)) + \sum_{i=0}^{p-1}\sum_{j=0}^{q}(-1)^{p+i+j}\cdot$$

$$\cdot\lambda([X_i,Y_j],X_0,\dots,\hat{X}_i,\dots,X_{p-1},Y_0,\dots,\hat{Y}_j,\dots,Y_q)+$$

$$\left.+\sum_{i<j=0}^{q}(-1)^{i+j}\lambda(X_0,\dots,X_{p-1},[Y_i,Y_j],Y_0,\dots,\hat{Y}_i,\dots,\hat{Y}_j,\dots,Y_q)\right\}+$$

$$+\sum_{i=0}^{p-1}\sum_{j=0}^{q}(-1)^{p+i+j}\lambda([(L_{Y_j}G^*)(X_i,\cdot)]^{\#_G},X_0,\dots,\hat{X}_i,\dots,X_{p-1},$$

$$Y_0,\dots,\hat{Y}_j,\dots,Y_q),$$

and this is exactly the requested result. The fact that the first large parenthesis is exactly d'' follows if we establish the formula of this operator in exactly the same way as we did for σ''. (See [Va2] if you need more details.)

4.12. Remarks. 1) A similar formula is obtained if G is replaced by a symplectic metric on $\nu\mathcal{S}$.

2) The operator σ' is not related to the decomposition of d, and using (4.29) one gets

$$(\tilde{\#}^{-1}\sigma'\lambda^{\tilde{\#}})(X_0,\ldots,X_{p-2},Y_0,\ldots,Y_{q+1}) =$$
$$= \sum_{i<j=0}^{q+1}(-1)^{i+j+1}\lambda(\#_G\{\#_w^{-1}Y_i,\#_w^{-1}Y_j\}',X_0\ldots \tag{4.42}$$
$$\ldots,X_{p-2},Y_0,\ldots,\hat{Y}_i,\ldots,\hat{Y}_j,\ldots,Y_{q+1}).$$

In this context, we shall also add

4.13. Proposition. *If (M,w) is a regular Poisson manifold as above, the Poisson codifferential δ_w has a decomposition*

$$\delta_w = \delta'_w + \delta''_w, \tag{4.43}$$

where type $\delta' = (1,-2)$ and type $\delta'' = (0,-1)$.

Proof. Let us take the form λ to be of type (p,q) $(p+q=k)$ in formula (4.22′). We can write like for any form

$$\delta\lambda = \sum_{u+v=k-1}(\delta\lambda)_{uv}, \tag{4.44}$$

and the (u,v)-component $(\delta\lambda)_{uv}$ will be computed if we compute $i(Q)\delta\lambda$ for $Q\in\mathcal{V}^{uv}(M)$, which can be done by (4.22′).

Namely, using (4.27) and (4.34), (4.22′) becomes

$$\begin{aligned} i(Q)\delta\lambda = i(\sigma'Q)\lambda + i(\sigma''Q)\lambda - (-1)^k i(w)d'(i(Q)\lambda) - \\ - (-1)^k i(w)d''(i(Q)\lambda) - (-1)^k i(w)d_{2,-1}(i(Q)\lambda). \end{aligned} \tag{4.45}$$

Since $i(Q)\lambda$ has either the type (1,0) or (0,1), and w has the type (0,2), the 3rd and 5th terms of the result of (4.45) vanish, and the other terms have respectively the types $(p-u+1,q-v-2)=(0,0)$, $(p-u,q-v-1)=(0,0)$ and again $(p-u,q-v-1)=(0,0)$. Hence nonzero results can be obtained only if either $(u,v)=(p+1,q-2)$ or $(u,v)=(p,q-1)$, and the respective components are

$$\delta'\lambda \overset{\text{def}}{=} (\delta\lambda)_{p+1,q-2}, \quad \delta''\lambda \overset{\text{def}}{=} (\delta\lambda)_{p,q-1}, \tag{4.46}$$

where these operators will be determined by the formulas

$$i(Q^{p+1,q-2})(\delta'\lambda) = i(\sigma'Q)\lambda \tag{4.47}$$
$$i(Q^{p,q-1})(\delta''\lambda) = i(\sigma''Q)\lambda - (-1)^k i(w)d''(i(Q)\lambda). \tag{4.48}$$

Q.e.d.

4.4 Cofoliations

Now we come back to an arbitrary Poisson manifold (M, w). An interesting utilization of the operator σ, inspired by the classical Frobenius theorem is indicated in

4.14. Definition. Let L be a rank k subbundle of TM (i.e., a regular k-dimensional distribution). We shall say that L is a k-dimensional *cofoliation of* (M, w) if, for every local cross-section X of L, $\sigma X = 0$(modulo L).

In other words, if (X_a) $(a = 1, \ldots, k)$ is a local basis of L, we have

$$\sigma X_a = -L_{X_a} w = X_b \wedge Y_a^b,$$

where Y_a^b are local vector fields on M. Using also a complementary distribution L' of L, with a basis (Y_u), and using the expression (4.8) of σ, it is easy to check

4.15. Proposition. *L is a cofoliation iff its annihilator* Ann L *is closed with respect to the bracket* $\{\ ,\ \}$ *of* 1*-forms.*

Following is a number of examples:

a) Any $(n-1)$-dimensional subbundle L of TM is a cofoliation. Indeed, then Ann L is 1-dimensional.

b) If M is a symplectic manifold, L is a cofoliation iff its symplectic-orthogonal bundle L' is a foliation. This follows by applying $\#^{-1}$ (which exists in the symplectic case) to σX_a, where X_a is a local basis of L. If L is also a foliation, we are in the case of a complete foliation of P. Libermann [Lb3]. Hence, it is natural to define a *Libermann foliation* of a Poisson manifold as a regular foliation which is also a cofoliation.

c) The orbits of a locally free action of a Lie group G on M by Poisson automorphisms define a Libermann foliation of M in the above mentioned sense. Indeed, these orbits have a basis (X_a) such that $\sigma X_a = -L_{X_a} w = 0$. Hence their tangent bundle is a cofoliation.

d) The symplectic distribution $\mathcal{S}(M)$ of a regular Poisson manifold (M, w) is a cofoliation. Indeed, during the proof of Proposition 4.5 we saw that if $\alpha, \beta \in$ Ann $\mathcal{S}(M)$ then $\{\alpha, \beta\} = 0$.

e) Let $\varphi : (M, w) \to (P, \pi)$ be a submersion which is a Poisson mapping (Definition 1.17). Then, if L is a cofoliation of P, $\varphi_*^{-1}(L)$ is a cofoliation of M. Indeed, we have $\text{Ann}(\varphi_*^{-1}(L)) = \varphi^*(\text{Ann } L)$, and, on the other hand, (1.41), (4.1) and (4.2) allow us to establish that $\{\varphi^*\alpha, \varphi^*\beta\} = \varphi^*\{\alpha, \beta\}$ by a local coordinate computation. Via Proposition 4.15, we get the desired result.

4.16. Remark. If L is a cofoliation of (M, w), $L_0 = \#\text{Ann}\, L$ is a general distribution on M, and (4.4) shows that L_0 is involutive. Moreover, by Proposition 4.15 $(\text{Ann}\, L, \{\ ,\ \}, \#)$ is a *Lie algebroid* as mentioned in Remark 4.2, and the general theory of Lie algebroids (e.g., [DS1]) tells us that, in fact, L_0 is completely integrable.

4.5 Contravariant derivatives on vector bundles

The next element of the Poisson calculus will be the notion of a *contravariant derivative* [Va10].

4.17. Definition. Let $\pi : E \to M$ be a vector bundle over the Poisson manifold (M, w). A *contravariant derivative* D on E consists of $\mathbb{R}$-linear operators

$$D_\alpha : \Gamma(E) \to \Gamma(E) \qquad (\alpha \in \Lambda^1 M), \tag{4.49}$$

where $\Gamma(E)$ is the space of the cross section of E, which are also $\mathbb{R}$-linear with respect to α, and satisfy the conditions

$$\begin{gathered} D_{f\alpha}s = fD_\alpha s, \quad D_\alpha(fs) = fD_\alpha s + \alpha(\sigma f)s \\ (f \in C^\infty(M),\ s \in \Gamma(E),\ \alpha(\sigma f) = \alpha^\# f). \end{gathered} \tag{4.50}$$

For instance if ∇ is a usual covariant derivative on E, and we put $D_\alpha = \nabla_{\alpha^\#}$, we obtain a contravariant derivative. This remark shows that contravariant derivatives always exist, but they may be more general; in fact, the equality $D_\alpha = \nabla_{\alpha^\#}$ of the example shows that contravariant derivatives can be identified with *partial connections* on E i.e., covariant derivatives along the leaves of the symplectic foliation $\mathcal{S}(M)$.

We shall also define a corresponding *curvature*:

$$C(\alpha, \beta)s = D_\alpha D_\beta s - D_\beta D_\alpha s - D_{\{\alpha,\beta\}}s \tag{4.51}$$

$(\alpha, \beta \in \Lambda^1 M)$, $s \in \Gamma(E))$, which satisfies $C(\alpha, \beta) = -C(\beta, \alpha)$, and

$$C(f\alpha, \beta) = fC(\alpha, \beta). \tag{4.52}$$

As in the classical case, we can use D_α to define a contravariant differential of *E-valued multivectors* i.e., sections $\mathbf{Q}$ of $\mathcal{V}^k(M) \otimes E$, by

$$\begin{gathered} (\sigma\mathbf{Q})(\alpha_0, \ldots, \alpha_k) = \sum_{i=0}^{k}(-1)^i D_{\alpha_i}(\mathbf{Q}(\alpha_0, \ldots, \hat{\alpha}_i, \ldots, \alpha_k) + \\ + \sum_{i<j=0}^{k}(-1)^{i+j}\mathbf{Q}(\{\alpha_i, \alpha_j\}, \alpha_0, \ldots, \hat{\alpha}_i, \ldots, \hat{\alpha}_j, \ldots, \alpha_k). \end{gathered} \tag{4.53}$$

Furthermore, we can extend D_α to E^*, End (E), etc. by the usual formulas

$$< D_\alpha s^*, s > = \alpha^\# < s^*, s > - < s^*, D_\alpha s >, \tag{4.54}$$

$$(D_\alpha\varphi)(s) = D_\alpha(\varphi s) - \varphi(D_\alpha s), \tag{4.55}$$

where $s \in \Gamma(E)$, $s^* \in \Gamma(E^*)$, $\varphi \in \Gamma(\text{End } E)$. Of course, the curvature $C \in \Gamma(\mathcal{V}^2(M) \otimes (\text{End } E))$, and, by applying (4.53), we obtain after a straightforward calculation (which, in fact, is not needed because it will be exactly the same as for the classical curvature of a connection on E)

$$\sigma C = 0. \tag{4.56}$$

This is the *contravariant Bianchi identity.*

Like for connections ∇, (4.50) ensures the local character of the operator D_α. If (e_i) $(i = 1, \ldots, \text{rank } E = r)$ is a local basis of cross sections of E, we shall have some expressions of the form

$$D_\alpha e_i = t_i^j(\alpha) e_j, \tag{4.57}$$

where t_i^j are local vector fields. A change of the basis, $\tilde{e}_i = a_i^j e_j$, implies

$$\tilde{t}_i^j = a_i^h b_k^j t_h^k + b_k^j (\sigma a_i^k) \qquad (a_k^p b_p^i = \delta_k^i). \tag{4.58}$$

Furthermore, we shall have some expressions

$$C(\alpha, \beta) e_i = C_i^j(\alpha, \beta) e_j, \tag{4.59}$$

where

$$C_i^j = \sigma t_i^j - t_i^k \wedge t_k^j, \tag{4.60}$$

and, if the basis is changed as above, these local bivectors satisfy

$$\tilde{C}_i^j = a_i^h b_k^j C_h^k. \tag{4.61}$$

Now, if $\mathbf{Q}$ of (4.53) is $Q^i e_i$, and we use (4.57) it follows from (4.53) that $\sigma \mathbf{Q}$ has the coefficients

$$(\sigma \mathbf{Q})^i = \sigma Q^i + t_j^i \wedge Q^j. \tag{4.62}$$

Furthermore, if e_*^i is the dual basis in E^*, (4.54) yields

$$D e_*^i = -t_j^i e_*^j \tag{4.63}$$

(the argument α of D_α is usually omitted), and then, from $C = C_j^i e_i \otimes e_*^j$, we get the following form of the Bianchi identity (4.56)

$$\sigma C_j^i + t_h^i \wedge C_j^h - t_j^h \wedge C_h^i = 0. \tag{4.64}$$

4.18. Exercise. Let $\mathbf{F}$ be an (End E)-valued multivector field and $\mathbf{F} = \varphi_i^j e_j \otimes e_*^i$. Then, a computation which uses (4.55), (4.53) and (4.60) gives the matrices of the components of $\sigma\mathbf{F}$ and $\sigma^2\mathbf{F}$:

$$(\sigma\mathbf{F}) = \sigma\varphi - [t, \varphi], \quad (\sigma^2\mathbf{F}) = [\varphi, C], \tag{4.65}$$

where $\varphi = (\varphi_i^j)$, $t = (t_i^j)$, $C = (C_i^j)$ are the corresponding matrices, and the bracket is the exterior product commutant of matrices:

$$[\lambda, \mu] = \lambda \wedge \mu + (-1)^{\deg \lambda . \deg \mu} \mu \wedge \lambda.$$

Another interesting result is

4.19. Proposition. *Every cofoliation L of the Poisson manifold (M, w) admits contravariant derivatives D whose curvature $C(\alpha, \beta)$ vanishes for $\alpha, \beta \in$ Ann L.*

Proof. This is just the contravariant version of a well known theorem of Bott in foliation theory [Bt]. The desired operator D will be obtained from a contravariant derivative, also denoted by D, on the vector bundle L^*, dual to L, which is, of course a quotient bundle of T^*M. Namely, we shall use a splitting $TM = L \oplus L'$, and an arbitrary $\tilde{D}$ on L^*. Then put

$$D_\lambda \theta = \tilde{D}_\lambda \theta, \quad D_\alpha \theta = pr_L * \{\alpha, \theta\}$$

($\alpha \in L'^* =$ Ann L; $\lambda, \theta \in L^*$). Then, the conditions of Definition 4.17 are easily verified, and the fact that Ann L is closed by the bracket $\{\ ,\ \}$ gives us, after an easy computation, $C(\alpha, \beta) = 0$ for $\alpha, \beta \in$ Ann L.

Furthermore, $\forall X \in L$, $\forall \gamma \in T^*M$, and α, θ as above, put

$$(D'_\gamma X)(\alpha) = 0, \quad (D'_\gamma X)(\theta) = \gamma^\#(X(\theta)) - X(D_\gamma \theta).$$

Again, the conditions of Definition 4.17 are satisfied, and the curvature C' of the new derivative is such that

$$(C'(\alpha, \beta)(X))(\theta) = -(C(\alpha, \beta)\theta)(X).$$

Q.e.d.

In the symplectic case, the required contravariant derivative comes from the Bott connection of the symplectic-orthogonal foliation of L.

4.6 More brackets

On a Poisson manifold, several other interesting bracket operations can be built as extensions of the Poisson bracket to differential forms and multivectors. This extension was done by Michor [Mh] for the symplectic manifolds, and by Koszul [Kz] and Cabras and Vinogradov [CV] for the Poisson manifolds (see also [Kr] and [KSM1,2]).

4.20. Definition. A *graded Lie algebra* is a graded vector space $\mathcal{A} = \oplus_k \mathcal{A}^k$, endowed with a bracket multiplication [,] that satisfies the conditions

i) $[\lambda, \mu] \in \mathcal{A}^{p+m}$

ii) $[\lambda, \mu] = -(-1)^{pm}[\mu, \lambda]$

iii) $(-1)^{ph}[\lambda, [\mu, \nu]] + (-1)^{mp}[\mu, [\nu, \lambda]] + (-1)^{hm}[\nu, [\lambda, \mu]] = 0,$

where $\lambda \in \mathcal{A}^p$, $\mu \in \mathcal{A}^m$, $\nu \in \mathcal{A}^h$.

These properties mean that the multiplication is *graded*, and *graded-skew-commutative*, and it satisfies the *graded Jacobi identity*.

This is an interesting algebraic structure, and it turns out that a Poisson manifold (M, w) bears several such graded Lie algebras. First, formulas (1.9) and (1.11) mean

4.21. Proposition [Kz]. *Put* $\mathcal{A}^k = \mathcal{V}^{k+1}(M)$, and $[P, Q] = (-1)^{p+1}[P, Q]$, *where* $P \in \mathcal{V}^p(M)$, $Q \in \mathcal{V}^q(M)$, *and we have the Schouten-Nijenhuis bracket in the right hand side. Then* $(\oplus \mathcal{A}^k, [\ ,\ ])$ *is a graded Lie algebra.*

Proof. The same P, Q belong respectively to $\mathcal{A}^{p-1}, \mathcal{A}^{q-1}$, hence, i) of Definition 4.20 holds, and it is straightforward to check that (1.9) means ii), and (1.11) means iii) of Definition 4.20. Q.e.d.

4.22. Remark. Formula (1.10) can be accommodated in this structure as follows. If a graded vector space $\oplus_k A^k$ is endowed with an associative product denoted $\wedge$ which makes it into an algebra, and if $\lambda \wedge \mu \in \mathcal{A}^{p+m+k}$ for a fixed $k(\lambda \in \mathcal{A}^p, \mu \in \mathcal{A}^m)$, $\wedge$ is said to be a *k-graduated product*, and it is k-graded-commutative if

$$\lambda \wedge \mu = (-1)^{(p+k)(m+k)} \mu \wedge \lambda. \tag{4.65}$$

Now, a graded Lie algebra will be said to be given the structure of a *graded Poisson algebra of index k* if it is given a product $\wedge$ as above, with (4.65), such that

$$[\lambda \wedge \mu, \nu] = \lambda \wedge [\mu, \nu] + (-1)^{(p+k)(m+k)} \mu \wedge [\lambda, \nu]. \tag{4.66}$$

(Compare this with the definition of a graded Poisson algebra in [Hu2].) The comparison of (1.10) and (4.66) shows that the exterior product of multivectors makes the graded Lie algebra of Proposition 4.21 into a graded Poisson algebra of index 1. Notice that the algebra of functions of a Poisson manifold can be seen as a graded Poisson algebra of index 0, if we take $\mathcal{A}^0 = C^\infty(M)$, and $\mathcal{A}^k = \{0\}$ for $k \neq 0$.

Now, essentially like in [CV], we can get another graded Lie algebra, where the grades are the "natural" ones. Take $\mathcal{V}(M) = \oplus_k \mathcal{V}^k(M)$, and define the σ-*bracket*

$$[P, Q]_\sigma = -[\sigma P, Q] = [[w, P], Q]. \tag{4.67}$$

The degree of the result is $p + q$, and the use of (1.11) gives

$$\sigma([P, Q]) = -[\sigma P, Q] - (-1)^p[P, \sigma Q], \tag{4.68}$$

whence

$$[P,Q]_\sigma = -(-1)^{pq}[Q,P]_\sigma + \sigma([P,Q]). \tag{4.69}$$

This formula suggests looking at the graded vector space

$$\mathcal{V}_\sigma(M) \stackrel{\text{def}}{=} \mathcal{V}(M)/_{\sigma(\mathcal{V}(M))}. \tag{4.70}$$

The σ-bracket induces a σ-bracket in $\mathcal{V}_\sigma(M)$ because of $\sigma^2 = 0$, and the latter is graded skew-commutative because of (4.69). Notice that, in view of (4.68), the bracket of $\mathcal{V}_\sigma(M)$ is also induced by

$$[P,Q]'_\sigma = (-1)^p[P,\sigma Q]. \tag{4.71}$$

Now, using again (1.11), we can make the following computation:

$$\begin{aligned}(-1)^{pr}[P,[Q,R]_\sigma]_\sigma &= (-1)^{pr}[\sigma P,[\sigma Q,R]] = \\ &= (-1)^{p+r}((-1)^{p(q+1)}[\sigma Q,[R,\sigma P]] + (-1)^{rq}[R,[\sigma P,\sigma Q]]) = \\ &= -(-1)^{qp}[Q,[R,P]'_\sigma]_\sigma + (-1)^{qr+p+r}[R,[\sigma P,\sigma Q]].\end{aligned} \tag{4.72}$$

For the last term here, we get using (4.68)

$$[R,[\sigma P,\sigma Q]] = -(-1)^{p+1}[R,\sigma[\sigma P,Q]] = (-1)^{r+p+1}[R,[P,Q]_\sigma]'_\sigma.$$

If this is replaced in (4.72) we get

$$(-1)^{pr}[P,[Q,R]_\sigma]_\sigma + (-1)^{qp}[Q,[R,P]'_\sigma]_\sigma + (-1)^{rq}[R,[P,Q]_\sigma]'_\sigma = 0. \tag{4.73}$$

In the quotient $\mathcal{V}_\sigma(M)$, this yields the graded Jacobi identity, hence, we proved ([CV], etc.)

4.23. Proposition. *$\mathcal{V}_\sigma(M)$ with the σ-bracket is a graded Lie algebra.*

Further graded Lie algebras will be obtained by constructing covariant versions of the algebras given by Propositions 4.21 and 4.23. That a *covariant Schouten-Nijenhuis bracket* should exist is suggested by formula (1.12) which can be reformulated for 1-forms, using Theorem 4.1. But, to avoid problems of invariance, we shall proceed in a different way suggested itself by formula (1.26). First, notice that, for $\alpha, \beta \in \Lambda^1 M$, one has

$$\{\alpha,\beta\} = (\delta\alpha)\beta - \alpha(\delta\beta) - \delta(\alpha\wedge\beta). \tag{4.74}$$

This can be easily verified for $\alpha = f_1 dg_1$, $\beta = f_2 dg_2$ using (4.21), and the formulas of Theorem 4.1, and, then, the general case is a straightforward consequence. Accordingly, if $\lambda \in \Lambda^k M$, and $\mu \in \Lambda^h M$,we shall define the *covariant Schouten-Nijenhuis bracket* defined and studied in [Kz] and [KSM2]

$$\{\lambda,\mu\} = (\delta\lambda)\wedge\mu + (-1)^{\mathbf{k}}\lambda\wedge(\delta\mu) - \delta(\lambda\wedge\mu), \tag{4.75}$$

and, of course, this is an invariant operation. (If λ and μ are functions, the result is 0 and not the Poisson bracket of the functions!)

The definition implies easily

$$\{\lambda, \mu\} = (-1)^{kh}\{\mu, \lambda\}. \tag{4.76}$$

Now, if we take in (4.75)

$$\lambda = f_0 df_1 \wedge \ldots \wedge df_k, \quad \mu = g_0 dg_1 \wedge \ldots \wedge dg_h,$$

and apply (4.21), we obtain

$$\{\lambda, \mu\} = \sum_{i=1}^{k}(-1)^i f_0\{g_0, f_i\} df_1 \wedge \ldots \wedge d\hat{f}_i \wedge \ldots \wedge df_k \wedge \tag{4.77}$$

$$\wedge\, dg_1 \wedge \ldots \wedge dg_h + \sum_{j=1}^{h}(-1)^{k+j} g_0\{f_0, g_j\} df_1 \wedge \ldots \wedge df_k \wedge dg_1 \wedge$$

$$\ldots \wedge d\hat{g}_j \wedge \ldots \wedge dg_h - \sum_{i=1}^{k}\sum_{j=1}^{h}(-1)^{i+j+k} f_0 g_0 d\{f_i, g_j\} \wedge$$

$$\wedge\, df_1 \wedge \ldots \wedge d\hat{f}_i \wedge \ldots \wedge df_k \wedge dg_1 \wedge \ldots \wedge d\hat{g}_j \wedge \ldots \wedge dg_h.$$

A first consequence of (4.77) is the coordinate expression of this operation. If

$$\lambda = \frac{1}{k!}\lambda_{i_1 \ldots i_k} dx^{i_1} \wedge \ldots \wedge dx^{i_k}, \quad \mu = \frac{1}{h!}\mu_{j_1 \ldots j_h} dx^{j_1} \wedge \ldots \wedge dx^{j_h},$$

the components of $\{\lambda, \mu\}$ as given by (4.77) are

$$\begin{aligned} \{\lambda, \mu\}_{\alpha_1 \ldots \alpha_{k+h-1}} &= \frac{-1}{(k-1)!h!}\delta^{i_2 \ldots i_k j_1 \ldots j_h}_{\alpha_1 \ldots \ldots \alpha_{k+h-1}} w^{su} \lambda_{u i_2 \ldots i_k} \cdot (\partial_s \mu_{j_1 \ldots j_h}) - \\ &- \frac{(-1)^k}{(h-1)!k!}\delta^{i_1 \ldots i_k j_2 \ldots j_h}_{\alpha_1 \ldots \ldots \alpha_{k+h-1}} w^{su} (\partial_s \lambda_{i_1 \ldots i_k}) \cdot \mu_{u j_2 \ldots j_h} - \\ &- \frac{(-1)^k}{(h-1)!(k-1)!}\delta^{s i_2 \ldots i_k j_2 \ldots j_h}_{\alpha_1 \ldots \ldots \alpha_{k+h-1}} (\partial_s w^{uv}) \lambda_{u i_2 \ldots i_k} \mu_{v j_2 \ldots j_h}. \end{aligned} \tag{4.78}$$

Since the invariance of the operation is already known, one may replace in (4.78) the partial derivatives by covariant derivatives with respect to any torsionless connection of M.

A second consequence is a formula corresponding to (1.12). In order to prove it, notice first, by applying (4.77) for $\lambda = f$, $\mu = g_0 dg_1 \wedge \ldots \wedge dg_h$, that one has the general formula

$$\{f, \mu\} = -i(X_f)\mu, \tag{4.79}$$

whence, with (4.75)

$$\delta(f\mu) = f\delta\mu + i(X_f)\mu, \tag{4.80}$$

and, after that

$$\{\lambda, f\mu\} = f\{\lambda,\mu\} - (i(X_f)\lambda)\wedge\mu. \tag{4.81}$$

(If, instead, λ is replaced by $f\lambda$, the corresponding result will follow by using (4.76).)

Now, let us define a bracket of decomposable forms by formula (1.12) i.e., for $\alpha_i, \beta_j \in \Lambda^1 M$ put

$$\{\alpha_1\wedge\ldots\wedge\alpha_k, \beta_1\wedge\ldots\wedge\beta_h\}' = (-1)^{k+1}\sum_{i=1}^{k}\sum_{j=1}^{h}(-1)^{i+j}\{\alpha_i,\beta_j\}\wedge \tag{4.82}$$
$$\wedge\,\alpha_1\wedge\ldots\wedge\hat{\alpha}_i\wedge\ldots\wedge\alpha_k\wedge\beta_1\wedge\ldots\wedge\hat{\beta}_j\wedge\ldots\wedge\beta_h.$$

Then, the following facts are immediate: a) the bracket $\{\ ,\ \}'$ satisfies (4.76) and (4.81); b) if $\alpha_i = df_i$ and $\beta_j = dg_j$, $\{\ ,\ \}' = \{\ ,\ \}$ as given by (4.77). Hence $\{\ ,\ \}' = \{\ ,\ \}$ is always true, and we may see (4.82) as a true formula for the bracket defined by (4.75).

At this moment, we may write down

4.24. Proposition. *The bracket* $\{\ ,\ \} : \Lambda^k M \times \Lambda^h M \to \Lambda^{k+h-1}M$, *which satisfies the commutation equality* (4.76) *also satisfies the equalities*

$$(-1)^{k(p-1)}\{\lambda,\{\mu,\nu\}\} + (-1)^{h(k-1)}\{\mu,\{\nu,\lambda\}\} + \tag{4.83}$$
$$+\,(-1)^{p(h-1)}\{\nu,\{\lambda,\mu\}\} = 0,$$

$$\{\lambda,\mu\wedge\nu\} = \{\lambda,\mu\}\wedge\nu + (-1)^{kh+h}\mu\wedge\{\lambda,\nu\}, \tag{4.84}$$

$$\{\lambda,\mu\}^\# = [\lambda^\#,\mu^\#], \tag{4.85}$$

where $\lambda \in \Lambda^k M$, $\mu \in \Lambda^h M$, $\nu \in \Lambda^p M$.

Proof. If we look back to Theorem 1.1, it is clear there that the formulas (1.11) and (1.10) are algebraic consequences of (1.12) (in spite of the fact that (1.10) was proven otherwise). Hence, in exactly the same way, (4.83) and (4.84) will follow from (4.82). Finally, in (4.85), # is the one defined in (4.11), and it is clearly compatible with the exterior product. Hence, if # is applied to (4.82) one gets (1.12), and this clearly proves (4.85). Q.e.d.

4.25. Remark. By a differentiation of (4.75), and using $d\delta + \delta d = 0$, one gets the interesting formula [Kz]

$$d\{\lambda,\mu\} = -\{d\lambda,\mu\} - (-1)^{\deg\lambda}\{\lambda, d\mu\}, \tag{4.86}$$

which is to be compared with (4.68). This formula shows that $\{\lambda, \mu\}$ is closed if λ and μ are closed. However, the induced product in the de Rham cohomology is trivial [Kz] since, if we put $\delta = i(w)d - di(w)$ in (4.75) we see easily that $d\lambda = d\mu = 0$ imply $\{\lambda, \mu\} = d$-exact.

Now, since Propositions 4.21 and 4.23 followed from the formulas of Theorem 1.1, similar results will follow for differential forms from the formulas of Proposition 4.24, and we get

4.26. Proposition. *Put* $\Theta^k M = \Lambda^{k+1}M$, *and* $\{\lambda, \mu\} = (-1)^{\deg \lambda + 1}\{\lambda, \mu\}$. *Then* $(\oplus_k(\Theta^k M), \{\ ,\ \})$ *is a graded Lie algebra, and the exterior product of forms makes it into a graded Poisson algebra of index* 1. *The mapping* # *is a homomorphism from these structures to those of Proposition* 4.21, *and Remark* 4.22.

Proof. See the explanation given above. The last assertion follows from (4.85). Q.e.d.

4.27. Proposition ([CV], etc.). *The d-bracket operation*

$$\{\alpha, \beta\}_d \overset{\text{def}}{=} - \{d\alpha, \beta\} \tag{4.87}$$

induces in $\Lambda_0 M = \Lambda M/_{d(\Lambda M)}$, *with its natural graduation, the structure of a graded Lie algebra, and* $-\#$ *sends this structure homomorphically into that of Proposition* 4.23.

Proof. The same computations as in the proof of Proposition 4.23 hold good now for differential forms. The last assertion follows from (4.85) and (4.12). Q.e.d.

4.28. Remark. From (4.87) and (4.84) one gets easily

$$\{\lambda \wedge \mu, \nu\}_d = \lambda \wedge \{\mu, \nu\}_d + (-1)^{(\deg \lambda)(\deg \mu)} \mu \wedge \{\lambda, \nu\}_d, \tag{4.88}$$

which is (4.66) for $k = 0$. However, $\Lambda_0 M$ is not a graded Poisson algebra because the exterior multiplication does not go over to this quotient. A similar result can be obtained for the σ-bracket (4.68).

Another interesting subject of Poisson calculus is an algorithm for building compatible Poisson structures by the use of fields of endomorphisms of the tangent bundle, known as *Nijenhuis tensors*. This led to the notion of a *Poisson-Nijenhuis structure*, and we send the reader to [KSM2] for the corresponding theory.

Finally, let us indicate that the Poisson calculus can also be developed in the purely algebraic context of Poisson algebras. The interested reader is referred to the papers by Huebschmann [Hu1,2,3,4].

5 Poisson Cohomology

5.1 Definition and general properties

The subject of this chapter is related to that of the previous one in an obvious way. Just as the usual calculus on manifolds leads to the de Rham cohomology, the Poisson calculus leads to the *Lichnerowicz-Poisson cohomology* first studied in [Lh2]. These new cohomology spaces are far from being as interrelated with the topology of the manifold as the de Rham spaces are, and, moreover, they are "too big", and their actual computation is both more complicated and less significant than in the case of the de Rham cohomology. However, they are interesting because they allow us to describe various results concerning Poisson structures, in particular, one important result about the *quantization* of the manifold

5.1. Definition. $\mathcal{V}(M) = \oplus_k \mathcal{V}^k(M)$ with the coboundary operator σ ((4.8), (4.10), (4.18)) is called the *Lichnerowicz-Poisson cochain complex* of the Poisson manifold (M, w), and

$$H^k_{LP}(M, w) = H^k_{LP}(M) \stackrel{\text{def}}{=} \frac{\ker(\sigma : \mathcal{V}^k(M) \to \mathcal{V}^{k+1}(M))}{\operatorname{im}(\sigma : \mathcal{V}^{k-1}(M) \to \mathcal{V}^k(M))} \tag{5.1}$$

are the *Lichnerowicz-Poisson* **or** *LP-cohomology spaces* of (M, w).

5.2. Remark. The LP-cohomology can also be defined for Poisson algebras, and its study in this sense was done by Huebschmann [Hu1].

From (5.1), and known facts about σ we get straightforwardly the following facts:

a) $H^0_{LP}(M, w)$ is the center of the Poisson algebra $(C^\infty(M), \{\ ,\ \})$;

b) $H^1_{LP}(M, w) = \mathcal{V}^1_w(M)/\chi_w(M)$, where $\mathcal{V}^1_w(M)$ is the space of the infinitesimal automorphisms of the Poisson structure w (i.e., vector fields X such that $L_X w = [w, X] = -\sigma X = 0$), and $\chi_w(M)$ is the space of the Hamiltonian vector fields X_f, $f \in C^\infty(M)$.

c) $H^2_{LP}(M, w)$ has the distinguished element $[w]$[1] defined by the cocycle w ($\sigma w = -[w, w] = 0$). If $[w] = 0$ i.e., there exists a vector field A such that $w = \sigma A = -L_A w$, some authors call (M, w) a *homogeneous Poisson manifold* [DLM], [DS1]. We shall prefer the name of *exact Poisson manifold* (w is σ-exact). For instance, the Lie-Poisson structure of a coadjoint Lie algebra $\mathcal{G}^*$ is exact, since, with the notation of the formula (3.5), the choice $A = x_i(\partial/\partial x_i)$ does the job. This kind of Poisson manifolds play a very important role in the study of the Jacobi manifolds.

---

[1] Following a usual practice we denote the cohomology classes by brackets.

d) Because of (1.10) and (4.18) we have

$$\sigma(P \wedge Q) = (\sigma P) \wedge Q + (-1)^{\deg P} P \wedge \sigma Q, \tag{5.2}$$

and this implies that $\wedge$ induces an associative and graded commutative product in $H_{LP}(M) = \oplus_k H^k_{LP}(M)$, so that we may speak of the LP-cohomology algebra $H_{LP}(M, w)$. The operation

$$[P] \wedge [Q] = [P \wedge Q] \tag{5.3}$$

will be called the *LP-cup-product.* On the other hand, formula (4.68) shows that the Schouten-Nijenhuis bracket also induces a product in $H_{LP}(M)$ by

$$[[P], [Q]] = [[P, Q]] \tag{5.4}$$

which, however, is not graded in the same sense. (See results on this product in [Kz].)

e) Let X be a nowhere vanishing vector field on M which defines a 1-dimensional cofoliation i.e., $\sigma X = X \wedge Y$ for another vector field Y (see Section 4.4). Then, using (5.2), we see that $X \wedge \sigma Y = 0$ i.e., $\sigma Y = \lambda X$, and, then, that $\sigma(Y \wedge \sigma Y) = 0$. Hence, X has an associated cohomology class $[Y \wedge \sigma Y] \in H^3_{LP}(M, w)$. If X is replaced by fX ($f \neq 0$ everywhere), Y must be replaced by $Y + \sigma \ln | f |$, and $[Y \wedge \sigma Y]$ is unchanged. Similarly, if Y is replaced by $Y + \varphi X$, $Y \wedge \sigma Y$ changes to $Y \wedge \sigma Y + \sigma(\varphi \sigma X)$, and, again, $[Y \wedge \sigma Y]$ is unchanged. Hence, this LP-cohomology class depends only on the cofoliation spanned by X. We call it the *Poisson-Godbillon-Vey class* of the cofoliation, since the above computation is just the contravariant version of the computation that leads to the Godbillon-Vey class of a foliation of codimension 1 (e.g., [Bt]).

Now, we have the following important result [Lh2], [Kz]

5.3. Proposition. *The mapping* $\#$ *of* (4.11) *induces a homomorphism of graded Lie algebras*

$$\# : H^*_{\mathrm{deR}}(M) \to H^*_{LP}(M, w), \tag{5.5}$$

which is an isomorphism if w *comes from a symplectic structure of* M.

Proof. Of course, H^*_{deR} denotes the de Rham cohomology (i.e., $H^*_{\mathrm{deR}}(M) = H^*(M, \mathbb{R})$). The homomorphism is $\#[\lambda] = [\lambda^\#]$ and it is well defined because of (4.12). Moreover, the preservation of the cup-product follows from (4.11). Finally, in the symplectic case the inverse mapping $\#^{-1}$ exists at the level of cochains, and, therefore, it also exists on cohomology. Q.e.d.

For instance, if we come back to the Poisson-Godbillon-Vey class of a cofoliation, it is clear that, in the case of a symplectic manifold, $[Y \wedge \sigma Y]$ is the $\#$-image of the Godbillon-Vey class of the symplectic orthogonal foliation.

Unlike for the de Rham cohomology, one cannot speak conveniently of a functorial character of the LP-cohomology. The best remark which we can make (and which is obvious) is that, if (M_a, w_a) are Poisson manifolds ($a = 1, 2$) and $\varphi : M_1 \to M_2$ is a Poisson mapping which is a local diffeomorphism, then one has an induced homomorphism

$$\varphi^* : H^k_{LP}(M_2) \to H^k_{LP}(M_1) \qquad (k = 0, 1, 2, \ldots) \tag{5.6}$$

and if φ is a diffeomorphism, φ^* will be an isomorphism.

Particularly if $\varphi : U \to (M, w)$ is the inclusion of an open set U into M, and if we restrict w to U, we get an induced homomorphism $H^*_{LP}(M, w) \to H^*_{LP}(U, w/_U)$. This shows that one has the necessary ingredients for a Mayer-Vietoris exact sequence, and, indeed, the usual proof of the corresponding theorem for the de Rham cohomology (e.g., [BT]) can be used to prove

5.4. Proposition. *Let (M, w) be a Poisson manifold, and let us endow all its open subsets by the restrictions of the Poisson structure w. Then, for every pair (U, V) of open subsets of M, the sequence*

$$0 \to \mathcal{V}((U \cup V) \xrightarrow{\alpha} \mathcal{V}(U) \oplus \mathcal{V}(V) \xrightarrow{\beta} \mathcal{V}(U \cap V) \to 0, \tag{5.7}$$

where α restricts the fields of multivectors to U and V and, $\beta(P, Q) = Q/_{U \cap V} - P/_{U \cap V}$, is an exact sequence of LP-cochain complexes, and its corresponding exact cohomology sequence has the form

$$\begin{gathered} \ldots \to H^k_{LP}(U \cup V) \xrightarrow{\alpha_*} H^k_{LP}(U) \oplus H^k_{LP}(V) \xrightarrow{\beta_*} \\ \xrightarrow{\beta_*} H^k_{LP}(U \cap V) \to H^{k+1}_{LP}(U \cup V) \to \ldots \end{gathered} \tag{5.8}$$

Proof. Same as in [BT] pp. 22–23. The sequence (5.8) is called the Mayer-Vietoris exact sequence for the LP-cohomology. Q.e.d.

Furthermore, let us say that h is a *filtration degree* of $Q \in \mathcal{V}^k(M)$ if $Q(\alpha_1, \ldots, \alpha_k) = 0$ as soon as $\geq k - h + 1$ arguments α_i belong to $\ker \#_w$. Then we get

5.5. Proposition. *The spaces*

$$F_h\mathcal{V}(M) = \{Q \in \mathcal{V}(M)/h \text{ is } a \text{ filtration degree of } Q\} \tag{5.9}$$

define a differential regular filtration of the LP-cochain complex $\mathcal{V}(M)$ hence, the corresponding spectral sequence ${}_{LP}E^{pq}_r(M, w)$ converges to $H^*_{LP}(M, w)$.

Proof. (See e.g. [Va2] for basics on spectral sequences.) It is obvious that $F_{h+1}\mathcal{V}(M) \subseteq F_h\mathcal{V}(M)$ i.e., (5.9) is indeed a filtration. Also $Q \in \mathcal{V}^k M$ cannot have a filtration degree $> k$, unless $Q = 0$, and this means that the filtration is regular. Then, if we look at the expression (4.8), and if we remember that $\ker\#$ is an abelian ideal of the Lie algebra $(\Lambda^1 M, \{\ ,\ \})$ (because of $\{\alpha,\beta\}^\# = [\alpha^\#,\beta^\#]$) we discover immediately that $\sigma(F_h) \subseteq F_h$ i.e., we have a differential filtration, and we can construct a spectral sequence to be denoted by ${}_{LP}E_r^{pq}(M,w)$ or $E_r^{pq}(M)$ shortly. Since the filtration is regular, the inductive limit of $E_r^{pq}(M)$ is $E_\infty^{pq}(M)$ which (as always) is

$$E_\infty^{pq}(M) = (H_{LP}^{p+q}(M))_p/(H_{LP}^{p+q}(M))_{p+1},$$

where $(H_{LP}^k(M))_h = [H_{LP}^k(M)] \cap [\mathrm{im}H(F_h) \text{ by } F_h \subseteq \mathcal{V}(M)]$. Q.e.d.

5.6. Remark. The construction of the spectral sequence ${}_{LP}E_r^{pq}(M,w)$ is the usual Serre-Hochschild construction of the cohomology theory of Lie algebras. Hence we shall call this sequence the *LP-SH spectral sequence of* (M,w).

Another general idea which can be used in studying the LP-cohomology is that of going over to differential forms by means of the choice of an auxiliary Riemannian metric g on M, as we did in Chapter 1. Then, if we denote $\pi = \#_g^{-1}w$, and if we use the formula (1.34), we get

$$\sigma(\#_g\lambda) = \#_g(d_\pi\lambda) \tag{5.10}$$

where

$$d_\pi = \delta_g e(\pi) - e(\pi)\delta_g - e(\delta_g\pi), \tag{5.11}$$

e denoting the exterior product by the respective form. Correspondingly, if we denote by $\Pi(M)$ the cochain complex $(\Lambda^* M, d_\pi)$ (that $d_\pi^2 = 0$ follows from (5.10)), we have

$$H_{LP}^k(M,w) = H^k(\Pi(M)). \tag{5.12}$$

(Notice that the Poisson condition (1.36) is now $d_\pi\pi = 0$.)

For instance, this remark yields

5.7. Proposition. *Let (M,w) be a Poisson manifold endowed with a Riemannian metric g. Let $\varphi : N \to M$ be a mapping such that $\delta_{\varphi *_g}(\varphi^*\lambda) = \varphi^*(\delta_g\lambda)$ for any $\lambda \in \Lambda M$. Then N has a well defined induced Poisson structure $\varphi^{-1}w$, and there are induced homomorphisms*

$$\varphi^* : H_{LP}^k(M,w) \to H_{LP}^k(N,\varphi^{-1}w). \tag{5.13}$$

Proof. Put $\pi = \#_g^{-1}w$, and define $\varphi^{-1}w$ as the bivector $\#_{\varphi^*g}(\varphi^*\pi)$. Then, it follows easily that $d_{\varphi *_\pi} \circ \varphi^* = \varphi^* \circ d_\pi$. Q.e.d.

5.2 Straightforward and inductive computations

We already said that computation of LP-cohomology is a difficult problem. This fact becomes apparent even in the following simple case

5.8. Example. Let us discuss the LP-cohomology of $(\mathbb{R}^2, w)$ where

$$w = (x^2 + y^2)\frac{\partial}{\partial x} \wedge \frac{\partial}{\partial y}, \tag{5.14}$$

whose unique singular point is 0.

For $f \in C^\infty(\mathbb{R}^2)$, we have then

$$\sigma f = -X_f = (x^2 + y^2)\left(\frac{\partial f}{\partial y}\frac{\partial}{\partial x} - \frac{\partial f}{\partial x}\frac{\partial}{\partial y}\right), \tag{5.15}$$

and $\sigma f = 0$ implies $f/_{\mathbb{R}^2\backslash\{0\}} = \text{const}$. Therefore, by continuity, $f = \text{const.}$ on $\mathbb{R}^2$, and we get $H^0_{LP}(\mathbb{R}^2, w) = \mathbb{R}$.

Furthermore, if we consider a vector field

$$A = a(x, y)\frac{\partial}{\partial x} + b(x, y)\frac{\partial}{\partial y}, \tag{5.16}$$

we can compute $\sigma A = -[w, A] = -L_A w$, and we get

$$\sigma A = \varphi(a, b)\frac{\partial}{\partial x} \wedge \frac{\partial}{\partial y}, \tag{5.17}$$

where

$$\varphi(a, b) \stackrel{\text{def}}{=} (x^2 + y^2)\left(\frac{\partial a}{\partial x} + \frac{\partial b}{\partial y}\right) - 2(xa + yb). \tag{5.18}$$

Now, the inclusion ι of $\mathbb{R}^2\backslash\{0\}$ into $\mathbb{R}^2$, yields a restriction as defined in (5.6)

$$\iota^* : H^k_{LP}(\mathbb{R}^2, w) \to H^k_{LP}(\mathbb{R}^2\backslash\{0\}, w) \approx H^k_{\text{deR}}(\mathbb{R}^2\backslash\{0\}), \tag{5.19}$$

where the last isomorphism exists since $w/_{\mathbb{R}^2\backslash\{0\}}$ is a symplectic structure, and we may apply Proposition 5.3. Particularly, for $k = 1$ this $\#$-isomorphism is given by

$$a\frac{\partial}{\partial x} + b\frac{\partial}{\partial y} \mapsto \frac{bdx - ady}{x^2 + y^2}. \tag{5.20}$$

But if we look at the classical deformation retraction of $\mathbb{R}^2\backslash\{0\}$ onto S^1, we see that $H^1_{\text{deR}}(\mathbb{R}^2\backslash\{0\}) = \mathbb{R}$, and every cohomology class in this space has a unique representative of the form

$$\rho d\theta = \rho\frac{xdy - ydx}{x^2 + y^2} \quad (\rho \in \mathbb{R}, \theta = \text{“the polar angle”}). \tag{5.21}$$

Since the form (5.21) comes from (5.20) applied to the vector field $-\rho[x(\partial/\partial x)+y(\partial/\partial y)]$ which extends to the whole of $\mathbb{R}^2$, we conclude that, for $k=1$, ι^* of (5.19) is an epimorphism, and $H^1_{LP}(\mathbb{R}^2,w)\neq 0$ (while, of course, $H^1_{\mathrm{deR}}(\mathbb{R}^2)=0$).

However, it is not true that ι^* of (5.19) is injective for $k=1$. Indeed, if we take $f=(1/2)\ln(x^2+y^2)$ on $\mathbb{R}^2\backslash\{0\}$, (5.15) yields $X\stackrel{\mathrm{def}}{=}\sigma f=y(\partial/\partial x)-x(\partial/\partial y)$, and we see that ι^* sends $[X]\in H^1_{LP}(\mathbb{R}^2,w)$ to zero. But, $[X]\neq 0$ since if $X=\sigma\varphi$, $\varphi=f+$ const. on $\mathbb{R}^2\backslash\{0\}$, and this is impossible because f does not extend to $\mathbb{R}^2$.

Hence, it is not quite clear what $H^1_{LP}(\mathbb{R}^2,w)$ is.

Furthermore, any bivector $Q(x,y)(\partial/\partial x)\wedge(\partial/\partial y)$ is a 2-cocycle, and it is a coboundary iff there are a and b such that $\varphi(a,b)=Q$, where φ is defined by (5.18). If this happens, and if we see $\varphi=Q$ as an equation for a, it is just a linear differential equation with respect to x, that yields

$$a/_{\mathbb{R}^2\backslash\{0\}}=(x^2+y^2)\left(C(y)+\int\frac{Q+2yb-(x^2+y^2)(\partial b/\partial y)}{(x^2+y^2)^2}dx\right), \tag{5.22}$$

and Q defines a coboundary iff there is a function b such that a extends to $\mathbb{R}^2$.

Again, it is not clear what $H^2_{LP}(\mathbb{R}^2,w)$ really looks like. But, in any case, $H^2_{LP}(\mathbb{R}^2,w)\neq 0$, since there are no solutions a,b, on $\mathbb{R}^2$ for $Q=x^2+y^2$ (i.e., w is not an exact structure) [Pt]. Indeed, if $\varphi(a,b)=x^2+y^2$ for some $a,b\in C^\infty(\mathbb{R}^2)$ then, on $\mathbb{R}^2\backslash\{0\}$, one has

$$a_x(x,y)+b_y(x,y)-1=2\frac{xa(x,y)+yb(x,y)}{x^2+y^2},$$

where the indices denote partial derivatives. From this, and using Taylor's formula, we get in a neighborhood of the origin

$$a_x(x,y)+b_y(x,y)-1=\frac{2}{x^2+y^2}\{xa(0,0)+yb(0,0)+$$

$$+x^2a_x(0,0)+(a_y(0,0)+b_x(0,0))xy+y^2b_y(0,0)+o(3)\}.$$

Since the left-hand side of this formula has a well defined finite value at 0, while the limit of the right-hand side is either infinite or depends on the ray, if we approach 0 along rays, the previous relation is contradictory if at least one of the coefficients of the quadratic terms of its right-hand side is nonzero. If all these coefficients are zero, the same relation leads to the limit contradiction $-1=0$.

Since a general Poisson manifold may have many kinds of singularities, the task of computing the LP-cohomology seems to be hopeless, generally. However, for instance, if $\mathcal{G}$ is the Lie algebra of a compact Lie group, and w

is the Lie-Poisson structure of $\mathcal{G}^*$ defined in Section 3.1, it follows from more general results of [Lu1] and [GW] that $H^k_{LP}(\mathcal{G}^*, w) = H^k(\mathcal{G}) \otimes$ {Casimirs of $(\mathcal{G}^*, w)$}, where $H^k(\mathcal{G})$ is the cohomology of the Lie algebra $\mathcal{G}$.

The situation is a little bit better in the regular case, and we shall discuss this case hereafter following [Va9].

Let (M^m, w) be a regular Poisson manifold, with the symplectic foliation $\mathcal{S}(M)$ ($\dim \mathcal{S} = 2n$; $m = 2n + s$), let us choose a transversal distribution $\nu\mathcal{S}$, and use types of multivectors, and the decomposition $\sigma = \sigma' + \sigma''$ given in Proposition 4.5. Then, as we shall see in what follows, one has, in principle, a recurrent computational process for the computation of the LP-cohomology. (Such a process was used first in [VK] for the case where $\mathcal{S}$ is a fibration.)

Any $Q \in \mathcal{V}^k(M)$ has a decomposition

$$Q = Q^{k,0} + Q^{k-1,1} + \ldots + Q^{0,k}, \tag{5.23}$$

where the indices denote the type of the components, and $\sigma Q = 0$ is equivalent to

$$\sigma'' Q^{i,k-i} + \sigma' Q^{i+1,k-i-1} = 0 \qquad (i = 0, \ldots, k). \tag{5.24}$$

In particular, if $i = k$ we get $\sigma'' Q^{k,0} = 0$, and, on the other hand, $\forall \tilde{Q} \in \mathcal{V}^{k-1}(M)$, $(\sigma\tilde{Q})^{k,0} = 0$. Hence, if we denote by $\mathcal{V}^{k,0}_0(M)$ the space of the σ''-closed k-vector fields of type $(k, 0)$, it follows that the mapping $[Q] \mapsto Q^{k,0}$ is a homomorphism

$$p_{k,0} : H^k_{LP}(M, w) \to \mathcal{V}^{k,0}_0(M). \tag{5.25}$$

The image of $p_{k,0}$ consists of k-vector fields that can be extended to a cocycle (5.23), and (5.24) shows us that this is true iff $\sigma' Q^{k,0}$ satisfies the following k σ''-exactness conditions

$$\begin{array}{lll}
(c_1) & \sigma' Q^{k,0} & = \sigma''\text{-exact} \stackrel{\text{def}}{=} -\sigma'' Q^{k-1,1}, \\
(c_2) & \sigma' Q^{k-1,1} & = \sigma''\text{-exact} \stackrel{\text{def}}{=} -\sigma'' Q^{k-2,2}, \\
\dots & \dots & \dots \\
(c_k) & \sigma' Q^{1,k-1} & = \sigma''\text{-exact} \stackrel{\text{def}}{=} -\sigma'' Q^{0,k}.
\end{array}$$

Let us denote $\mathrm{im} p_{k,0} = \mathcal{V}^{k,0}_{0(k)}(M)$, and $\ker p_{k,0} = {}^0H^k_{LP}(M, w) = \{[Q] \in H^k_{LP}(M, w) / Q^{k,0} = 0\}$. Then we may write down the result of this first step as

$$H^k_{LP}(M, w) \approx {}^0H^k_{LP}(M, w) \oplus \mathcal{V}^{k,0}_{0(k)}(M). \tag{5.26}$$

In the next step, we must analyse ${}^0H^k_{LP}(M, w)$, and, for this purpose, we take the subcomplex ${}^0\mathcal{V}(M) \subseteq \mathcal{V}(M)$ of the multivectors (5.23) with $Q^{k,0} = 0$, and we denote by $H^k({}^0\mathcal{V}(M))$ its cohomology spaces. Of course, ${}^0H^k_{LP}(M, w)$

is the image of $H^k({}^0\mathcal{V}(M))$ with respect to the inclusion $\iota : {}^0\mathcal{V}(M) \subseteq \mathcal{V}(M)$, and, since the quotient $\mathcal{V}(M)/{}^0\mathcal{V}(M)$ has the 0 coboundary operator i.e., $H^k(\mathcal{V}/{}^0\mathcal{V}) = (\mathcal{V}/{}^0\mathcal{V})^k = \mathcal{V}^{k,0}(M)$, we get the following form of the exact cohomology sequence of these complexes

$$\mathcal{V}^{k-1,0} \xrightarrow{\sigma_*} H^k({}^0\mathcal{V}(M)) \xrightarrow{\iota_*} H^k(\mathcal{V}(M)), \tag{5.27}$$

which implies

$${}^0H^k_{LP}(M,w) \approx H^k({}^0\mathcal{V}(M))/\sigma_*(\mathcal{V}^{k-1,0}(M)). \tag{5.28}$$

Therefore, in order to complete this second step, we have to compute $H^k({}^0\mathcal{V}(M))$. But, clearly, this can be done, following the procedure of the first step, which will lead us to a formula of the type (5.26) for $H^k({}^0\mathcal{V}(M))$, and so on.

One may object that this procedure is not very efficient and practical but, at least, it exists. In small dimensions it gives the following results.

a) If $k = 1$, then

$${}^0H^1_{LP}(M,w) = \{X \in \mathcal{V}^{0,1}(M)/\sigma''X = 0\}/\sigma''(\mathcal{V}^0(M)).$$

If we look at the formula (4.37), and notice that the operator $\mathcal{L}_G$ always vanishes for $(0,q)$-forms it follows that

$${}^0H^1_{LP}(M,w) \approx \{\lambda \in \Lambda^{01}M/d''\lambda = 0\}/d''(C^\infty(M)),$$

and this space is known from foliation theory. Namely, (e.g., [Va2]) the sheaf $\Phi^p(\mathcal{S})$ of *projectable* p-forms (i.e., induced by forms on the space of leaves) has the fine resolution

$$0 \to \Phi^p(\mathcal{S}) \subseteq \underline{\Lambda^{p,0}M} \xrightarrow{d''} \underline{\Lambda^{p,1}M} \xrightarrow{d''} \dots, \tag{5.29}$$

where underlining denotes sheaves of germs. Therefore what we have is in fact

$${}^0H^1_{LP}(M,w) \approx H^1(M,\Phi^0(\mathcal{S})), \tag{5.30}$$

where $\Phi^0(\mathcal{S})$ is the sheaf of the germs of functions of M that are constant along the leaves of $\mathcal{S}$.

Hence, by using (5.26) we can write

$$H^1_{LP}(M,w) \approx H^1(M,\Phi^0(\mathcal{S})) \oplus \mathcal{V}^{1,0}_{0(1)}(M). \tag{5.31}$$

A further analysis of this result depends on the possibility to make its terms more explicit which, generally, is not very simple. In order to give an example let us establish first

5.9. Proposition. *Let $\mathcal{F}$ be the foliation of $M = F \times N$ by the leaves $F \times \{x\}$ ($x \in N$), and assume that F has finite Betti numbers. Then one has*

$$H^q(M, \Phi^p(\mathcal{F})) = H^q(F, \mathbb{R}) \otimes \Lambda^p N. \tag{5.32}$$

Proof. Let x^i be local coordinates of N and y^u be local coordinates of F. Then $\lambda \in \Lambda^{p,q}M$ is given by

$$\lambda = \frac{1}{p!q!}\lambda_{i_1\dots i_p\alpha_1\dots\alpha_q}(x,y)dx^{i_1}\wedge\dots\wedge dx^{i_p}\wedge dy^{\alpha_1}\wedge\dots\wedge dy^{\alpha_q},$$

and for the calculation of

$$H^q(M, \Phi^p(\mathcal{F})) = \frac{\ker(d'' : \Lambda^{p,q}M \to \Lambda^{p,q+1}M)}{\operatorname{im}(d'' : \Lambda^{p,q-1}M \to \Lambda^{p,q}M)} \tag{5.33}$$

x^i are just parameters. Now, let $\mu_1 \dots \mu_s$ be closed forms of F which represent a basis of $H^q(F, \mathbb{R})$ via de Rham's theorem. Then, if $d''\lambda = 0$, we must have

$$\lambda = \alpha_1 \wedge \mu_1 + \dots + \alpha_s \wedge \mu_s + d''\beta,$$

where $\alpha_i \in \Lambda^p N$, $\beta \in \Lambda^{p,q-1}M$. Formula (5.32) is a straightforward consequence of this expression of λ. Q.e.d.

Now, if we come back to (5.31), and if the Poisson manifold is a product of a symplectic manifold S of finite Betti numbers by another manifold N, then the first term of (5.31) is $C^\infty(N) \otimes_{\mathbb{R}} H^1(S, \mathbb{R})$.

Let us simplify even more our case, and ask the Poisson structure of $S \times N$ to be transversally constant (Definition 4.8) with respect to the transversal distribution TN i.e., the symplectic structure of all the leaves $S \times \{x\}$ is the same. Then $\sigma' = 0$ (Proposition 4.9), and $\mathcal{V}^{k,0}_{0(k)} = \mathcal{V}^{k,0}_0$. If we also use (4.37) for a metric G^* of N, we may go to differential forms, and we see that $\mathcal{V}^{1,0}_{0(1)} \approx \Lambda^1 N$. Overall, we get

5.10. Proposition. *Let S be a symplectic manifold with a fixed symplectic structure, and finite Betti numbers. Take $M = S \times N$ with the Poisson structure defined by the fixed symplectic structure of S. Then one has*

$$H^1_{LP}(M) \approx (\Lambda^1 N) \oplus (H^1(S, \mathbb{R}) \otimes C^\infty(N)), \tag{5.34}$$

where the vector space operations are over $\mathbb{R}$.

This is, in fact, the result established in [VK].

b) For $k = 2$, we have first

$$H^2({}^0\mathcal{V}(M)) = \frac{\{Q^{1,1} + Q^{0,2} / \sigma'' Q^{1,1} = 0, \sigma'' Q^{0,2} + \sigma' Q^{1,1} = 0\}}{\{\sigma'' X^{0,1}\}} \tag{5.35}$$

and similar to (5.26) we get now

$$H^2({}^0\mathcal{V}(M)) \approx H^2(\oplus_k \mathcal{V}^{0,k}(M), \sigma'') \oplus \mathcal{V}^{1,1}_{0(1)}(M), \tag{5.36}$$

where

$$\mathcal{V}^{1,1}_{0(1)}(M) = \{Q^{1,1}/\sigma'' Q^{1,1} = 0 \text{ and } \sigma' Q^{1,1} = \sigma''\text{-exact}\}. \tag{5.37}$$

For the same reasons as in the case $k = 1$, we have

$$H^2(\oplus_k \mathcal{V}^{0,k}(M), \sigma'') \approx H^2(M, \Phi^0(\mathcal{S})), \tag{5.38}$$

and, if we sum up the results of (5.26), (5.28), (5.36), and (5.38) we get

$$H^2_{LP}(M, w) \approx \left[\left(H^2(M, \Phi^0(\mathcal{S})) \oplus \mathcal{V}^{1,1}_{0(1)}(M)\right)/\sigma(V^{1,0}(M))\right] \oplus \mathcal{V}^{2,0}_{0(2)}(M). \tag{5.39}$$

If w is a transversally constant structure, $\sigma = \sigma''$, and

$$\mathcal{V}^{1,1}_{0(1)} = \{Q^{1,1}/\sigma'' Q^{1,1} = 0\}, \mathcal{V}^{2,0}_{0(2)}(M) = \mathcal{V}^{2,0}_0(M).$$

The result is nicely interpretable if (4.37) holds with $\mathcal{L}_G = 0$, for instance if $\mathcal{S}(M)$ has a *bundle-like metric* (e.g., [Mn1], [Va2]). In this case (5.39) becomes

$$H^2_{LP}(M, w) \approx H^2(M, \Phi^0(\mathcal{S})) \oplus H^1(M, \Phi^1(\mathcal{S})) \oplus \Lambda^2_{\text{proj}}(M), \tag{5.40}$$

where the index "proj" denotes projectable forms.

Particularly, using (5.32) we have

5.11. Proposition. *Under the same hypotheses as in Proposition* 5.10, *we have*

$$H^2_{LP}(S \times N) = \sum_{i=0}^{2} H^i(S, \mathbb{R}) \otimes \Lambda^{2-i} N. \tag{5.41}$$

Compare again with [VK]. Formula (5.41) will be extended later on for all $H^k_{LP}(S \times N)$. The calculation of $H^k_{LP}(S \times N)$ for more general Poisson structures, that are not transversally constant, has been done in [Xu1] using a different method which we shall indicate later, in Chapter 9.

5.3 The spectral sequence of Poisson cohomology

Further interesting information on the LP-cohomology of a regular Poisson manifold can be obtained from its LP-SH spectral sequence. Namely, one has [Va9].

5.12. Theorem. *The first terms of the LP-SH spectral sequence of a regular Poisson manifold (M, w) are given by*

$$
{}_{LP}E_0^{pq}(M,w) = {}_{LP}E_1^{pq}(M,w) = V^{q,p}(M),
$$
$$
{}_{LP}E_2^{pq}(M,w) = H^p(\oplus_k V^{q,k}, \sigma''). \tag{5.42}
$$

Moreover, if the symplectic foliation $\mathcal{S}$ of w has either a transversally Riemannian or a transversally symplectic structure, then

$$
{}_{LP}E_2^{pq}(M,w) = E_1^{pq}(\mathcal{S}) = H^p(M, \Phi^q(\mathcal{S})). \tag{5.43}
$$

Proof. By looking at the definition of the filtration degree, and at the formula (5.9), we see that, now

$$
F_h\mathcal{V}(M) = \bigoplus_{i \geq h} \bigoplus_{p} \mathcal{V}^{p,i}(M). \tag{5.44}
$$

Remember that, following the general definition (e.g. [Va2]), the spectral sequence is given by the following formula

$$
E_r^{pq} = Z_r^{pq}/(B_{r-1}^{pq} + Z_{r-1}^{p+1,q-1}), \tag{5.45}
$$

where

$$
Z_r^{pq} = \{Q \in F_p \cap \mathcal{V}^{p+q} / \sigma Q \in F_{p+r}\},
$$
$$
B_r^{pq} = \{Q \in F_p \cap \mathcal{V}^{p+q} / Q = \sigma P, P \in F_{p-r}\} \subseteq Z_r^{pq}. \tag{5.46}
$$

From (5.44), it is easy to see that $F_p \cap \mathcal{V}^{p+q} = \bigoplus_{s=0}^{q} \mathcal{V}^{s,p+q-s}$, $Z_0^{pq} = F_p \cap \mathcal{V}^{p+q}$, $Z_{-1}^{p+1,q-1} = \bigoplus_{s=0}^{q-1} \mathcal{V}^{s,p+q-s}$. Hence, $E_0^{pq} = \mathcal{V}^{q,p}(M)$ as requested.

Furthermore, every E_r^{pq} has a differential

$$
d_r : E_r^{pq} \to E_r^{p+r,q-r+1}, \tag{5.47}
$$

which in our case is induced by σ, and E_{r+1}^{pq} is determined by the formula

$$
H^n(E_r) = E_{r+1}^n \stackrel{\text{def}}{=} \bigoplus_{p+q=n} E_{r+1}^{pq}. \tag{5.48}
$$

Taking into account the values of E_0^{pq}, we see that d_0 must be the part of σ that sends $\mathcal{V}^{q,p}$ into $\mathcal{V}^{q+1,p}$. But, by Proposition 4.5, σ has no component of type (1,0). Hence $d_0 = 0$, and $E_1^{pq} = E_0^{pq}$, again as requested.

Similarly, we must have $d_1 = \sigma''$, and this justifies the last one of the formulas (5.42).

In (5.43), $E_r^{pq}(\mathcal{S})$ denotes the spectral sequence of the foliation $\mathcal{S}$, which is defined by filtering the complex $(\Lambda M, d)$ by

$$F_p \Lambda M = \bigoplus_{i \geq p} \bigoplus_q \Lambda^{iq} M \tag{5.49}$$

(e.g. [Va2]). As above, it follows that $E_0^{pq} = \Lambda^{pg}(M)$, and $d_0 = d''$ of (4.34). Hence, in view of the resolution (5.29) of Φ^p we have the justification of the last equality in (5.43). The first equality in this formula then follows because of (4.37) where G is either the transversally Riemannian metric of $\mathcal{S}$ ([Va2], [Mn1]) or its transversal symplectic form, a choice that, obviously, makes $\mathcal{L}_G = 0$. Q.e.d.

The previous theorem leads to a number of interesting corollaries.

5.13. Corollary. *If w is a transversally constant regular Poisson structure, then*

$$H^k_{LP}(M, w) = \bigoplus_{q=0}^{k} E_2^{k-q,q}(M, w). \tag{5.50}$$

If, moreover, $\mathcal{S}(M)$ is either transversally Riemannian or transversally symplectic one has

$$H^k_{LP}(M, w) = \bigoplus_{q=0}^{k} H^q(M, \Phi^{k-q}(\mathcal{S})). \tag{5.51}$$

Proof. Straightforward from (5.42), (4.33) and (5.43) since, in this case, $\sigma = \sigma''$. Q.e.d.

5.14. Corollary. *Let w be the Poisson structure of $M = S \times N$, that is defined by a fixed symplectic structure of S, and assume that S has finite Betti numbers. Then, one has*

$$H^k_{LP}(M, w) = \bigoplus_{q=0}^{k} [H^q(S, \mathbb{R}) \otimes \Lambda^{k-q} N]. \tag{5.52}$$

Proof. Straightforward from (5.51) and (5.32). Q.e.d.

This result is the promised generalization of (5.34) and (5.41).

5.15. Corollary. *Let (M, w) be an arbitrary regular Poisson manifold. Then, $\forall x \in M$ there is a connected open neighbourhood U such that*

$$H^k_{LP}(U, w/_U) = \Gamma(\Phi^k(\mathcal{S}/_U)), \tag{5.53}$$

i.e., the space of projectable k-forms over U.

Proof. Use Weinstein's splitting theorem 2.16, and take $U = S \times N$ with the canonical coordinates of Definition 2.18. (In the regular case, these coordinates were also obtained earlier, in [Lh2].) Besides, take S to be contractible. For this U, the result follows obviously from (5.52), and we shall say that U is an *LP-simple neighbourhood.* Q.e.d.

Finally, we can obtain here as a corollary an *LP-Poincaré lemma*, a straightforward proof of which was given in [Lh2].

5.16. Corollary [Lh2]. *Let (M, w) be a regular Poisson manifold, and $x \in M$. Then, x has an open neighbourhood U in M such that, if $Q \in \mathcal{V}^k(U)$, and $\sigma Q = 0$, then $Q = A + \sigma B$ for some $B \in \mathcal{V}^{k-1}(U)$, and for a k-vector field A on U that is projectable to a k-vector field of a local transversal submanifold of $\mathcal{S}$ in U.*

Proof. Take the neighbourhood U to be LP-simple, and endowed with the w-canonical coordinates $(p_a, q^a) = (x^a)$, and (x^σ) of Definition 2.18. Then, $\mathcal{S}/_U$ has leafwise coordinates (x^a), and transversal coordinates (x^σ), and, with this structure, $w/_U$ is obviously transversally Riemannian and transversally constant. Hence, $\sigma = \sigma''$, and $\mathcal{L}_G$ of (4.37) vanishes.

For the given Q, we can write

$$Q = \sum_{p=0}^{k} (\lambda^{p,k-p})^{\tilde{\#}}, \tag{5.54}$$

where $\lambda^{p,k-p} \in \Lambda^{p,k-p}U$, and $\tilde{\#}$ is like in (4.37). Furthermore, $\sigma Q = 0$ is equivalent to

$$d''\lambda^{p,k-p} = 0 \qquad (k = 0, \ldots, p),$$

and we see that $\lambda^{k,0}$ is a projectable k-form, and $\lambda^{p,k-p} (k \neq p)$ are d''-exact, since d'' satisfies a Poincaré lemma (e.g., [Va2]). If these facts are introduced in (5.54), and (4.37) is taken once more into account, we are done. Q.e.d.

While the *LP-SH* spectral sequence of a Poisson manifold certainly offers important information, it doesn't give us the explicit values of the LP-cohomology spaces, usually. To balance this kind of remarks, let me make here the following comment.

Although the LP-cohomology is hardly computable, it represents an important theory since it offers a natural place where the obstructions to the solutions of many problems can be found. Then, in principle, one can discuss the vanishing of these obstructions individually, even if it is not possible to compute the whole LP-cohomology. Let us give some examples.

We already know that, if X is an infinitesimal automorphism of w, then $[X] \in H^1(M, w)$ is the obstruction to the existence of $f \in C^\infty(M)$ such that $X = X_f$.

Furthermore, if a function $f \in C^\infty(M)$ is seen as a Hamiltonian in the sense of classical mechanics, then the corresponding Hamiltonian dynamical system is

$$\frac{dx}{dt} = X_f(x) = i(df)w/_x \qquad (x \in M), \tag{5.55}$$

and a *perturbation* of this system can be seen as

$$\frac{dx}{dt} = i(df)(w + \epsilon u) \qquad (u \in \mathcal{V}^2(M)), \tag{5.56}$$

where ϵ is a small parameter. Such a perturbation will be said to be *Hamiltonian*, if (5.56) can be put under the form

$$\frac{dx}{dt} = i(df_\epsilon)w \tag{5.57}$$

for a *perturbed Hamiltonian*

$$f_\epsilon = f + \epsilon\tilde{f}. \tag{5.58}$$

Obviously, the condition for this to happen is the existence of $\tilde{f} \in C^\infty(M)$ such that $i(d\tilde{f})w = i(df)u$ i.e., $\sigma\tilde{f} = -i(df)u$. Hence, one must ask first that $\sigma(i(df)u) = 0$, and then $[-i(df)u] \in H^1_{LP}(M, w)$ is the obstruction to the existence of $\tilde{f}$.

In the same vein, we may ask the following question: does there exist an infinitesimal transformation X (i.e., $\tilde{x} = x + \epsilon X + o(2)$) on M such that, $\forall f \in C^\infty(M)$, the perturbed Hamiltonian system (5.56), is transformed by X into (5.57), up to $o(\epsilon^2)$, where f_ϵ is obtained from f by applying X i.e., $\tilde{f} = Xf$? In order to answer, notice that X sends (5.56) into

$$\frac{dx}{dt} = i(df)w + \epsilon(i(df)u + L_X(i(df)w)) + o(\epsilon^2), \tag{5.59}$$

and this is (5.57) iff, $\forall f$,

$$i(d(Xf))w = L_X(i(df)w) + i(df)u. \tag{5.60}$$

But,

$$L_X(i(df)w) = i(L_X df)w + i(df)L_X w = i(d(Xf)w - i(df)(\sigma X).$$

Hence X satisfies the desired condition iff $\sigma X = $ u. Therefore, we must ask that $\sigma u = 0$, and then $[u] \in H^2_{LP}(M, w)$ is the obstruction to the existence of the desired infinitesimal transformations.

The previous example is taken from [VK], and we refer the reader to that paper for several more examples of obstructions that are 2- and 3-dimensional LP-cohomology classes.

5.4 Poisson homology

We end this chapter by the discussion of a related subject, namely, the *canonical homology* of a Poisson manifold [Kz], [Br], also called *Poisson homology* [Hu3, 4]. This is natural since we saw in Chapter 4 that the LP-coboundary σ is related to the Koszul-Brylinski boundary $\delta_w = \delta$ (Proposition 4.4).

5.17. Definition. The homology spaces of the chain complex $(\Lambda^k M, \delta_w)$ will be denoted by $H_k^\pi(M, w)$ and called the *(Koszul-Brylinski)-Poisson homology spaces of* (M, w).

In other words, we have

$$H_k^\pi(M,w) = \frac{\ker(\delta : \Lambda^k M \to \Lambda^{k-1} M)}{\operatorname{im}(\delta : \Lambda^{k+1} M \to \Lambda^k M)}. \tag{5.61}$$

One of the main facts concerning the Poisson homology is

5.18. Theorem [Br]. *If (M^m, w) is a symplectic manifold, then*

$$H_k^\pi(M,w) \approx H^{m-k}(M, \mathbb{R})(\approx H_{LP}^{m-k}(M,w)). \tag{5.62}$$

Proof. In the comments that followed Proposition 4.4 it was noticed (fact a)) that the Poisson codifferential is equal to the symplectic codifferential defined by (1.31), with the symplectic significance instead of the Riemannian one. But, then one has a relation of the form $\delta = -\tilde{*} d \tilde{*}$, where $\tilde{*} : \Lambda^k M \to \Lambda^{m-k} M$ is like the Riemannian $*$-isomorphism, defined with the symplectic volume form instead of the Riemannian one [Lb1], [LM]. It follows easily that $\tilde{*}$ induces the isomorphism (5.62). Q.e.d.

Now, we give an example which shows that, generally, $H_k^\pi(M,w) \neq H^{m-k}(M,\mathbb{R})$, and also $H_k^\pi(M,w) \neq H_{LP}^{m-k}(M,w)$.

5.19. Example. Take again the Poisson manifold $(\mathbb{R}^2, w)$ of Example 5.8, where w is (5.14). Remember that we had $H_{LP}^0(\mathbb{R}^2, w) = \mathbb{R}$. Now, to compute $H_2^\pi(\mathbb{R}^2, w)$, we look at a 2-form $\alpha = A(x,y) dx \wedge dy$, and use (4.21) and (5.15) to get

$$\delta\alpha = -d[(x^2+y^2)A]. \tag{5.63}$$

Hence, $\delta\alpha = 0$ implies $A/_{\mathbb{R}^2 \setminus \{0\}} = k/(x^2+y^2)$ ($k =$ const.), and A extends to $\mathbb{R}^2$ only if $k = 0$. Thus, the only 2-cycle is 0, and since there are no boundaries either, we get $H_2^\pi(\mathbb{R}^2, w) = 0 \neq H^0(\mathbb{R}^2, \mathbb{R}) = H_{LP}^0(\mathbb{R}^2, w) = \mathbb{R}$.

In the same example, we get easily

$$H_0^\pi(\mathbb{R}^2, w) = C^\infty(\mathbb{R}^2)/\{(x^2+y^2)\left(\frac{\partial b}{\partial x} - \frac{\partial a}{\partial y}\right) / a, b \in C^\infty(\mathbb{R}^2)\}, \tag{5.64}$$

and this space is not zero. Indeed, $\forall\varphi \in C^\infty(\mathbb{R}^2), \varphi/_{\mathbb{R}^2\backslash\{0\}}$ must be of the form indicated in the denominator of (5.64), because this means $\varphi = \delta(adx + bdy)$, and $H_0^\pi(\mathbb{R}^2\backslash\{0\}, w) = H^2(\mathbb{R}^2\backslash\{0\}, \mathbb{R}) = 0$ by Theorem 5.18. But, for instance, for $\varphi = (x^2 + y^2)^{1/2}(x - y)$, we have $\varphi = \delta[(x^2 + y^2)^{1/2}(dx + dy)]$, and the corresponding Pfaff form is continuous, but not differentiable on $\mathbb{R}^2$. Moreover, we cannot have $(x^2 + y^2)^{1/2}(x - y) = \delta(adx + bdy)$ with $a, b \in C^\infty(\mathbb{R}^2)$ since this would give

$$\frac{\partial b}{\partial x} - \frac{\partial a}{\partial y} = \frac{x - y}{(x^2 + y^2)^{1/2}},$$

on $\mathbb{R}^2\backslash\{0\}$, and the right-hand side has no limit for $(x, y) \to (0, 0)$. Hence the homology class of φ in $H_0^\pi(\mathbb{R}^2, w)$ is not zero.

Finally, the formulas (5.63) and (5.64) also give us

$$H_1^\pi(\mathbb{R}^2, w) = \frac{\{\lambda \in \Lambda^1(\mathbb{R}^2)/d\lambda = 0\}}{\{d[(x^2 + y^2)A]/A \in C^\infty(M)\}}. \tag{5.65}$$

Notice that $d[(x^2+y^2)A_1] = d[(x^2+y^2)A_2]$ iff $A_1 = A_2+k/(x^2+y^2)$ on $\mathbb{R}^2\backslash\{0\}$, and taking $(x, y) \to (0, 0)$ we get $k = 0$ necessarily. Hence, the denominator of (5.65) is isomorphic to $C^\infty(\mathbb{R}^2)$.

5.20. Remark [BV1]. The natural pairing of $\lambda \in \lambda^k M$ with $P \in \mathcal{V}^k(M)$ by $i(P)\lambda$ yields a pairing $H_{LP}^k(M, w) \times H_k^\pi(M, w) \to H_0^\pi(M, w)$. This follows straightforwardly from formula (4.22) which shows that, if $P \mapsto P + \sigma Q$, $\lambda \mapsto \lambda + \delta\mu$ $(\sigma P = \delta\lambda = 0)$, then $i(P)\lambda \mapsto i(P)\lambda + (-1)^k\delta[i(Q)\lambda - i(P)\mu + i(Q)\delta\mu]$.

Now, let us consider the case of a regular Poisson manifold (M, w). Then, we shall decompose $TM = T\mathcal{S} \oplus \nu\mathcal{S}$, and use type computations, as we did for the LP-cohomology. The operator δ decomposes as shown by Proposition 4.13 i.e., $\delta = \delta'_{(1,-2)} + \delta''_{0,-1)}$. A first simple consequence is that every form $\lambda \in \Lambda^{k,0}M$ is a δ-cycle, therefore, we have the exact homology sequence

$$\ldots \to \Lambda^{k,0}M \to H_k^\pi(M, w) \to H_k(\Lambda M/_{\Lambda^{*,0}M}) \to \Lambda^{k-1,0}M \to \ldots \tag{5.66}$$

but, in fact, it is not simpler to discuss $H_k(\Lambda M/_{\Lambda^{*,0}M})$ than $H_k^\pi(M, w)$.

An arbitrary k-form λ can be written as

$$\lambda = \lambda_{k,0} + \lambda_{k-1,1} + \ldots + \lambda_{0,k}, \tag{5.67}$$

where, as usual, indices denote type. This form is a δ-cycle iff

$$\delta''\lambda_{0,k} = 0, \quad \delta''\lambda_{k-i,i} + \delta'\lambda_{k-i-1,i+1} = 0 \tag{5.68}$$

$(i = 1, \ldots, k - 1)$. Similarly, λ is a δ-boundary iff there exists $\mu \in \Lambda^{k+1}M$ such that the following relations hold for its bihomogeneous components

$$\lambda_{0,k} = \delta''\mu_{0,k+1}, \quad \lambda_{k-i+1,i-1} = \delta''\mu_{k-i+1,i} + \delta'\mu_{k-i,i+1} \tag{5.69}$$

$$(i - 1, \ldots, k).$$

The relations (5.68) and (5.69) should be the starting point for a recursive procedure of computing $H^{\pi}_{*}(M,w)$, but we see that this will be more complicated than in the case of the LP-cohomology. For instance, this time $[\lambda] \mapsto [\lambda_{0,k}]$ yields a homomorphism

$$H^{\pi}_{k}(M,w) \to {}''H^{\pi}_{k}(M,w), \tag{5.70}$$

where $''H^{\pi}_{k}(M,w)$ is the homology of the chain complex $(\Lambda^{0k}M,\delta'')$ ($\delta''^2 = 0$ is an easy consequence of $\delta^2 = 0$), and this mapping is more complicated than its LP-equivalent (5.25). Notice however the

5.21. Remark. For $k = 0$, (5.70) is an isomorphism, and for $k = 1$ it is an epimorphism.

In the general (regular) case the best thing that we can do is look at a homology spectral sequence similar to the one of Theorem 5.12. Namely, similar to (5.44), the formula

$$F_h\Lambda M = \bigoplus_{i\geq h}\bigoplus_{p}\Lambda^{pi}M, \tag{5.71}$$

defines an increasing filtration ($F_h \subseteq F_{h+1}$), compatible with the boundary δ (i.e., $\delta F_h \subseteq F_h$), and if we remember that a homology spectral sequence is defined by $E^r_{pq} = E^{-p,-q}_r$ as given by (5.45) (e.g., [Sp]) we obtain

5.22. Theorem. *The Poisson homology $H^{\pi}_{k}(M,w)$ of a regular Poisson manifold (M,w) has an associated homology spectral sequence $E^r_{pq}(M,w)$ determined by the filtration* (5.71), *and whose first terms are*

$$E^0_{pq}(M,w) = E^1_{pq}(M,w) = \Lambda^{qp}M,$$
$$E^2_{pq}(M,w) = H_p(\bigoplus_{k}\Lambda^{qk}M,\delta'') \overset{\text{def}}{=} {}''H^{(q)}_{p}(M,w). \tag{5.72}$$

Proof. Same as for Theorem 5.12. Q.e.d.

Furthermore, the same idea as in the proof of Theorem 5.18 yields

5.23. Proposition. *(i) For a regular Poisson manifold (M,w) with* rank $w = 2n$, *and with the symplectic foliation $\mathcal{S}$, one has*

$$''H^{(q)}_{p}(M,w) = H^{2n-p}(M,\Phi^q(\mathcal{S})), \tag{5.73}$$

where $\Phi^q(\mathcal{S})$ is the sheaf of germs of $\mathcal{S}$-projectable q-forms on M. (ii) If moreover, $\mathcal{S}$ has either a transversally Riemannian or a transversally symplectic structure, then

$$E^2_{pq}(M,w) =_{LP} E^{2n-p,q}_2(M,w). \tag{5.74}$$

Proof. (i) Practically, the proof is the same as that of Theorem 5.18 but along the symplectic leaves of S only, which requires the checking of a number of details.

In order to do this, we use the coordinates and the bases encountered in the formulas (3.20), (3.21), and the components of the tensors will be with respect to these bases. In particular, $w^{ab} = w^{au} = 0$, and $\det(w^{uv}) \neq 0$, such that the symplectic structure of the leaves is induced by a (0,2)-form ω, where $\omega_{ab} = \omega_{au} = 0$, and $\omega_{uv}w^{us} = \delta^s_v$.

Now, ω defines an associated "volume form" on $T\mathcal{S}$, that can be used to define a $*$-operator, which is an isomorphism

$$\tilde{*}_w : \Lambda^{pq}M \to \Lambda^{p,2n-q}M, \tag{5.75}$$

acting on $\lambda \in \Lambda^{pq}M$ by the same formula as the symplectic $\tilde{*}$, but on the indices u only.

The next step is to reproduce the computation of [Lh1 §83], in order to calculate $-\tilde{*}_w d'' \tilde{*}_w$. This is possible because the componentwise expression of d'' is just like that of d, but on the indices u only, and it includes the usual partial derivatives (e.g., [Va2]).

Then we may replace the latter by covariant derivatives with respect to a Poisson connection of (M, w) (see Theorem (2.20), since the relevant connection coefficients satisfy

$$\Gamma^w_{uv} = \Gamma^w_{vu}, \Gamma^b_{av} = 0. \tag{5.76}$$

Indeed, if ∇ is a Poisson connection, $\nabla w = 0$ (hence $\nabla\omega = 0$ too), and $T\mathcal{S}$ is a parallel distribution i.e., with the notation of (3.20), and the usual definition of the connection coefficients,

$$\nabla_{\frac{\partial}{\partial y^u}} \frac{\partial}{\partial y^v} = \Gamma^w_{vu} \frac{\partial}{\partial y^w}, \quad \nabla_{X_a} \frac{\partial}{\partial y^u} = \Gamma^w_{ua} \frac{\partial}{\partial y^w}. \tag{5.77}$$

The first relation (5.76) follows from the first relation (5.77) since ∇ has no torsion, and the second relation (5.76) follows from the second relation (5.77) and from

$$\nabla_{\frac{\partial}{\partial y^u}} X_a = \Gamma^b_{au} X_b + \Gamma^v_{au} \frac{\partial}{\partial y^v} = \nabla_{X_a} \frac{\partial}{\partial y^u} + [\frac{\partial}{\partial y^u}, X_a],$$

since, by (3.20), $[\partial/\partial y^u, X_a] \in T\mathcal{S}$. The final result will be

$$(-1)^{p+1} \tilde{*}_w d'' \tilde{*}_w = w^{uv} \nabla_u \lambda_{a_1 \ldots a_p v u_1 \ldots u_{q-1}}, \tag{5.78}$$

and, by looking at the formula (4.23′), we shall obtain

$$-\tilde{*}_w d'' \tilde{*}_w = \delta''. \tag{5.79}$$

This formula, together with the exact resolution (5.29) of $\Phi^q(\mathcal{S})$, yields precisely (5.73).

(ii) (5.74) is a straightforward consequence of (5.73) and (5.43). Q.e.d.

5.24. Corollary. *(i) If the Poisson structure w is transversally constant then*

$$H_k^\pi(M,w) = \bigoplus_{q=0}^{k} H^{2n-k+q}(M,\Phi^q(\mathcal{S})). \tag{5.80}$$

(ii) If $M = S \times N$, and w is given by a fixed symplectic structure of S, and if S has finite Betti numbers, then

$$H_k^\pi(M,w) = \bigoplus_{q=0}^{k} [H^{2n-k+q}(S,\mathbb{R}) \otimes \Lambda^q N]. \tag{5.81}$$

Proof. (i) If w is transversally constant, formula (4.47) and Proposition 4.9 show that $\delta' = 0$, hence the Poisson homology is defined by δ'' alone. Then, (5.80) follows straightforwardly from (5.73). (ii) Now, (5.81) is a consequence of (5.80) and of Proposition 5.9. Q.e.d.

In connection with Poisson homology in general, let us also make the

5.25. Remark. In opposition to the case of Poisson cohomology, Poisson homology has a functorial character as follows. If $\varphi : (M_1, w_1) \to (M_2, w_2)$ is a Poisson morphism (Definition 1.17), it follows from formulas (1.41) and (4.21) that $\varphi^* : \Lambda M_2 \to \Lambda M_1$ commutes with the operator δ and, therefore, we have induced homomorphisms in homology

$$\varphi^* : H_k^\pi(M_2, w_2) \to H_k^\pi(M_1, w_1).$$

Finally, we just mention two other questions that have been raised in the study of the Poisson homology.

The first [Br] is to study the spectral sequence ${}_\pi\mathcal{E}_r^{pq}(M,w)$ of the double cochain complex

$$C_\pi^{pq}(M,w) = \Lambda^{p-q}M, \tag{5.82}$$

with the operators d, seen of the type (1,0), and δ seen of the type (0,1). Of course, this is based on the known relation $d\delta + \delta d = 0$ (Chapter 4). For this sequence one has

$${}_\pi\mathcal{E}_1^{pq}(M,w) = H_{p-q}^\pi(M,w), \tag{5.83}$$

and Brylinski [Br] shows that ${}_\pi\mathcal{E}_r^{pq}(M,w)$ degenerates at $\mathcal{E}_1$ if (M,w) is a compact symplectic manifold. In view of the proof of Proposition 5.23 Brylinski's proof should give a similar result for a similar spectral sequence constructed with forms of the type $(0,q)$, and with the operators d'', δ'', in the case of a compact regular Poisson manifold. One just has to use the sheaf $\Phi^0(\mathcal{S})$ instead of $\mathbb{R}$.

The second question (Y. Kosmann-Schwarzbach [Hu3]) is that of obtaining some kind of dual of the mapping $\# : H^*_{\mathrm{deR}}(M) \to H^*_{LP}(M, w)$ given by Proposition 5.3, which should, therefore, look as a mapping

$$H^{\pi}_*(M, w) \to ?, \tag{5.84}$$

and reduce to an isomorphism in the symplectic case. We refer the reader to [Hu3] for Huebschmann's answer, and indicate another answer here.

Before we do this, notice that the question should be understood in the sense that ? is a certain cohomology theory of the graded vector space $\oplus_k \mathcal{V}^k(M)$ since otherwise it has a trivial answer. Namely, if we look at the space of the Grassmann algebra $\oplus_k \mathcal{V}^k_S(M)$ generated by im$\#_w$, the relation $\rho \circ \# = \# \circ \delta$ defines, obviously a boundary $\rho : \mathcal{V}^k_S(M) \to \mathcal{V}^{k-1}_S(M)$, and the corresponding homology may replace the question mark in (5.84), making the latter an isomorphism always.

A nontrivial answer to the previous question is given by

5.26. Proposition. *(i) If the Poisson manifold (M, w) has a Poisson connection ∇, then, for any k-form λ on M, one has $(\delta\lambda)^\# = -D_\nabla(\lambda^\#)$. (ii) (See also* [Kz].*) If, moreover, ∇ has a symmetric Ricci curvature, or, equivalently, it preserves a volume form, then $\mathcal{V}_\nabla(M) \overset{\mathrm{def}}{=} (\oplus \mathcal{V}^k(M), D_\nabla)$ is a chain complex, and there is a well defined induced mapping*

$$\# : H^{\pi}_k(M, w) \to H_k(\mathcal{V}_\nabla(M)), \tag{5.85}$$

which is an isomorphism in the symplectic case.

Proof. The operator D_∇ above was defined by formula (1.25), and (i) is a straightforward computational consequence of (1.25), (4.11), (4.23′) and $\nabla w = 0$. Furthermore, the classical commutation formulas of the covariant derivatives, and the Bianchi identity, yield

$$(D^2_\nabla Q)^{j_3 \cdots j_k} = -R_{uv} Q^{uvj_3 \cdots j_k} \qquad (Q \in \mathcal{V}^k(M)), \tag{5.86}$$

where R is the Ricci curvature of ∇. This gives us the chain complex requested, and we get (5.85). In the symplectic case, the Ricci tensor is always symmetric (e.g., [Va6]), and (5.85) is an isomorphism since $\#^{-1}$ also exists. Q.e.d.

5.27. Remark. If ∇ is an arbitrary Poisson connection, we also have the mapping (5.85), if we take $H_k(\mathcal{V}_\nabla(M))$ to be the *twisted homology groups*

$$H_k(\mathcal{V}_\nabla(M)) = \frac{\ker(D_\nabla : \mathcal{V}^p \to \mathcal{V}^{p-1})}{\mathrm{im}(D_\nabla : \mathcal{V}^{p+1} \to \mathcal{V}^p) \cap \ker(D_\nabla : \mathcal{V}^p \to \mathcal{V}^{p-1})} \tag{5.87}$$

defined in [HL] and [Va1].

6 An Introduction to Quantization

6.1 Prequantization

The present chapter is intended to provide some further important motivation for the study of the LP-cohomology of Poisson manifolds. Namely, LP-cohomological obstructions appear in the problem of the *quantization* of Poisson manifolds.

The general idea is to give a mathematical procedure able to produce a structure that physicists would agree to call the quantum system associated with the classical system represented by the Poisson manifold. Such a theory is far from being unique, and different people give different meanings to the word *quantization*. Here, we make a brief study of the quantization theory known as *geometric quantization*. This theory is very interesting from the mathematical viewpoint at least, since they can be used in the theory of Lie group representations [Kt]. We shall also give some indications, mainly references to the literature, for another theory: *deformation quantization*.

The geometric quantization theory of Kostant [Kt] and Souriau [So] was first developed for symplectic manifolds, and it has important applications. It was amply developed by many authors, and interested readers can use corresponding books such as [Wd], [Sn2], [Hr], etc. Here, we refer only to some basic facts in the Poisson case [Hu1,4], [Va10].

The theory starts from a principle of Dirac which states that the quantization of a classical system is the process of associating Hermitian operators on a Hilbert space to the *classical observables* i.e., functions $f \in C^\infty(M)$, where M is the *phase space* i.e., a symplectic or a Poisson manifold, in such a way that, up to a constant factor depending on the Planck constant, the Poisson bracket on M is associated with the commutator of the operators. The operator $\hat{f}$ associated with the function f is called the *quantization* of f.

Then, a second principle is that for simple physical examples, this quantization should agree with the one accepted by physics.

This second principle can be seen to be related to the demand that the obtained representation of the Poisson algebra $(C^\infty(M), \{\ ,\ \})$ is irreducible. But a theorem of van Hove (e.g., [AM], [Gy]) shows that, with this principle, the quantization problem has no solution.

The exit from this situation consists in looking at the problem of quantization in two steps: *prequantization* which gives a linear representation of the whole algebra $(C^\infty(M), \{\ ,\ \})$ by operators on a complex vector space, and, then, *quantization* where one restricts the problem to a convenient subalgebra of $(C^\infty(M), \{\ ,\ \})$, and one represents it irreducibly, in a way which is compatible with physics, on a Hilbert space built out of the prequantization space.

It was the discovery of Kostant and Souriau (independently) that, modulo a certain obstruction, the prequantization problem can be solved by means of the space of cross-sections of a complex line bundle $\pi : K \to M$, the *prequantization bundle*, and by a well-chosen *prequantization formula* which, in the symplectic case, involved the covariant derivative with respect to a connection of K.

In the symplectic theory, the condition for the existence of K was that the cohomology class of the symplectic form be an *integral cohomology class* (i.e., the image of a class in $H^2(M, \mathbb{Z})$). As we shall see soon, in the Poisson case, one must ask the same condition for $[w] \in H^2_{LP}(M, w)$.

Now, let us go to some details. For the sake of simplicity, and following [Kt], we forget about the Planck constant, and we just ask that prequantization satisfies the condition

$$\widehat{\{f, g\}} = [\hat{f}, \hat{g}] \overset{\text{def}}{=} \hat{f} \circ \hat{g} - \hat{g} \circ \hat{f}. \tag{6.1}$$

Now, let $\Gamma(K)$ be the complex linear space of the global cross-sections s of the prequantization bundle K (if K exists). The Kostant-Souriau prequantization formula adapted to the Poisson case is [Va10]

$$\hat{f}s = D_{df}s + 2\pi \mathrm{ifs} \ , \tag{6.2}$$

where D is a contravariant derivative of K (Definition 4.17). Accordingly, (6.1) is equivalent to

$$C_D = -2\pi i w, \tag{6.3}$$

where C_D is the curvature of D (formula (4.51)).

Hence, the prequantization problem has the Kostant-Souriau solution iff there is a complex line bundle K that possesses a contravariant derivative satisfying (6.3).

Particularly, we must have C_D purely imaginary, and this suggests looking for K together with a Hermitian metric h on it, and to ask D to be compatible with h in the sense that

$$\sigma(h(s_1, s_2)) = h(Ds_1, s_2) + h(s_1, Ds_2). \tag{6.4}$$

Indeed, then, if e is an h-orthonormal local basis of K, D is given, as in (4.57) with the writing of α omitted, by

$$De = te, \tag{6.5}$$

and, as shown by (4.60), C_D identifies with the bivector given locally by σt. This bivector is global, since a change of basis by $\tilde{e} = ae$ leads to $\tilde{t} = t + \sigma \ln a$, and $\sigma\tilde{t} = \sigma t$. If we take $s_1 = s_2 = e$ in (6.4), we have $t + \bar{t} = 0$ hence, $C_D + \bar{C}_D = 0$, as requested.

In order to understand the significance of (6.3), we notice first that the usual Chern-Weil theory of characteristic classes of a vector bundle $\pi : E \to M$ extends in a straightforward manner to contravariant derivatives. We reref the reader to [KN] or [Va2] for the usual theory. If f is an invariant polynomial of degree k on the Lie algebra $gl(n, \mathbb{C})$ and if D is a contravariant derivative on E with the curvature matrix (C_i^j) of (4.60), then, because of (4.61), $f(C)$ is a well defined global $(2k)$-vector field, and it is σ-closed, in view of the Bianchi identity (4.56). The cohomology class $[f(C)] \in H^{2k}_{LP}(M, w)$ is the *Poisson characteristic class* defined by f. By the same computation as in the classical case, it follows that $[f(C)]$ is independent of the choice of the contravariant derivative D. Hence, if we take $D_\alpha = \nabla_{\alpha^\#}$ for a connection ∇ on E we see that the Poisson characteristic class of f is the image of the usual real characteristic class of f by the #-homomorphism (5.5).

It is worth noticing that similar considerations can be made in connection with the secondary (exotic) characteristic classes of foliation theory (e.g., [Bt], [Va8]). Namely, if L is a cofoliation of the Poisson manifold (M, w), the *Bott contravariant derivative* of Proposition 4.19 can be used just as one uses the Bott connection of a foliation in order to get, now, *Poisson secondary characteristic classes.* In the symplectic case, these will be the #-image of the secondary classes of the symplectic orthogonal foliation of L, but there is no reason for the Poisson secondary classes to be images of real classes in the general case of a Poisson manifold.

In particular, we can talk of the *Poisson-Chern classes* of E, and these will be the images of the real Chern classes. Moreover, if we look at the usual formula of the Chern classes (e.g., [KN]) we see that the first Poisson-Chern class of a complex line bundle K is

$$ {}_Pc_1(K) = -\frac{1}{2\pi i}[C_D] \in H^2_{LP}(M, w), \tag{6.6} $$

where D is a Hermitian contravariant derivative on K i.e., D satisfies (6.4).

Now, we can formulate the main result

6.1. Theorem. *The Poisson manifold (M, w) has prequantization bundles iff $[w] \in H^2_{LP}(M, w)$ is the image of an integral cohomology class.*

Proof. If K exists, then, by (6.6) and (6.3), $[w]$ is the image of the real class $c_1(K)$ which is well known to be an integral Chern class ([KN], [Va2]).

Conversely, the fact that $[w]$ is *integral* is to be understood as follows: there exists $a \in H^2(M, \mathbb{Z})$ which is sent by the inclusion $\mathbb{Z} \subseteq \mathbb{R}$ to $[\alpha] \in H^2_{\mathrm{deR}}(M)$, and $\#[\alpha] = [w]$ ($\alpha \in \Lambda^2 M$; $d\alpha = 0$). Then, by a classical result of A. Weil [Wl] and S. Kobayashi [Kb] (see also [Kt]), on M there exist complex line bundles

K, endowed with Hermitian metrics, such that $[\alpha]$ is the first real Chern class of K and, more exactly, $-(2\pi i)\alpha$ is the curvature form of a certain Hermitian connection ∇ of K.

Briefly, this result can be seen as follows. Take an open covering $M = \cup_{h\in S}U_h$ such that $\alpha/U_h = d\varphi_h$ ($\varphi_h \in \Lambda^1 U_h$). Then $d(\varphi_h - \varphi_k) = 0$, and we may assume that

$$\varphi_k - \varphi_h = \frac{1}{2\pi i}\ln c_{kh} \qquad (h, k \in S)$$

for some complex valued functions c_{kh} over $U_h \cap U_k$. Then (e.g., [Va2]) $[\alpha]$ is represented in Čech cohomology by the Čech 2-cocycle

$$\gamma_{khs} = \frac{1}{2\pi i}\{\ln c_{kh} + \ln c_{hs} + \ln c_{sk}\},$$

and, since $[\alpha]$ is the image of $a \in H^2(M, \mathbb{Z})$, we have $\gamma_{khs} \in \mathbb{Z}$ or, equivalently, $c_{kh}c_{hs}c_{sk} = 1$. Accordingly, $\{c_{kh}\}$ are transition functions that define a complex line bundle K on M (up to an isomorphism), $\{-2\pi i\varphi_h\}$ are the local connection forms of a Hermitian connection on K (since they are purely imaginary), and $-2\pi i\alpha$ is the curvature of this connection. (Again, see [Wl], [Kb] or [Kt] for details.)

Furthermore, $\#[\alpha] = [w]$ means

$$w = \alpha^{\#} + \sigma A$$

for a certain vector field A on M, and then we get that

$$D_\lambda s = \nabla_{\lambda^{\#}} s - (2\pi i)\lambda(A)s \qquad (s \in \Gamma(K))$$

is the Hermitian contravariant derivative of K with the curvature $C_D = -2\pi i w$, as requested by (6.3). Q.e.d.

6.2. Remarks. 1) K is not unique. A unique K is associated with the integral first Chern class [Kb].

2) The condition of Theorem 6.1 can also be expressed as follows: (M, w) has a prequantization bundle iff there exists a vector field A, and a closed 2-form α that represents an integral cohomology class of M, such that

$$w + L_A w = \alpha^{\#}. \tag{6.7}$$

3) A theorem similar to 6.1 holds for Poisson algebras [Hu1, 4].

4) If Theorem 6.1 is satisfied, we shall say that (M, w) is *quantizable*.

6.3. Examples. a) If (M, w) is an exact Poisson manifold (i.e., $w = \sigma A$), the trivial bundle $M \times \mathbb{C} \to M$ is a prequantization bundle. In particular, this is the case for the Lie-Poisson structure of a coadjoint Lie algebra. However, this does not imply that every coadjoint orbit is a quantizable symplectic manifold.
b) Let (M, ω) be a quantizable symplectic manifold (i.e., $[\omega]$ is an integral cohomology class), with a prequantization bundle K, and let $\mathcal{S}$ be a symplectic foliation of M. Then, the associated Dirac bracket $\{\ ,\ \}_{\mathcal{S}}$ (Definition 3.7) is quantizable with the same prequantization bundle K. Indeed, it is easy to see that the bivector w of this Dirac bracket satisfies the condition $w = \omega^{\#_w}$. Hence, if ∇ is a connection of K with the curvature $-2\pi i\omega$, $D_\alpha = \nabla_{\alpha^{\#}w}$ is a contravariant derivative with the curvature $C = -2\pi \mathrm{i}w$. Q.e.d.

6.2 Quantization

We do not intend to develop in detail the second step: *quantization*, which is not related to the LP-cohomology, and we refer the reader to the relevant literature ([Wd], [Sn2], etc.). Let us give a few hints, however. The basic notion to be introduced is that of a *polarization*, which we define as a subalgebra $\mathcal{Q}$ of $(\Lambda^1 M \otimes \mathbb{C}, \{\ ,\ \})$ such that $\forall \alpha, \beta \in \mathcal{Q}$, $w(\alpha, \beta) = 0$. (Usually, one takes a maximal subalgebra like this.) Then, the subalgebra of $(C^\infty(M), \{\ ,\ \})$ defined by

$$\mathcal{P}(\mathcal{Q}) = \{f \in C^\infty(M) / \forall \alpha \in \mathcal{Q}, \{df, \alpha\} \in \mathcal{Q}\} \tag{6.8}$$

is called the subalgebra of the *straightforwardly quantizable observables.*

Furthermore, let $\mathcal{D}$ be the complex line bundle of complex *half-densities* of M, which is defined by transition functions that are the square roots of the absolute values of the "covariant" Jacobians of the coordinate transformations $\tilde{x}^i = \tilde{x}^i(x^j)$ i.e., $\mid \partial x^j / \partial \tilde{x}^i \mid^{1/2}$. The cross-sections ρ of $\mathcal{D}$ are called *half-densities* on M, and the Lie derivative of such objects can be defined as for tensors (e.g. [Ya1]). (Remember the more popular notion of a *density* which is just the integrand of an integral on a manifold [Sg]. A half density is the "square root" of a density, and one can also use s-densities which are the sth power of a density [Ya1].)

Then, the prequantization formula (6.2) can be extended to a representation on $\Gamma(K \otimes \mathcal{D})$ by putting

$$\hat{f}(s \otimes \rho) = D_{df}(s \otimes \rho) + 2\pi \mathrm{i} f(s \otimes \rho), \tag{6.9}$$

where, $\forall \alpha \in \Lambda^1 M$, we define

$$D_\alpha(s \otimes \rho) = (D_\alpha s) \otimes \rho + s \otimes (L_{\alpha^{\#}} \rho). \tag{6.10}$$

It is easy to see that the Dirac condition (6.1) is still satisfied. Indeed, in view of (6.10), (6.9) becomes

$$\hat{f}(s \otimes \rho) = (\hat{f} s) \otimes \rho + s \otimes L_{X_f} \rho, \tag{6.9'}$$

and the result follows by a direct computation that uses (6.1) and

$$L_{[X_f,X_g]} = L_{X_f} \circ L_{X_g} - L_{X_g} \circ L_{X_f},$$

which is a known property of the Lie derivative.

On the other hand, these formulas imply

$$D_\alpha \hat{f}(s \otimes \rho) = \hat{f} D_\alpha (s \otimes \rho) - D_{\{df,\alpha\}}(s \otimes \rho). \tag{6.11}$$

Namely, in order to get (6.11) we just compute the commutant $[D_\alpha, \hat{f}] = D_\alpha \circ \hat{f} - \hat{f} \circ D_\alpha$ on $s \otimes \rho$ by using (6.9) and (6.10). After reductions one gets

$$[D_\alpha, \hat{f}](s \otimes \rho) = (C_D(\alpha, df)s) \otimes \rho + (D_{\{\alpha,df\}}s) \otimes \rho +$$

$$+ s \otimes L_{[\alpha^\#, X_f]}\rho + 2\pi i(\alpha^\# f)s \otimes \rho,$$

and then (4.4), (6.3) and (6.10) show that the previous relation is exactly (6.11).

Hence, if we denote

$$\mathcal{H}_0 = \{\nu \in \Gamma(K \otimes \mathcal{D}) / \forall \alpha \in \mathcal{Q}, D_\alpha \nu = 0\}, \tag{6.12}$$

it will follow that $\nu \in \mathcal{H}_0$ and $f \in \mathcal{P}(\mathcal{Q})$, imply $\hat{f}\nu \in \mathcal{H}_0$.

Now, the *quantization* of $f \in \mathcal{P}(\mathcal{Q})$ is defined as $\hat{f} \mid_{\mathcal{H}_0}$, where $\hat{f}$ is given by (6.9), and, again, (6.1) is satisfied.

A difficulty of this scheme is that it is unclear whether $\mathcal{H}_0 \neq 0$. This is a so-called *Bohr-Sommerfeld* condition [Sn2], and we suppose it holds. (If not, the quantization scheme becomes more complicated.)

Furthermore, if M is compact, it is easy to make $\mathcal{H}_0$ into a pre-Hilbert space, by taking

$$\langle s_1 \otimes \rho_1, s_2 \otimes \rho_2 \rangle = \int_M h(s_1, s_2)\rho_1 \bar{\rho}_2, \tag{6.13}$$

where h is the Hermitian metric of K, and the bar denotes complex conjugation.

Moreover, with respect to (6.13), the operators $\hat{f}$ of (6.9), (6.9′) are anti-Hermitian as it follows from the following computation

$$\begin{aligned}
&\langle \hat{f}(s_1 \otimes \rho_1), (s_2 \otimes \rho_2)\rangle + \langle s_1 \otimes \rho_1, \hat{f}(s_2 \otimes \rho_2)\rangle = \\
&\overset{(6.9')}{=} \langle (\hat{f}s_1) \otimes \rho_1 + s_1 \otimes L_{X_f}\rho_1, s_2 \otimes \rho_2 \rangle + \\
&+ \langle s_1 \otimes \rho_1, (\hat{f}s_2) \otimes \rho_2 + s_2 \otimes L_{X_f}\rho_2 \rangle = \\
&= \int_M \{[h(\hat{f}s_1, s_2) + h(s_1, \hat{f}s_2]\rho_1\bar{\rho}_2 + h(s_1, s_2)[(L_{X_f}\rho_1)\bar{\rho}_2 + \rho_1(L_{X_f}\bar{\rho}_2)]\} = \\
&\overset{(6.4)}{=} \int_M \{X_f[(h(s_1, s_2)]\rho_1\bar{\rho}_2 + h(s_1, s_2)[(L_{X_f}\rho_1)\bar{\rho}_2 + \rho_1(L_{X_f}\bar{\rho}_2)]\} = \\
&= \int_M L_{X_f}[h(s_1, s_2)\rho_1\bar{\rho}_2] = 0,
\end{aligned} \tag{6.14}$$

where the last result is implied by the density version of Stokes' theorem, $\int_M L_X \lambda = 0$ for any vector field X and any density λ (e.g., [Va3]).

Thus, if we want Hermitian operators we shall multiply $\hat{f}$ by i, for instance, and we shall remember that we have "forgotten" a constant factor in Dirac's condition (6.1). Finally, if we want a Hilbert space, we shall just take the completion of $\mathcal{H}_0$.

If M is not compact, we will require the subalgebra $\mathcal{Q}_0$ of $\Lambda^1 M$ whose complexification is $\mathcal{Q} \cap \bar{\mathcal{Q}}$ to be such that $\#\mathcal{Q}_0$ is tangent to a fibering of M over another manifold N. Then $\mathcal{H}_0$ will be replaced by its subspace $\mathcal{H}_0^c$ of sections that are projectable to N, and whose projection has a compact support. Replacing M by N in (6.13), and using the projected integrand, will then give a pre-Hilbert structure on $\mathcal{H}_0^c$, and one will use its complexification as the quantum Hilbert space. Formula (6.14) will still hold good.

Let us mention that the described solution of the quantization step is not complete, since there are important physical observables, such as energy, that are not straightforwardly quantizable. Therefore, the theory must be completed by methods of extending quantization to more functions f, and one such method is the *intertwining* of two polarizations by a so-called *BKS kernel*. Again, we must refer the interested reader to the already quoted literature [Wd], [Sn2], [Va4], [Va10], etc.

6.3 Prequantization representations

Following Urwin [Ur], if we have a Poisson manifold (M, w) and a complex line bundle K with a Hermitian metric h and a Hermitian contravariant derivative D, we shall define

6.4. Definition. A *prequantization representation* of $(C^\infty(M), \{\ ,\ \})$ is a homomorphism Q of this algebra into the algebra of the first order differential operators on $\Gamma(K)$, with the usual commutator as a product, such that $\forall f \in (C^\infty(M)$, symbol $Q(f) = X_f$.

Remember that an operator Q as requested is one which acts on $s \in \Gamma(K)$ as follows: if b is a local basis of cross sections of K then $s = \sigma b$ for a complex valued function σ, and $Qs = (X\sigma + \varphi\sigma)b$, where X is a vector field and φ is a function. The vector field X does not change if b is changed to $\tilde{b} = \beta b$ hence, X is global on M, and it is called the *symbol* of Q.

Since, because of (4.50), the symbol of D_{df} is exactly X_f, it follows that Q must be of the form

$$Q(f)s = D_{df}s + 2\pi \mathrm{i} m(f)s, \tag{6.15}$$

where $m(f)$ is given by an $\mathbb{R}$-linear mapping $m : C^\infty(M) \to C^\infty(M, \mathbb{C})$. (The factor $2\pi i$ is here for technical reasons.) Accordingly, the commutation condition $[Q(f), Q(g)] = Q(\{f, g\})$ becomes

$$\{f, m(g)\} - \{g, m(f)\} - m(\{f, g\}) = -\frac{1}{2\pi i} C_D(df, dg). \tag{6.16}$$

Now, the meaning of the left-hand side of this formula comes from another cohomology theory, whose study in the theory of Poisson manifolds is, partly, motivated by the problem of finding all the prequantization representations, namely, the *Chevalley-Eilenberg cohomology* (e.g., [Lh2]). This study will get an even stronger motivation in the theory of the deformation quantization.

Remember that, generally, if $(\mathcal{A}, [\ ,\ ])$ is a (e.g., real) Lie algebra (not necessarily finite dimensional), and $\mathcal{M}$ is a (real) linear space endowed with a bilinear multiplication $\mathcal{A} \times \mathcal{M} \to \mathcal{M}$ such that

$$[a_1, a_2]m = a_1(a_2 m) - a_2(a_1 m), \tag{6.17}$$

one says that $\mathcal{M}$ is an $\mathcal{A}$*-module*, or that one has an *action* of $\mathcal{A}$ on $\mathcal{M}$. Then, a k-linear skew symmetric mapping $\mathcal{A}^k \to \mathcal{M}$ is called an $\mathcal{M}$*-valued* k*-cochain*, and these cochains form a linear space $C^k(\mathcal{A}; \mathcal{M})$. The formula

$$\begin{aligned}(\partial c)(a_0, \ldots, a_k) = &\sum_{i=0}^{k} (-1)^i a_i . c(a_0, \ldots, \hat{a}_i, \ldots, a_k) + \\ &+ \sum_{i<j}^{k} (-1)^{i+j} c([a_i, a_j], a_0, \ldots, \hat{a}_i, \ldots, \hat{a}_j, \ldots, a_k)\end{aligned} \tag{6.18}$$

defines a coboundary since, as for the exterior differential d, $\partial^2 = 0$. Hence we have corresponding cohomology spaces $H^k(\mathcal{A}; \mathcal{M}) = \ker \partial_k / \mathrm{im} \partial_{k-1}$, called the *cohomology of the Lie algebra* $\mathcal{A}$ *with values (or coefficients) in* $\mathcal{M}$, or *relative to the given representation of* $\mathcal{A}$ *on* $\mathcal{M}$.

Because of the Jacobi identity, any Lie algebra $\mathcal{A}$ is an $\mathcal{A}$-module, to be denoted by ($\mathcal{A}$-ad), for the operation $a.a' = [a, a']$, and this justifies

6.5. Definition. The cochains of $\mathcal{A}$ with values in $\mathcal{A}$-ad are called *Chevalley-Eilenberg cochains*, and $\mathcal{H}^k(\mathcal{A}; \mathcal{A}\text{-ad})$ are the *Chevalley-Eilenberg cohomology spaces* of $\mathcal{A}$. In particular, if $\mathcal{A} = (C^\infty(M), \{\ ,\ \})$ one obtains the *Chevalley-Eilenberg cochains* $C^k_{CE}(M, w)$ and the *Chevalley-Eilenberg cohomology* of the Poisson manifold (M, w), to be denoted by $H^k_{CE}(M, w)$.

For instance, a 1-dimensional Chevalley-Eilenberg cocycle Δ is just what is called a *derivation* of $(\mathcal{A}, [\ ,\ ])$, since $\partial \Delta = 0$ means exactly

$$\Delta[a_1, a_2] = [\Delta a_1, a_2] + [a_1, \Delta a_2].$$

Lichnerowicz (e.g., [Lh2], [Lh4], etc.) has pointed out several important subcomplexes of $C^k_{CE}(M, w)$ namely,

6.6. Definition. An element $\ell \in C^k_{CE}(M,w)$ is *local* if $\ell(u_1,\dots,u_k)/_U = 0$ whenever some $u_i/_U = 0$, where U is any open subset of M. The same ℓ is *d-differentiable* if it is defined by a linear multi-differential operator of maximum order d in each argument. (The case $d = 1$ is the most important.)

The corresponding subspaces will be denoted by ${}_{CE}C^k_{\mathrm{loc}}(M,w)$ and ${}_{CE}C^k_{d-\mathrm{diff}}(M,w)$, respectively.

Formula (6.18) shows directly that if ℓ is local so is $\partial\ell$, and, by a computation of the terms with the highest order derivatives, that uses the skew symmetry of w, (6.18) also shows that, if ℓ is d-differentiable so is $\partial\ell$ [Lh2]. Hence, one has corresponding cohomologies ${}_{CH}H^*_{\mathrm{loc}}(M,w)$ and ${}_{CE}C^*_{d-\mathrm{diff}}(M,w)$. (The indices CE are usually omitted.)

6.7. Remark. The LP-cochain complex can also be seen as a subcomplex of the CE-cochain complex namely, it consists of the 1-differentiable cochains that are derivations with respect to each argument. Accordingly, one has the following homomorphisms induced by inclusions

$$H^k_{LP}(M,w) \xrightarrow{\iota_1^*} {}_{CE}\,H^k_{1-\mathrm{dif}}(M,w) \xrightarrow{\iota_2^*} {}_{CE}\,H^k_{\mathrm{loc}}(M,w) \xrightarrow{\iota_3^*} H^k_{CE}(M,w). \quad (6.19)$$

We shall denote by j^* the composition of all these homomorphisms. The existence of j^*shows the relevance of the LP-cohomology for the CE-cohomology. In fact, there is more to say about it than that. Namely, $H^k_{1-\mathrm{dif}}(M,w)$ can be computed in terms of the LP-cohomology of (M,w) [Lh2]. As for the whole CE-cohomology, its calculation is more difficult. We refer the reader to the recent thesis of Melotte [Ml] for this calculation in dimensions 1, 2, 3, in the case of regular Poisson manifolds, and Lie-Poisson structures.

Now, let us come back to the prequantization representation (6.15). From (6.16) we see that m can be seen as a complex valued 1-dimensional CE-cochain that must satisfy

$$\partial m = -\frac{1}{2\pi i} C_D. \quad (6.20)$$

Since the right-hand side of (6.20) is real, the imaginary part of m will be a 1-CE-cocycle, and we only have to find the real cochains m satisfying (6.20). The corresponding result is [Va10]

6.8. Theorem. *A complex line bundle $K \to (M,w)$ has pre-quantization representations iff its Poisson-Chern class ${}_{PC_1}(K)$ belongs to* $\ker j^*$. *Moreover, K admits 1-differentiable pre-quantization representations iff ${}_{PC_1}(K) = [\varphi w]$ for some Casimir function φ of M.*

Proof. The first assertion follows from (6.6) and (6.20). If we have a 1-differentiable representation we have

$$m(f) = Af + \varphi f \qquad (A \in \mathcal{V}^1(M), \varphi \in C^\infty(M)), \quad (6.21)$$

and then (6.16) becomes

$$\begin{aligned} \varphi w(df,dg) - (L_A w)(df,dg) + f(X_\varphi g) - g(X_\varphi f) = \\ = -\frac{1}{2\pi i} C_D(df,dg). \end{aligned} \tag{6.22}$$

For $f = 1$, and an arbitrary g, this shows that $X_\varphi = 0$ i.e., φ must be a Casimir function. Then, by looking again at the general form of (6.22), and since $L_A w = -\sigma A$, we see that ${}_P c_1(K) = [\varphi w]$.

Conversely, if this happens, and if we define A such that

$$\sigma A = -\frac{1}{2\pi i} C_D - \varphi w,$$

(6.22) will hold good, which means that (6.21) is a prequantization representation. Q.e.d.

The notion of a prequantization representation, and the previous theorem can conveniently be extended to Jacobi manifolds [Va10].

6.4 Deformation quantization

Now, let us say a few words about *deformation quantization.* This is a theory which has been developed by A. Lichnerowicz and his school (e.g., [BF-]), and it starts from the principle that a quantum system has to reduce to the corresponding classical system if $\hbar \to 0$ ($\hbar = h/2\pi$, $h =$ the Planck constant). Therefore, the quantization of a classical system should consist of a deformation of this classical system into a system that depends on a parameter $\hbar$.

Of course, in this principle, the word *system* should be given a precise meaning. If we remember what happened in geometric quantization, we can say that, from the classical viewpoint, *system* should mean the commutative algebra of *observables* $C^\infty(M) \otimes \mathbb{C}$ with its usual product on one hand, and with the Poisson bracket $\{\ ,\ \}$ on the other hand. Hence, the *quantization* of this system should be an associative, non-commutative deformation of the usual product into a new operation the $*$-*product* $f * g$ ($f, g \in C^\infty(M)$) that depends on $\hbar$, and is such that the commutant $[f * g] \stackrel{\text{def}}{=} (f * g - g * f)/i\hbar$ will be a deformation of the Lie algebra operation $\{f, g\}$. Accordingly, the main mathematical aim of deformation quantization is the construction of $*$-products, and, then, the corresponding physics will have to be constructed and interpreted.

A first operation of this kind was discovered by Moyal [My], [BF-] within the framework of H. Weyl's quantization theory, that has the usual quantum machinery of Hilbert spaces. Moyal's bracket covers the flat case, and it gives the basis needed for physical interpretations [BF-]. The further development of the theory went through the proof of the existence of $*$-products on symplectic manifolds, modulo the vanishing of an obstruction in $H^3(M, \mathbb{R})$, by J. Vey

[Ve], and then by the proof that this obstruction indeed vanishes for any symplectic manifold [DL], [OMY1,2], [Fd]. Recently, there are efforts to study the $*$-products on Poisson manifolds [LMR], [Rg], [Ri], etc.

For physical interpretations, as explained in [BF-], one has to consider in the classical system a finite subalgebra of *distinguished observables* $\mathcal{A} \subseteq (C^\infty(M), \{\ ,\ \})$. These observables will play the role of *preferred coordinates* of the system, and they have to satisfy the property

$$[a * f] = \{a, f\} \qquad (a \in \mathcal{A}).$$

Furthermore, once $\mathcal{A}$ is fixed, a good $*$-product for quantization is one that is *invariant* by $\mathcal{A}$ in the sense that, $\forall a \in \mathcal{A}$ one has

$$\{a, f * g\} = \{a, f\} * g + f * \{a, g\},$$

and only such $*$-products should be called *quantizations.* We refer the reader to [BF-] for details.

In what follows, we want to explain more exactly what are the *deformations* of the algebraic structures that we mentioned above, namely that of an associative algebra, and that of a Lie algebra, and see that the study of such deformations is related to cohomological theories.

6.9. Definition. i) Let B be a $C^\infty(M)$-valued $\mathbb{R}$-bilinear form on the vector space $C^\infty(M)$ of differentiable functions on a manifold M, and let ν be an indeterminate. Then, a *(formal) deformation* of B is a formal power series

$$B_\nu(f, g) = \sum_{r=0}^{\infty} \nu^r C_r(f, g), \tag{6.23}$$

where $C_0(f, g) = B(f, g)$ and $C_r(f, g)$ are again $\mathbb{R}$-bilinear forms on $C^\infty(M)$ with values in $C^\infty(M)$.
ii) If $B(f, g) = fg$ (usual product), one denotes $B_\nu(f, g) = f *_\nu g$, and, if the extension of this operation to formal power series is associative, it is said to be a *(formal) associative deformation* of the algebra $C^\infty(M)$. Furthermore, if M is a Poisson manifold, and $B(f, g) = \{f, g\}$, we shall denote $B_\nu(f, g) = \{f, g\}_\nu$, and if this extends to a Lie algebra structure on formal power series, we call it a *(formal) Lie deformation* of the Poisson bracket $\{f, g\}$.

In order to write down the conditions for an associative deformation, we denote (e.g., [Lh4])

$$(f_1 *_\nu f_2) *_\nu f_3 - f_1 *_\nu (f_2 *_\nu f_3) = \sum_{k=1}^{\infty} \nu^k D_k(f_1, f_2, f_3),$$

where

$$D_k(f_1,f_2,f_3)\sum_{\substack{r+s=k\\(r,s\geq 0)}} C_r(C_s(f_1,f_2),f_3)-C_r(f_1,C_s(f_2,f_3))], \tag{6.24}$$

and then the mentioned conditions are $D_k=0$ for all $k=1,2,\ldots$.

Let us denote the sum of the terms with $r,s\geq 1$ in (6.24) by $E_k(f_1,f_2,f_3)$. Then we get

$$D_k(f_1,f_2,f_3)=E_k(f_1,f_2,f_3)-(\tilde{\partial}C_k)(f_1,f_2,f_3), \tag{6.25}$$

where the last term is a notation which comes from the fact that algebraists have recognized in the corresponding expression a coboundary that we must explain.

Let $\mathcal{A}$ be an associative algebra. Define p-cochains as p-linear mappings $\mathcal{A}^p\to\mathcal{A}$ (and elements of $\mathcal{A}$ if $p=0$), and put

$$\begin{gathered}(\tilde{\partial}C)(a_0,\ldots,a_p)=a_0.C(a_1,\ldots,a_p)+\\ +\sum_{i=0}^{p-1}(-1)^{i+1}C(a_0,\ldots,a_i.a_{i+1},\ldots,a_p)+(-1)^{p+1}C(a_0,\ldots,a_{p-1}).a_p.\end{gathered} \tag{6.26}$$

Then associativity implies $\tilde{\partial}^2=0$, and the corresponding cohomology $H^p_{\mathrm{Hoch}}(\mathcal{A})$ is called the *Hochschild cohomology* of $\mathcal{A}$.

The coefficients $C_k(f,g)$ of (6.23) can be seen as Hochschild 2-cochains of $C^\infty(M)$, D_k and E_k of (6.25) are 3-cochains, and the last term of (6.25) is, precisely, a Hochschild coboundary.

Now, the construction of an associative deformation of $C^\infty(M)$ can be seen as a recurrent process. Namely, assume that we have succeeded in finding some 2-cochains $C_i(f,g) i\leq k$ such that $D_i=0$ for $i\leq k$, and that we want to find C_{k+1} such that $D_{k+1}=0$ too. E_{k+1} depends only on C_r with $r\leq k$, and it has been proven by Gerstenhaber [Gh] that $D_i=0$ for $i\leq k$ implies $\tilde{\partial}E_{k+1}=0$. In other words E_{k+1} defines a cohomology class $[E_{k+1}]\in H^3_{\mathrm{Hoch}}(C^\infty(M))$, and, iff this class vanishes, a good choice of C_{k+1} is possible (see again (6.25)).

Therefore, the obstructions to the existence of an associative deformation lie in $H^3_{\mathrm{Hoch}}(C^\infty(M))$, and a computation of this space is essential in deformation quantization. If $H^3_{\mathrm{Hoch}}(C^\infty(M))=0$, an associative deformation of $C^\infty(M)$ is possible.

6.10. Remark. It is possible to look at various versions of Hochschild cohomology for instance, at $\tilde{H}^p_{\mathrm{Hoch}}(C^\infty(M))$ defined by cochains given by differential operators, and null on constants (e.g. [Lh4]). J. Vey [Ve] has proven that

$$\tilde{H}^p_{\mathrm{Hoch}}(C^\infty(M))=\mathcal{V}^p(M). \tag{6.27}$$

In fact, it makes good sense to ask $C_r(f,g)$ in (6.23) to vanish on constants, since the product by constants should rather not be deformed.

Now, let us suppose that (6.23) is a Lie deformation, and, to have a different notation, write it as

$$\{f,g\}_\nu = \sum_{r=0}^{\infty} \nu^r \Gamma_r(f,g) \qquad (\Gamma_0(f,g) = \{f,g\}), \tag{6.28}$$

where $\Gamma_r(f,g) + \Gamma_r(g,f) = 0$, i.e., Γ_r are Chevalley-Eilenberg 2-cochains. Accordingly, we get

$$\begin{aligned} \{f_1,\{f_2,f_3\}\} + \{f_2,\{f_3,f_1\}\} + \{f_3,\{f_1,f_2\}\} = \\ = \sum_{k=1}^{\infty} \nu^k \Delta_k(f_1,f_2,f_3), \end{aligned} \tag{6.29}$$

where

$$\Delta_k(f_1,f_2,f_3) = (\text{Cyclic} \sum_{r+s=k} \Gamma_r(f_1,\Gamma_s(f_2,f_3)) \quad (r,s \geq 0), \tag{6.30}$$

and (6.28) is a Lie deformation iff all $\Delta_k = 0$ $(k = 1,2,\ldots)$.

As in the previous case, if we put

$$\Xi_k(f_1,f_2,f_3) = (\text{Cyclic} \sum_{r+s=k} \Gamma_r(f_1,\Gamma_s(f_2,f_3)) \quad (r,s \geq 1) \tag{6.31}$$

the relation (6.30) becomes

$$\Delta_k = \Xi_k + \partial\Gamma_k, \tag{6.32}$$

where ∂ is the coboundary of the Chevalley-Eilenberg cohomology (Definition 6.5).

Now, again, the construction of a Lie deformation can be approached recursively. Assume that we already have Γ_r such that $\Delta_r = 0$ for $r \leq k-1$, and we want Γ_k with $\Delta_k = 0$.

Ξ_k contains only Γ_r for $r \leq k-1$, and, again, the computations of Gerstenhaber [Gh], show that Ξ_k is a CE-cocycle. Iff $[\Xi_k] \in H^3_{CE}(M,w)$ vanishes we can find Γ_k such that $\Delta_k = 0$, and the construction of the deformation may go on. In particular, if $H^3_{CE}(M,w) = 0$, the formal Lie deformation can be achieved.

The concrete application of this deformation machinery for quantization will use complex-valued functions, and the purely imaginary parameter $\nu = i\hbar$. Then, one has to find an associative deformation

$$f * g = fg + i\hbar C_1(f,g) + (i\hbar)^2 C_2(f,g) + \ldots, \tag{6.33}$$

such that

$$C_1(f,g) - C_1(g,f) = \{f,g\}. \tag{6.34}$$

For instance, we can take $C_1(f,g) = (1/2)\{f,g\}$, and proceed by looking for C_2, and so on.

The other way around would be to look for a Lie deformation

$$\{f,g\}_* = \{f,g\} + (i\hbar)\Gamma_1(f,g) + (i\hbar)^2\Gamma_2(f,g) + \dots, \tag{6.35}$$

and then for the $*$-product that makes

$$\{f,g\}_* = (f * g - g * f)/(i\hbar). \tag{6.36}$$

We refer the reader to such papers as [FLS], [BF-], [Lh4], [DL], [LMR], [Rg], [Rw], [MV] etc., for an actual description of $*$-products and deformations of Poisson algebras with various conditions on the cocycles of the deformations.

Finally, notice that everything in the above description referred to *formal* deformations. A more profound analysis should also regard the problem of the *convergence* of the series (6.33), (6.35). This is not an easy problem. A recent effort in this direction is [Ri]. A convergence result also appears in [LMR].

7 Poisson Morphisms, Coinduced Structures, Reduction

7.1 Properties of Poisson mappings

We defined in Chapter 1 (Definition 1.17) the notion of a Poisson mapping or morphism (as well as that of an automorphism or equivalence) for reasons of reference. Now, we shall discuss this notion in more detail, and use it in the study of the process of defining a Poisson structure on a manifold N from a mapping $\varphi : M \to N$, if M is endowed with a Poisson structure w. Such a process is possible in some interesting cases. Furthermore, this process is generalized in a procedure known as the reduction of a Poisson structure and, particularly, reductions appear in connection with group actions and momentum mappings.

Let us recall that a mapping $\varphi : (M_1, w_1) \to (M_2, w_2)$ between two Poisson manifolds is a Poisson mapping (morphism) if $\forall f, g \in C^\infty(M_2)$ one has

$$\{f \circ \varphi, g \circ \varphi\}_1 = \{f, g\}_2 \circ \varphi. \tag{7.1}$$

Obviously, this relation is equivalent to the fact that the two Poisson bivectors w_1, w_2 are φ-related i.e., $\forall x \in M_1$, $\forall \alpha, \beta \in T^*_{\varphi(x)}M_2$, one has

$$w_{1,x}(\varphi^*\alpha, \varphi^*\beta) = w_{2,\varphi(x)}(\alpha, \beta). \tag{7.2}$$

In fact, (7.1) and (7.2) are exactly the same thing if $\alpha = df$, $\beta = dg$.

Furthermore, if the Poisson brackets of (7.1) are expressed by using Hamiltonian vector fields, it follows that (7.1) is equivalent to the fact that the vector fields $X_{f\circ\varphi}$ and X_f are φ-related i.e., $\varphi_* X_{f\circ\varphi} = X_f$. Finally, since $X_f = \#_{w_2}df$ and $X_{f\circ\varphi} = \#_{w_1}d(f \circ \varphi)$, this φ-relatedness is equivalent to

$$\#_{w_2,\varphi(x)} = \varphi_{*,x} \circ \#_{w_1,x} \circ \varphi^*_{\varphi(x)}, \quad \forall x \in M, \tag{7.3}$$

The φ-relatedness of $X_{f\circ\varphi}$ and X_f also shows that the spaces of the characteristic distribution $\mathcal{S}(M_2)$ (see Chapter 2) are images of subspaces of $\mathcal{S}(M_1)$ by φ_*. Hence, $\mathrm{rank}_x w_1 \geq \mathrm{rank}_{\varphi(x)} w_2$, and, in particular, if M_2 is a symplectic manifold, φ must be a submersion.

It is also clear from (7.2) that if φ is a Poisson morphism and a diffeomorphism, then φ^{-1} is a Poisson morphism too and, in this case, we call φ a *Poisson isomorphism* (or, *equivalence* or, if $M_1 = M_2$, a *Poisson automorphism*).

The following are a few elementary properties [LM].

7.1. Proposition. *Consider the mappings $\varphi : (M_1, w_1) \to (M_2, w_2)$ and $\psi : (M_2, w_2) \to (M_3, w_3)$. (i) If φ, ψ are Poisson mappings so is $\psi \circ \varphi$. (ii) If φ is a surjective Poisson mapping, and if $\psi \circ \varphi$ is Poisson, then ψ is a Poisson mapping too.*

Proof. (i) follows straightforwardly from (7.1). (ii) Take $y \in M_2$, $z = \psi(y) \in M_3$. There exists $x \in M_1$ with $\varphi(x) = y$, and the following computation holds good

$$w_{3,z}(\alpha, \beta) = w_{1,x}((\psi \circ \varphi)^*\alpha, (\psi \circ \varphi)^*\beta) =$$
$$= w_{1,x}(\varphi^*(\psi^*\alpha), \varphi^*(\psi^*\beta)) = w_{2,y}(\psi^*\alpha, \psi^*\beta).$$

Q.e.d.

If we look again at (7.2), we see that a surjective mapping $\varphi : (M_1, w_1) \to M_2$ can only be a Poisson mapping for at most one Poisson structure w_2 of M_2. This remark suggests giving the

7.2. Definition. Let (M_1, w_1) be a Poisson manifold, and $\varphi : M_1 \to M_2$ be a surjective differentiable mapping. Then, if there is a Poisson structure w_2 on M_2 that makes φ into a Poisson mapping, we shall say that w_2 is *coinduced* by the mapping φ.

7.3. Proposition. *If $\varphi : (M_1, w_1) \to M_2$ is a surjective mapping, M_2 has a coinduced Poisson structure iff, $\forall f, g \in C^\infty(M_2)$, $\{f \circ \varphi, g \circ \varphi\}$ is constant along the fibers of φ.*

Proof. The condition is obviously a sufficient one since it allows us to define $\{f, g\}$ by (7.1). Conversely, if a coinduced structure exists we have

$$\{f \circ \varphi, g \circ \varphi\}(\varphi^{-1}(y)) = \{f, g\}(y) \qquad (y \in M_2),$$

and this means that the result is constant along the fibers. Q.e.d.

7.4. Corollary. *Let $\varphi : (M_1, w_1) \to M_2$ be a surjective submersion with connected fibers. Assume that, $\forall x \in M_1, \ker \varphi_*(x)$ is spanned by local Hamiltonian vector fields (i.e., $\ker \varphi_*(x) \subseteq S_x(M_1)$). Then, M_2 has a Poisson structure coinduced by φ.*

Proof. We take $f, g \in C^\infty(M_2)$, and prove that $\{f \circ \varphi, g \circ \varphi\}$ is constant along the fibers of φ. Since φ is a submersion, the fibers of φ are submanifolds of M_1, and, because of the spanning hypothesis, it will be enough to prove that $\forall \lambda \in C^\infty(M_1)$ such that $X_\lambda \in \ker \varphi_*$ one has $X_\lambda(\{f \circ \varphi, g \circ \varphi\}) = 0$. This follows by the Jacobi identity:

$$X_\lambda(\{f \circ \varphi, g \circ \varphi\}) = \{\lambda, \{f \circ \varphi, g \circ \varphi\}\} =$$
$$= \{f \circ \varphi, \{\lambda, g \circ \varphi\}\} + \{\{\lambda, f \circ \varphi\}, g \circ \varphi\} =$$
$$= \{f \circ \varphi, X_\lambda(g \circ \varphi)\} + \{X_\lambda(f \circ \varphi), g \circ \varphi\} = 0,$$

since $f \circ \varphi$ and $g \circ \varphi$ are constant along the fibers. Q.e.d.

Now, we want to discuss another interesting characterization of Poisson morphisms, given by Weinstein [We6], which is based on the notion of a coiso tropic submanifold that is itself an important geometric notion.

7.5. Definition. A submanifold N of a Poisson manifold (M, w) is called *coisotropic* if $w/_{\mathrm{Ann}TN} \equiv 0$, where Ann $T_xN = \{\alpha \in T_x^*M/\alpha(v) = 0, \forall v \in T_xN\}$ is the *annihilator* of $T_xN(x \in N)$, or, equivalently

$$\#_w(\text{Ann } TN) \subseteq TN. \tag{7.4}$$

Notice that, for any submanifold N of M, $\#(\text{Ann } T_xN)$ $(x \in N)$ is just the symplectic orthogonal space of $(T_xN) \cap \mathcal{S}_x(M)$ with respect to the symplectic structure of $\mathcal{S}_xM$, since the symplectic scalar product $\omega_{\mathcal{S}}(v_1, v_2) = w(\alpha_1, \alpha_2)$ for $\#\alpha_1 = v_1$, $\#\alpha_2 = v_2$. Hence, if $(TN) \cap \mathcal{S}(M)$ is a differentiable distribution on N, $\#$ Ann(TN), may not have a lower semi-continuous dimension i.e., it may not be differentiable.

The notion of coisotropy is meaningful for immersed submanifolds, but it is interesting, particularly, in the case of a locally closed submanifold because of

7.6. Proposition. *A locally closed submanifold N of the Poisson manifold (M, w) is coisotropic iff one of the following equivalent conditions is satisfied: a) $\forall f, g \in C^\infty(M)$ such that $f/_N = 0, g/_N = 0$, one has $\{f, g\}/_N = 0$; b) $\forall f \in C^\infty(M)$ such that $f/_N = 0$, the Hamiltonian vector field X_f/N is tangent to N.*

Proof. $f/_N = 0$ obviously implies $df/N \in \text{Ann}(TN)$. Hence, N coisotropic implies a) and b) because of Definition 7.5. Conversely, since N is locally closed in M, every $\alpha \in \text{Ann}_x TN$ extends to a local 1-form df where $f/_N = 0$ (use local coordinates x^i such that N has the local equations $x^a = 0$, $i = 1, \ldots, \dim M$, $a = 1, \ldots, \dim M - \dim N$). This makes clear the fact that a) implies the condition of Definition 7.5, and that b) implies a). Q.e.d.

7.7. Remark. The name coisotropic comes from symplectic geometry. Indeed, if the Poisson structure w comes from a symplectic one, $\#_w(\text{Ann } TN)$ is the symplectic orthogonal of TN, and (7.4) means that N is a coisotropic submanifold of the symplectic manifold M. Moreover, since, generally, $\#(\text{Ann } TN)$ is the symplectic orthogonal space of $T_xN \cap \mathcal{S}_xM$ in $\mathcal{S}_xM$, (7.4) shows that N is a coisotropic submanifold of M iff, $\forall x \in N$, $(T_xN) \cap \mathcal{S}_x(M)$ is a coisotropic vector subspace of the symplectic vector space $\mathcal{S}_x(M)$. In case N has a *clean intersection* [AM] with the symplectic leaves of (M, w) (i.e. $N\cap$ (leaf) is a submanifold with the tangent space $TN \cap \mathcal{S}(M)$), N is coisotropic in M iff $N\cap$ (leaf of $\mathcal{S}$) is a coisotropic submanifold of leaf $\mathcal{S}$, for every leaf.

7.8. Remark. The coisotropy condition (7.4) has two extreme cases namely,

$$\#(\text{Ann } TN) = 0. \tag{7.5}$$

Then, again because $\#(\text{Ann } TN)$ is the symplectic orthogonal of $(T_xN) \cap \mathcal{S}_x(M)$, the latter is just $\mathcal{S}_x(M)$, and N is foliated by subsets of the symplectic leaves of (M, w). Therefore N is a Poisson submanifold of (M, w) as defined in Chapter 1.

The second extreme case is

$$\#(\mathrm{Ann}\ T_xN) = (T_xN) \cap \mathcal{S}_x(M) \tag{7.6}$$

Then $(T_xN) \cap \mathcal{S}_xM$ is a Lagrangian subspace of $\mathcal{S}_x(M)$, and N will be called a *Lagrangian submanifold* of M because, in the symplectic case, one has usual Lagrangian submanifolds. In the Poisson case, this notion is much less important than it was in the symplectic case.

Now, we shall prove [We6]

7.9. Theorem. *A mapping $\varphi : (M_1, w_1) \to (M_2, w_2)$ is a Poisson morphism iff* Graph φ *is a coisotropic submanifold of $M_1 \times M_2^-$.*

Proof. In this theorem M_2^- is the Poisson manifold $(M_2, -w_2)$, and the product is that defined by Definition 1.18. Furthermore, Graph $\varphi = \{(x, \varphi(x))/x \in M_1\}$, and obviously, it is an embedded closed submanifold of $M_1 \times M_2$.

Now, a tangent vector of Graph φ is a pair $(v_x, \varphi_*(v_x))$ $(x \in M_1)$. Therefore, Ann T(Graph φ) consists of the pairs of covectors $(-\varphi^*\lambda, \lambda)$, where $\lambda \in T^*_{\varphi(x)}M_2$, and (7.4) will be satisfied iff

$$\varphi_*(\#_{w_1}(-\varphi^*\lambda)) = -\#_{w_2}(\lambda), \tag{7.7}$$

which is exactly (7.3). Q.e.d.

In view of this result, it is natural to extend the notion of a Poisson mapping to that of a *Poisson relation*, defined as a subset R of $M_1 \times M_2$ which is a coisotropic submanifold of $M_1 \times M_2^-$ [We6]. It turns out that the Poisson relations have interesting categorical properties. We refer the reader to [We6] for the study of these properties in the Poisson case, and to [We1] for the symplectic case.

7.10. Proposition [We6]. *Let $\varphi : (M_1, w_1) \to M_2$ be a surjective submersion. Then φ coinduces a Poisson structure w_2 on M_2 iff the equivalence relation ρ determined on M_1 by the fibers of φ as equivalence classes is a Poisson relation.*

Proof. The relation ρ is given by the subset

$$\rho = \{(x, y) \in M_1 \times M_1/\varphi(x) = \varphi(y)\}, \tag{7.8}$$

and, if we use coordinates (x^i), (y^i) on the two copies of M_1, and z^α on M_2 $(i = 1, \ldots, n_1 = \dim M_1;\ \alpha = 1, \ldots, n_2 = \dim M_2 \leq n_1)$ such that φ has the canonical equations $z^\alpha = x^\alpha$ $(z^\alpha = y^\alpha)$, the subset (7.8) has the local equations $x^\alpha = y^\alpha$. This shows that ρ is an embedded submanifold of $M_1 \times M_1$, with the local coordinates $(x^{\alpha+1}, \ldots, x^{n_1}, y^1, \ldots, y^{n_1})$.

If ρ is a coisotropic submanifold of $M_1 \times M_1^-$, and if $f, g \in C^\infty(M_2)$, since the functions $F(x,y) = f\circ\varphi(x) - f\circ\varphi(y)$, $G(x,y) = g\circ\varphi(x) - g\circ\varphi(y)$ vanish on ρ, their bracket

$$\{F,G\}(x,y) = \{f\circ\varphi, g\circ\varphi\}_1(x) - \{f\circ\varphi, g\circ\varphi\}_1(y)$$

must also vanish on ρ. This means that $\{f\circ\varphi, g\circ\varphi\}_1$ is constant on the fibers of φ, and w_2 exists by Proposition 7.3.

Conversely, if w_2 exists, Proposition 7.3 shows that $\{f\circ\varphi, g\circ\varphi\}$ is constant along the fibers of φ. Now, $\mathrm{Ann}(T\rho) = \mathrm{span}\ \{dx^\alpha - dy^\alpha\}$, and we have

$$(w_1 \oplus (-w_1))(dx^\alpha - dy^\alpha, dx^\beta - dy^\beta) = \{x^\alpha, x^\beta\}_1 - \{y^\alpha, y^\beta\}_1 =$$

$$= \{z^\alpha\circ\varphi, z^\beta\circ\varphi\}_1 - \{z^\alpha\circ\varphi, z^\beta\circ\varphi\}_1 = \{z^\alpha, z^\beta\}_2\circ\varphi - \{z^\alpha, z^\beta\}_2\circ\varphi = 0.$$

Hence ρ is a coisotropic submanifold. Q.e.d.

7.11. Corollary. *If $\varphi : (M_1, w) \to M_2$ is a surjective submersion, then $\varphi^*(C^\infty(M_2))$ is a Lie subalgebra of $(C^\infty(M_1), \{\ ,\ \})$ iff the equivalence relation defined by the fibers of φ is a Poisson relation on M_1.*

Proof. It follows from Propositions 7.3 and 7.10. Q.e.d.

7.12. Corollary. *Let $\mathcal{F}$ be an arbitrary regular foliation of a Poisson manifold (M, w). Then, the sheaf $\Phi^0(\mathcal{F})$ of germs of functions of M that are constant along the leaves of $\mathcal{F}$ is closed under the Poisson bracket iff the equivalence relation ρ whose equivalence classes are the leaves of $\mathcal{F}$ is a Poisson relation (i.e., ρ is a (possibly immersed) coisotropic submanifold of $M \times M^-$).*

Proof. Use Proposition 7.10 on $\mathcal{F}$-adapted neighborhoods. Q.e.d.

7.2 Reduction of Poisson structures

In this section, we shall discuss a number of properties which are important since they extend the famous reduction method of symplectic geometry (e.g. [AM], [LM]). First we state

7.13. Definition. Let $i : N \to (M, w)$ be the inclusion of a submanifold, and $\varphi : M \to P$ be a surjective submersion. Then, a Poisson structure π on P is called a *reduction of w via (N, φ)* if $\forall f, g \in C^\infty(P)$ one has

$$\{f\circ\varphi, g\circ\varphi\}_w \circ i = \{f, g\}_\pi \circ \varphi \circ i. \tag{7.9}$$

If w and π are defined by symplectic forms ω_M, ω_P respectively, (7.9) means $(\varphi\circ i)^*\omega_P = i^*\omega_M$, as it is usually requested in symplectic geometry [LM]. The name *reduction* comes from the fact that the Poisson brackets on P are obtained by a pullback to M followed by restriction to N of the bracket of M.

7.14. Proposition [We6]. *Let N, M, P, φ be as in Definition 7.13, and assume that N is coisotropic. Then P has a reduced Poisson structure via (N, φ) iff the relation θ defined on M by the intersections of the fibers of φ with N is a Poisson relation.*

Proof. We have $\theta = \{(x, y) \in N \times N / \varphi(x) = \varphi(y)\} \subseteq M \times M$, and, by using the same local coordinates as in the proof of Proposition 7.10, we see that the local equations of θ are $x^\alpha = y^\alpha$, $F^\sigma(x^i) = 0$, $F^\sigma(y^i) = 0$, where $F^\sigma = 0$ are the local equations of N in M. Now, the same proof as for Proposition 7.10 gives the present result (i.e., a bracket $\{f, g\}_\pi$ satisfying (7.9)) since the bracket $\{F, G\}$ there vanishes modulo $F^\sigma = 0$. In the opposite direction, Ann $(T\theta)$ now also includes dF^σ, but w vanishes on these differentials since N was coisotropic in M. Q.e.d.

The previous result may be turned around to say that θ is a Poisson relation iff N is coisotropic, and (7.9) holds good [We6].

Proposition 7.14 is not the only way to extend symplectic reduction to Poisson geometry, and the subject is so important that it deserves a more thorough analysis.

A general definition of reduction in the Poisson context is that given by Marsden and Ratiu [MR] namely,

7.15. Definition. Let (M, w) be a Poisson manifold. Then, a pair (N, E) that consists of a submanifold $i : N \subseteq M$, and a vector subbundle E of $TM/_N$ will be called a *reductive structure* of (M, w) if the following conditions are satisfied: i) $E \cap TN$ is tangent to a foliation $\mathcal{F}$ of N whose leaves are the fibers of a submersion $\pi : N \to P$; ii) $\forall \varphi, \psi \in C^\infty(M)$ such that $d\varphi$ and $d\psi$ vanish on E, $d\{\varphi, \psi\}$ also vanishes on E. Furthermore, if P above has a Poisson structure λ such that for any local C^∞ functions f, g on P, and any local extensions φ, ψ of $f \circ \pi$, $g \circ \pi$, with $d\varphi = d\psi = 0$ on E, the relation

$$\{\varphi, \psi\}_w \circ i = \{f, g\}_\lambda \circ \pi \tag{7.10}$$

holds good, we say that (M, N, E) is a *reducible triple*, and (P, λ) is the *reduced Poisson manifold of* (M, w) *via* (N, E).

The difference between this situation and the one of Definition 7.13 is, mainly, in the fact that φ of Definition 7.13 yielded a precise way of extending the functions $f \circ \pi$, while in Definition 7.15 we must take care that the results of (7.10) be independent of the "E-controlled" extension. This is done in

7.16. Theorem [MR]. *Let (N, E) be a reductive structure of the Poisson manifold (M, w). Then (M, N, E) is a reducible triple iff*

$$\#(\text{Ann } E) \subseteq TN + E \tag{7.11}$$

Proof. We shall say that (7.11) is the *reducibility condition*, and Ann E is a subspace of T^*M. Let (M, N, E) be reducible to (P, λ), and take $x \in N$, $\alpha \in$ Ann E_x, and φ smooth around x such that $d\varphi = 0$ on E and $d_x\varphi = \alpha$. Then, take $\beta \in \mathrm{Ann}(E_x + T_xN) = (\mathrm{Ann}\ E_x) \cap (\mathrm{Ann}\ T_xN)$, and ψ smooth around x, that vanishes on N, and satisfies $d\psi = 0$ on E, $d_x\psi = \beta$. (The existence of φ and ψ is ensured by condition i) of Definition 7.15, and since E is defined only along N.) For these data we have

$$\beta(\#_w\alpha) = (d_x\psi)(X_\varphi(x)) = \{\varphi, \psi\}_w(x) \overset{(7.10)}{=} \{f, 0\}(\pi(x)) = 0,$$

where f is a function on P such that φ is an extension of $f \circ \pi$. Hence any such β vanishes on $\#(\mathrm{Ann}\ E)$, and we get (7.11).

Conversely, assume that (7.11) holds good, and let $f, g \in C^\infty(P)$, and φ, ψ be extensions of $f \circ \pi, g \circ \pi$, respectively, such that $d\varphi, d\psi$ vanish on E. Because of condition ii) of Definition 7.15, $\{\varphi, \psi\}_w$ is constant along the leaves of $\mathcal{F}$, and it projects to a function on P that we shall take by definition to be $\{f, g\}_\lambda$. This definition ensures (7.10), but, we must check the independence of $\{f, g\}_\lambda$ on the choice of the extensions φ, ψ, and it suffices to discuss only the change of φ because of the skew-symmetry of the Poisson bracket.

Therefore, let us assume that φ' also extends $f \circ \pi$ and $d\varphi' = 0$ on E. Then $(\varphi - \varphi')/_N = 0$, hence $d\varphi = d\varphi'$ along $E + TN$ and, in view of (7.11),

$$(d\varphi)(\#_w(d\psi))_x = (d\varphi')(\#_w(d\psi))_x \qquad (x \in N).$$

This relation means exactly $\{\varphi, \psi\}(x) = \{\varphi', \psi\}(x)$ as requested.

Accordingly, λ is well defined, and it satisfies (7.10), whence it follows easily that λ is a Poisson structure on P. Q.e.d.

Several examples of applications of Theorem 7.16 are given in [MR], and one of them, that concerns momentum maps, will be described later on in this chapter. Here, let us just look at the situation that one studies in symplectic geometry. Namely, let (M, ω_M) be a symplectic manifold, and w its natural Poisson structure. Let N be a submanifold of M such that $\mathrm{rank}(i^*\omega_M) = \mathrm{const.}$ $(i : N \subseteq M)$. Then, it is well known that, if $T'N$ is the ω-orthogonal space of TN, $(T'N) \cap (TN)$ is a foliation of N, and it has a transversal symplectic structure (e.g., [LM], [Va8]). Hence, if the leaves of this foliation are the fibers of a submersion $\pi : N \to P$, we get a symplectic structure ω_P on P such that $\pi^*\omega_P = i^*\omega_M$. Furthermore, since we are on a symplectic manifold, $d\varphi$ vanishes on $T'N$ iff the Hamiltonian field $X_\varphi \in \#\ \mathrm{Ann}(T'N) = (T'N)' = TN$. Then, $X_\varphi, X_\psi \in TN$ implies $[X_\varphi, X_\psi] = X_{\{\varphi,\psi\}} \in TN$. Hence, the conditions i) and ii) of Definition 7.15 hold, and $(N, T'N)$ is a reductive structure of M. Moreover, (7.11) obviously holds in this case, since $\#\mathrm{Ann}(T'N) = TN$. Hence, $(M, N, T'N)$ is a reducible triple, and it is clear that the corresponding reduced structure λ on P is associated with the symplectic form ω_P.

On the other hand, if we take $E = (T'N) \cap (TN)$, we can see that (M, N, E) is a reducible triple iff N is a coisotropic submanifold of M. Of course, we still assume the existence of the submersion $\pi : N \to P$. Indeed, if N is coisotropic, $E = T'N$, and we may apply the previous result. Conversely, reducibility implies (7.11), and N must be coisotropic since $\#(\mathrm{Ann}\, E) = TN + T'N$.

This example shows the importance of the choice of E, and it extends to the Poisson case, where we have

7.17. Proposition. *Let N be a submanifold of the Poisson manifold (M, w), such that $\mathcal{C}(N) \stackrel{\text{def}}{=} (\#\mathrm{Ann}(TN)) \cap TN$ is a distribution of constant dimension along N. Then $\mathcal{C}(N)$ is differentiable and involutive. Furthermore, if N is transversal to the leaves of the symplectic foliation $\mathcal{S}(M)$ of (M, w), $\mathcal{C}(N)$ is a subfoliation of $\mathcal{S}(M) \cap TN$, which is transversally symplectic along each leaf of the latter.*

Proof. Locally, $\mathcal{C}(N)$ can be defined by Pfaff equations of the form $\alpha_i = 0$, $\beta_j = 0$, where $\alpha_i = 0$ define TN, and $\beta_j = 0$ define $\#\mathrm{Ann}(TN)$ i.e., β_j are solutions of $w(\alpha_i, \beta) = 0$. Since dimension $\mathcal{C}(N) =$ const., this is enough to justify the differentiability of $\mathcal{C}(N)$. Now, to check involutivity of $\mathcal{C}(N)$, let u, v be local vector fields along N in $\mathcal{C}(N)$, such that $u = \alpha^{\#}$, $v = \beta^{\#}$, where $\alpha, \beta \in \mathrm{Ann}\, TN$ are defined along N. Take local extensions $\tilde{\alpha}, \tilde{\beta}$ of α, β to an open neighbourhood in M, and put $\tilde{u} = \tilde{\alpha}^{\#}$, $\tilde{v} = \tilde{\beta}^{\#}$. Then, by using the bracket of 1-forms defined in Chapter 4 (formula (4.6)) we have $[u, v] = [\tilde{u}, \tilde{v}]/_N = \#\{\tilde{\alpha}, \tilde{\beta}\}/_N$, and we have to prove that $\{\tilde{\alpha}, \tilde{\beta}\}/_N \in \mathrm{Ann}\, TN$. This happens since

$$\{\tilde{\alpha}, \tilde{\beta}\} = d(w(\tilde{\alpha}, \tilde{\beta})) + i(\tilde{\alpha}^{\#})d\tilde{\beta} - i(\tilde{\beta}^{\#})d\tilde{\alpha},$$

and if this formula is pulled back by $i : N \to M$, and computed for $X \in TN$ it gives

$$\{\tilde{\alpha}, \tilde{\beta}\}/_N(X) = X(w(\alpha, \beta)) + (d\tilde{\beta})/_N(\alpha^{\#}, X) - d\tilde{\alpha}/_N(\beta^{\#}, X) = 0$$

as we see by making the differentials explicit, and by using $\alpha, \beta \in \mathrm{Ann}\, TN$.

Now, if N is transversal to the leaves of $\mathcal{S}(M)$, it will intersect cleanly each leaf along a submanifold of constant rank $= \dim(N \cap \text{leaf}) - \dim \mathcal{C}(N)$, and we have the situation described for the symplectic case. Q.e.d.

In what follows we shall call $\mathcal{C}(N)$ the *subcharacteristic distribution* (*foliation*) of N. If this foliation is given by the fibers of a submersion $\psi : N \to P$, then, $TN \cap \mathcal{S}(M)$ projects to a symplectic foliation $\psi(\mathcal{S})$ on P, and P has a well defined Poisson structure π whose symplectic foliation is $\psi(\mathcal{S})$. We shall say that π is the *leafwise reduction* of w via N, because π is obtained by reducing each leaf of $\mathcal{S}(M)$ by the standard procedure of symplectic geometry.

In this situation, it follows in exactly the same way as for the symplectic case discussed earlier that, if $\#\mathrm{Ann}(TN)$ has a constant dimension (which happens, for instance, if $\dim(\mathcal{S}(M) \cap TN) = \text{const.}$ since N is transversal to $\mathcal{S}(M)$, and $\#\text{Ann}(TN)$ is the symplectic orthogonal of $\mathcal{S}(M) \cap TN$ in the leaf of $\mathcal{S}(M)$) then $(N, \#\text{Ann}(TN))$ is a reductive structure, $(M, N, \#\text{ Ann}(TN))$ is a reducible triple, and the reduced Poisson structure π of P is exactly the one defined by the leafwise reduction. (The corresponding relation (7.11) follows since $\#\text{ Ann } \#\text{ Ann } (TN) \subseteq TN$, as we can easily check.) Similarly, $(M, N, \mathcal{C}(N))$ is reducible iff N is a coisotropic submanifold of M.

Somewhat in the same vein, i.e., reduction along the leaves of $\mathcal{S}(M)$, is the following generalization of a result of P. Libermann [Lb2,3]

7.18. Proposition. *Let $\mathcal{F}$ be a (possibly nonregular) foliation of the Poisson manifold (M, w), such that $\dim(T\mathcal{F} \cap \mathcal{S}(M)) = \text{const.}$ Then, if the sheaf $\Phi^0(\mathcal{F})$ of germs of projectable functions of M is closed by w-Poisson brackets, $\#\text{ Ann}(T\mathcal{F})$ is a completely integrable distribution on M. If, moreover, the leaves of $\mathcal{F}$ are the fibers of a surjective submersion $\pi : M \to N$ there is a coinduced Poisson structure on N, and the symplectic distribution of it is the π_*-image of $\#\text{ Ann}(T\mathcal{F})$ i.e., the structure of the symplectic leaves is obtained by reducing that of the leaves of $\mathcal{S}(M)$ via the leaves of $\#\text{ Ann}(T\mathcal{F})$.*

Proof. Since $\text{Ann}(T\mathcal{S}) = \ker \#$, we have

$$\#\text{Ann}(T\mathcal{F}) = \#(\ker \# + \text{ Ann}(T\mathcal{F})) = \#\text{Ann}(T\mathcal{S} \cap T\mathcal{F}),$$

and this is a differentiable distribution because $\dim(T\mathcal{S} \cap T\mathcal{F}) = \text{const.}$ Furthermore, $\#\text{ Ann}(T\mathcal{F})$ satisfies condition *b*) of Viflyantsev's Theorem 2.9, since if $v \in \#\text{ Ann}(T_{x_0}\mathcal{F})$ $(x_0 \in M)$, there is a local function f around x_0 with $v = X_f(x_0)$, and $f = \text{const.}$ on the leaves of $\mathcal{F}$, and, obviously, the integral path of X_f through x_0 consists of points where $\mathcal{S}(M)$, and, therefore, $\#\text{ Ann}(T\mathcal{F})$ have a constant dimension ($\dim \#\text{Ann}(T\mathcal{F}) = \dim \mathcal{S}(M) - \dim(T\mathcal{F} \cap \mathcal{S}(M))$ because of the known symplectic orthogonality of $\#\text{Ann}(T\mathcal{F})$ and $T\mathcal{F} \cap \mathcal{S}(M)$ in $\mathcal{S}(M)$). Hence, the only fact that we have to deal with is the involutivity of $\#\text{ Ann}(T\mathcal{F})$, and, to do this, we prove first

7.19. Lemma. *With the notation of Proposition 7.18, $\Phi^0(\mathcal{F})$ is closed by brackets iff $\text{Ann}(T\mathcal{F})$ is closed by the bracket of 1-forms.*

Proof. If $\Phi^0(\mathcal{F})$ is closed by brackets, then f, g constant on the leaves of $\mathcal{F}$ implies $\{f, g\}$ is constant on the leaves, which means that $df = 0$, $dg = 0$ on $T\mathcal{F}$ imply $\{df, dg\} = 0$ on $T\mathcal{F}$. Now, $\forall \alpha, \beta \in \text{Ann}(T\mathcal{F})$ we have local expressions

$$\alpha = \sum_\sigma \varphi_\sigma df_\sigma, \qquad \beta = \sum_\tau \psi_\tau dg_\tau$$

with $df_\sigma, dg_\tau \in \text{Ann}(T\mathcal{F})$, and, by (4.5)

$$\{\alpha, \beta\} = \sum_{\sigma,\tau} (\varphi_\sigma \psi_\tau \{df_\sigma, dg_\tau\} + \varphi_\sigma (X_{f_\sigma} \psi_\tau) dg_\tau -$$

$$- \psi_\tau (X_{g_\tau} \varphi_\sigma) df_\sigma)) \in \text{ Ann}(T\mathcal{F}).$$

Conversely, if Ann$(T\mathcal{F})$ is closed by brackets, and f, g are constant on the leaves of $\mathcal{F}$, we have $d\{f,g\} = \{df, dg\} = 0$ on $T\mathcal{F}$. Q.e.d.

Let us come back to Proposition 7.18. Now, we know that its hypotheses imply the closure of Ann $(T\mathcal{F})$ by brackets. If $u, v \in \#\text{Ann}(T\mathcal{F})$, we have $u = \alpha^{\#}$, $v = \beta^{\#}$, $\alpha, \beta \in$ Ann $(T\mathcal{F})$, and, by (4.4),

$$[u, v] = \{\alpha, \beta\}^{\#} \in \# \text{ Ann}(T\mathcal{F}),$$

which proves the first part of the Proposition.

Concerning the second part (where the foliation $\mathcal{F}$ is, necessarily regular) the existence of the coinduced structure on N follows from Proposition 7.3, and the other assertions are rather clear since $\pi_*(\mathcal{S}(M) \cap T\mathcal{F}) = 0$, and (# Ann$(T\mathcal{F})) \cap T\mathcal{F}$ is the subcharacteristic distribution along the leaves of $\mathcal{F}$. Q.e.d.

7.20. Remark. If $\mathcal{F}$ is a subfoliation of $\mathcal{S}(M)$ i.e., $T\mathcal{F} \subseteq \mathcal{S}(M)$, Ann$(T\mathcal{F})$ is closed by brackets iff # (Ann $T\mathcal{F}$) is involutive. The first implication was proven during the proof of Proposition 7.18. Conversely, if # (Ann $T\mathcal{F}$) is involutive, and $u = \alpha^{\#}$, $v = \beta^{\#}$ are vector fields with $\alpha, \beta \in$ Ann$(T\mathcal{F})$, we get $[u, v] = \lambda^{\#}$, $\lambda \in$ Ann$(T\mathcal{F})$, and we see that $\{\alpha, \beta\} = \lambda + \nu$, where $\nu \in \ker \# =$ Ann $\mathcal{S}(M) \subseteq$ Ann$(T\mathcal{F})$ i.e., $\{\alpha, \beta\} \in$ Ann$(T\mathcal{F})$ as requested. Hence, in this case the results of Proposition 7.18 become much stronger namely, $\Phi^0(\mathcal{F})$ will be bracket closed iff #Ann$(T\mathcal{F})$ is involutive, and, in the second part, N has the coinduced structure if #Ann$(T\mathcal{F})$ is integrable. In particular, if M is a symplectic manifold, and $\mathcal{F}$ is a regular foliation, we get precisely P. Libermann's theorem since, in this case, # Ann$(T\mathcal{F})$ is just the symplectic orthogonal of $T\mathcal{F}$.

We notice from the previous results on reduction that the Poisson structure w of a manifold M *reduces* if it induces a Poisson algebra structure on a conveniently defined algebra of differentiable functions that are constant along the classes of a certain partition of a submanifold N of M. In Proposition 7.14, we had the partition of N by the fibers of $\varphi \circ i$, and the differentiable functions on N constant along those fibers. In the reduction based on Proposition 7.17 the partition was by the leaves of the subcharacteristic foliation on N. Etc.

For geometry, the most interesting partitions are those defined by foliations, and the above considerations suggest that, instead of looking at submanifolds of a Poisson manifold we might just look at manifolds N, endowed with a foliation $\mathcal{F}$, and define the notion of a *reduced Poisson structure* on $(N, \mathcal{F})$ to be a structure of a Poisson algebra on $C^{\infty}_{\text{proj}}(N, \mathcal{F})$, or on the sheaf $\Phi^0(\mathcal{F})$ of projectable germs, if we want something of the local type. It might be interesting to make an independent study of such reduced Poisson structures. For instance, in view of Lemma 7.19, the full Poisson structure w of Proposition 7.18 defines a reduced one for the foliation $\mathcal{F}$ iff Ann$(T\mathcal{F})$ is closed by the bracket of the 1-forms.

7.3 Group actions and momenta

As in symplectic geometry, important reduction processes appear in connection with Lie group actions on a Poisson manifold. Let (M, w) be a Poisson manifold, G a Lie group, and $\Phi : G \times M \to Ma$ differentiable left action of G on M. We shall use the notation

$$\Phi(g,x) = \Phi_g(x) = g(x) = \Phi^x(g) = x(g) \quad (g \in G, x \in M), \tag{7.12}$$

that also defines the mappings $\Phi_g : M \to M$, $\Phi^x : G \to M$, if g, respectively x, are constant. Furthermore, remember again that if $\mathcal{G}$ is the Lie algebra of G, we shall denote by X_M the vector field on M associated to $X \in \mathcal{G}$ by

$$X_M(x) = -\Phi^x_*(e)(X) = \left.\frac{d}{dt}\right|_{t=0} \Phi_{\exp(-tX)}(x), \tag{7.13}$$

where the minus sign was inserted to ensure that $X \mapsto X_M$, is a Lie algebra homomorphism. (7.13) is the *infinitesimal action* associated with Φ.

7.21. Definition. Φ is a *Poisson action* if, $\forall g \in G, \Phi_g$ is a Poisson morphism. If, moreover, $\forall X \in \mathcal{G}$, there is a function $f_X \in C^\infty(M)$ such that X_M is precisely the Hamiltonian field of f_X, Φ is called a *Hamiltonian action.*

Clearly, if M is a symplectic manifold a Poisson action is just a symplectic action i.e., an action of G on M by symplectomorphisms. If G is connected, Φ is a Poisson action, iff the vector fields X_M are infinitesimal automorphisms of the Poisson structure w, as encountered in Chapter 4, i.e., vector fields V such that $L_V w = 0$, or V satisfies one of the equivalent conditions of

7.22. Proposition. *A vector field V of a Poisson manifold (M, w) is an infinitesimal automorphism of w iff one of the following equivalent conditions is satisfied:*

a) *V is a derivation of the Lie algebra $(C^\infty(M), \{\ ,\ \})$ i.e.,*

$$V\{f,g\} = \{Vf,g\} + \{f,Vg\};$$

b)
$$V\{f,g\} = [V,X_f](g) - [V,X_g](f).$$

c)
$$[V,X_f] = X_{V(f)};$$

Proof. Remember that the X in these formulas are Hamiltonian vector fields. a) is obtained from $(L_V w)\ (df,dg) = 0$, and b), c) are obtained from a) by writing the brackets as commutators. Q.e.d.

In particular, if G is connected, the second condition of Definition 7.21 alone implies the fact that one has a Poisson action. Furthermore, since the X_f are tangent to the leaves of the symplectic foliation $\mathcal{S}(M)$ of w, we see that a Hamiltonian action on M induces a Hamiltonian action on every symplectic leaf, in the sense of symplectic geometry. In this later case Hamiltonian actions have been studied at length, and there is a rich literature about them. See for instance [AM], [LM], [Mn2], etc. For the Poisson case, we quote [Tl2], [Mr], [Oz1] and [Mk].

7.23. Example. Take $M = \mathbb{R}^2$, and w given by

$$w = (x^2 + y^2)^{1/2} \frac{\partial}{\partial x} \wedge \frac{\partial}{\partial y}.$$

Then, the group S^1 of rotations acts on $\mathbb{R}^2$ with the preservation of w hence, we have a Poisson action. The generator of the Lie algebra of S^1 is sent by (7.13) to $X = y(\partial/\partial x) - x(\partial/\partial y)$, which is precisely the Hamiltonian field of $f_X = -(x^2+y^2)^{1/2}$. Hence, we have an example of a Hamiltonian group action. But the same action of S^1 is Poisson non-Hamiltonian for

$$\tilde{w} = (x^2 + y^2) \frac{\partial}{\partial x} \wedge \frac{\partial}{\partial y}$$

since the corresponding function would be $-\frac{1}{2} \ln(x^2 + y^2)$, which is not continuous at (0,0).

7.24. Proposition. *i) The orbits of a Poisson action of G on (M, w) define a generalized foliation $\mathcal{F}_G$ on M such that $\Phi^0(\mathcal{F}_G)$ is closed by w-brackets, and, if the action is Hamiltonian, $\mathcal{F}_G$ is a subfoliation of $\mathcal{S}(M)$. ii) If, moreover,* $\dim(T\mathcal{F}_G) \cap \mathcal{S}(M) = \text{const.}$, $\#\ \text{Ann}\ (T\mathcal{F}_G)$ *is integrable and, in case $M/\mathcal{F}_G$ is a manifold N, the latter gets a coinduced Poisson structure, with the symplectic distribution equal to the projection of* $\#\ \text{Ann} T\mathcal{F}_G$, *and a reduced symplectic structure on its symplectic leaves.*

Proof. i) is obvious from the definitions, and, then, ii) follows by applying Proposition 7.18 to the present situation. Notice that the tangent distribution of $\mathcal{F}_G$ is generated by $\{X_M / X \in \mathcal{G}\}$. Q.e.d.

It is clear that a Hamiltonian action of G on (M, w) is composed by Hamiltonian actions along the symplectic leaves of w, hence we may expect results of symplectic geometry to have Poisson analogs.

7.25. Proposition. *An action of a connected Lie group G on the Poisson manifold (M, w) is Hamiltonian iff there exists a differentiable mapping $J : M \to \mathcal{G}^*$= the dual of the Lie algebra $\mathcal{G}$ of G such that $\forall X \in \mathcal{G}$ the function $J(X) \in C^\infty(M)$ defined by*

$$J(X)(x) = J(x)(X) \tag{7.14}$$

satisfies

$$X_M = X_{J(X)}. \tag{7.15}$$

Proof. When convenient, we shall also denote J_X instead of $J(X)$. If J exists, we obviously have a Hamiltonian action. Conversely, if $\forall X \in \mathcal{G}$ we have some $f_X \in C^\infty(M)$, such that $X_M = X_{f_X}$ holds, let us take a basis $X_1, \ldots, X_s$ of $\mathcal{G}$, and define $J(x)$ by (7.14) where, if $X = \sum_{i=1}^s c^i X_i$, we take $J(X) = \sum_{i=1}^s c^i f_{X_i}$. Then $J : M \to \mathcal{G}^*$, and it satisfies all the properties requested. Q.e.d.

7.26. Remark. The above result is also true if G is nonconnected, provided that we have a Poisson action.

7.27. Definition. A mapping $J : M \to \mathcal{G}^*$ that satisfies (7.14) and (7.15) is called a *momentum map* (or a *moment map*, following other authors) of the Hamiltonian action of G on (M, w).

Clearly, such a momentum map induces a momentum map for the induced Hamiltonian action on the symplectic leaves. Furthermore, if J^a $(a = 1, 2)$ are two momentum maps, $J^1 - J^2$ is constant along these symplectic leaves. Finally, if J is a momentum map, we shall prove that one has

$$J(g(x)) = (\text{Coad } g)(J(x)) + \theta(g, x) \tag{7.16}$$

for a well defined function $\theta : G \times M \to \mathcal{G}^*$ which does not vary along the leaves of the symplectic foliation $\mathcal{S}(M)$. Indeed, with (7.14) for $X \in \mathcal{G}$ we have

$$J(g(x))(X) = J_X(g(x)) = (J_X \circ g)(x). \tag{7.17}$$

On the other hand, since the action is by Poisson morphisms, we have

$$\begin{aligned} X_{J(X)\circ g} &= g_*^{-1}(X_{J(X)} \circ g) = g_*^{-1}(X_M \circ g) = \\ &= \text{Ad}g^{-1}(X))_M = X_{J(\text{Ad}g^{-1}(X))}, \end{aligned} \tag{7.18}$$

where the third equality sign is a classical equality valid for an arbitrary action of a Lie group on a manifold (e.g., [LM]). Hence,

$$J(X) \circ g = J(\text{Ad } g^{-1}(X)) + \phi(g, X), \tag{7.19}$$

where $\phi(g, X)$ is a Casimir function on M, and (7.16) is just the combination of the results of (7.17) and (7.19), with the notation

$$\theta(g, x)(X) = \phi(g, X)(x). \tag{7.20}$$

Formula (7.16) suggests

7.28. Definition. A momentum map $J : M \to \mathcal{G}^*$ is called *equivariant* if

$$J(g(x)) = (\text{Coad } g)(J(x)) \tag{7.21}$$

i.e., θ of 7.16 vanishes.

7.29. Remark. Definition 7.28 introduces the most important type of momentum maps. As in the symplectic case (e.g. [LM]) θ can be given a cohomological interpretation that leads to an existence result for equivariant momentum maps [Mk].

7.30. Proposition. *An equivariant momentum map $J : M \to \mathcal{G}^*$ is a Poisson morphism, if $\mathcal{G}^*$ is endowed with its Lie-Poisson structure.*

Proof. The Lie-Poisson structure of $\mathcal{G}^*$ was defined in Chapter 3 (Example *b*)). We begin by proving that an equivariant momentum map satisfies

$$\{J_X, J_Y\} = J_{[X,Y]} \qquad (X, Y \in \mathcal{G}). \tag{7.22}$$

Indeed, for $x \in M$, we have

$$\{J_X, J_Y\}(x) = X_{J(X)}(x)(J_Y) = X_M(x)(J_Y) =$$
$$= -\left.\frac{d}{dt}\right|_{t=0} J_Y(\exp(tX)(x)) = -\left.\frac{d}{dt}\right|_{t=0} (\mathrm{Coad}(\exp tX))(J(x))(Y) =$$
$$= \left.\frac{d}{dt}\right|_{t=0} (J(x))(\mathrm{Ad}(\exp tX)Y) = (J(x))([X,Y]) = J_{[X,Y]}(x).$$

Now, it follows from (3.8) that in order to check that J is a Poisson morphism, it is enough to check that

$$\{\ell_1 \circ J, \ell_2 \circ J\} = \{\ell_1, \ell_2\} \circ J$$

for linear functions $\ell_1, \ell_2 : \mathcal{G}^* \to \mathbb{R}$, i.e., $\ell_1, \ell_2 \in \mathcal{G}$, and where the bracket of the right-hand side is the Lie-Poisson bracket (3.7) of $\mathcal{G}^*$. This result follows by a computation:

$$(\{\ell_1, \ell_2\} \circ J)(x) = \{\ell_1, \ell_2\}(J(x)) = (J(x))([\ell_1, \ell_2]) =$$
$$J_{[\ell_1,\ell_2]}(x) = \{J_{\ell_1}, J_{\ell_2}\}(x) = \{\ell_1 \circ J, \ell_2 \circ J\}(x)$$

($x \in M$). Q.e.d.

7.4 Group actions and reduction

Now, let us refer again to reduction. For the sake of simplicity, we consider only the case of an equivariant momentum map $J : (M, w) \to \mathcal{G}^*$. The reduction that is of interest in many applications is that of a level set $M_\gamma = J^{-1}(\gamma)$ for $\gamma \in \mathcal{G}^*$, and the simplest case is that where γ is a noncritical value for all the restrictions of J to the symplectic leaves of (M, w). In this case, M_γ is a submanifold of M, and, for any symplectic leaf S, $S_\gamma = M_\gamma \cap S$ is a submanifold of S. We shall also make the hypothesis that M_γ intersects S cleanly i.e., $TS_\gamma = TM_\gamma \cap$ TS. This ensures that TS_γ is a differentiable distribution on M_γ, and, therefore, that $\cup_S S_\gamma$ is a generalized foliation of M_γ.

Furthermore, let G_γ denote the isotropy subgroup of γ with respect to the coadjoint action of G on $\mathcal{G}^*$. G_γ is closed, hence it is a Lie subgroup of G, and, since J is equivariant, the action of G on M obviously induces an action of G_γ on M_γ. Moreover, since the action of G is Hamiltonian, the orbit $G(x)$ ($x \in M$) is contained in the symplectic leaf $S(x)$, and the orbit $G_\gamma(x)$ is contained in $S(x)_\gamma$; in fact, $S(x)_\gamma = (J/_{S(x)})^{-1}(\gamma)$, where $J/_{S(x)}$ is an equivariant momentum map of the induced action of G on the symplectic leaf $S(x)$. The following relations are obvious $\forall x \in M_\gamma$

$$M_\gamma \cap G(x) = S(x)_\gamma \cap G(x) = G_\gamma(x), \tag{7.23}$$

and prove that these intersections are submanifolds of M.

Under these circumstances the following reduction theorem holds good.

7.31. Theorem. *Let (M, w) be a Poisson manifold endowed with a Hamiltonian action of the Lie group G, and the equivariant momentum map J. Let $\gamma \in \mathcal{G}^*$ be such that: i) γ is a regular value for all the restrictions of J to the symplectic leaves of M; ii) the submanifold $M_\gamma = J^{-1}(\gamma)$ of M has a clean intersection with both the symplectic leaves of M, and the orbits of G in M. Let G_γ be the isotropy subgroup of γ with respect to the coadjoint action of G. Then, the subcharacteristic distribution of M_γ is a regular foliation of M_γ, and its leaves are the orbits of the connected component G_γ^0 of the unit in G_γ, for the action of G_γ on M_γ. Furthermore, if this foliation is defined by a submersion $\pi : M_\gamma \to N$, N has a well defined Poisson structure whose symplectic distribution is the projection of $S(M) \cap TM_\gamma$, where $S(M)$ is the symplectic distribution of (M, w). This is precisely the reduced Poisson manifold of (M, w) via $(M_\gamma, E = T(\text{orbits } G))$.*

Proof. We begin by recalling the following fact of linear algebra: let $\ell : V \to W$ be a linear map of finite dimensional vector spaces, and $\ell' : W^* \to V^*$ its transposed map; then one has

$$\mathrm{im}\ell' = \text{ Ann } \ker \ell. \tag{7.24}$$

Indeed, one checks easily that $\mathrm{im}\ell' \subseteq \text{Ann} \ker \ell$, and that the dimensions of the two spaces are equal.

We shall apply (7.24) to $\ell = J_*(x) : T_xM \to \mathcal{G}^*$, while noticing that $\ker J_*(x) = T_xM_\gamma$ if $x \in M_\gamma$. The transposed mapping is $J'_*(x) : \mathcal{G} \to T^*_xM$, where, for $v \in T_xM$, $X \in \mathcal{G}$, one has

$$(J'_*(x)(X))(v) = (J_*(x)(v))(X) = \left.\frac{d}{dt}\right|_{t=0} J(\sigma(\tau))(X),$$

$\sigma(\tau)$ being a path tangent to v at $t = 0$. If we go on using (7.14) we get

$$(J'_*(x)(X))(v) = v(J_X),$$

and we see that $(J'_*(x)(X) = d_x J_X$. Furthermore, with (7.15), this yields

$$\#_w(J'_*(x)(X)) = X_{J(X)}(x) = X_M(x). \tag{7.25}$$

The relation (7.25) obviously shows that we have

$$\#(\mathrm{im}J'_*(x)) = T_x(G(x)) \tag{7.26}$$

hence, in view of (7.24), it follows that

$$\#(\mathrm{Ann}(T_x M_\gamma)) = T_x(G(x)) \qquad (x \in M_\gamma). \tag{7.27}$$

As we already know, this space is the symplectic orthogonal of $(T_x M_\gamma) \cap T_x S(x)$ with respect to the symplectic structure of $S(x)$, and the plane of the subcharacteristic distribution at x (Proposition 7.17) is $\mathcal{C}_x M_\gamma = T_x(G(x)) \cap T_x M_\gamma = T_x(G_\gamma(x))$, because of (7.23) and of the clean intersection hypothesis.

In order to end the proof of the first part of the theorem, we must show that the orbits of G_γ in M_γ are of a constant dimension. To see this, we notice first that, since, $\forall y \in S(x)_\gamma$, $J/_{S(x)} : S(x) \to \mathcal{G}^*$ is a submersion at y, we have

$$\dim S(x)_\gamma = \dim S(x) - \dim\ \mathrm{G}. \tag{7.28}$$

On the other hand, using (7.27) for the manifold $S(x)$ instead of M, we get

$$\begin{aligned} \dim S(x)_\gamma &= \dim T_x S(x)_\gamma = \dim S(x) - \\ &- \dim\ \mathrm{Ann}\ T_x S(x)_\gamma = \dim S(x) - \dim(G(x)) = \\ &= \dim S(x) - (\dim G - \dim G_x), \end{aligned} \tag{7.29}$$

where G_x is the isotropy group of x in $S(x)$.

The comparison of (7.28) and (7.29) gives $\dim G_x = 0$, and, obviously, also $\dim(G_\gamma)_x = 0$, for $(G_\gamma)_x =$ the isotropy group of x for the action of G_γ. But then $\dim(G_\gamma(x)) = \dim G_\gamma =$ const., and we are done.

As for the second part of the theorem, its very formulation defines the requested structure of N. On the other hand, $E = T$(orbits G) is of a constant dimension, because we had $\dim G_x = 0$, and, obviously, it satisfies condition i) of Definition 7.15. It also satisfies ii) of Definition 7.15 because of the fact that the action of G preserves the Poisson structure w, and the reducibility condition (7.11) holds since (7.27) yields $\#(\mathrm{Ann}\ E) = \#\mathrm{Ann}\#\mathrm{Ann}(T_x M_\gamma) \subseteq TM_\gamma$. Q.e.d.

In fact, what happens in Theorem 7.31 is a reduction on each symplectic leaf $S(x)$, as described usually in symplectic geometry [AM], [LM]. Let us also notice that together with the reduced Poisson manifold $N = M_\gamma/G^0_\gamma$, we also have the Poisson manifold M_γ/G_γ that has N as a covering manifold.

Indeed, it is clear that the Poisson structure of N is invariant by the covering transformations.

The pointwise considerations of the proof of Theorem 7.31 also hold for nonequivariant momentum maps. But then, since the term θ of (7.16) depends on x, we have different groups $G_\gamma(S) =$ the isotropy group of γ for the action given by (7.16) on $\mathcal{G}^*$, for every symplectic leaf S, and the subcharacteristic distribution is no longer regular. Therefore, the situation is more complicated in this case.

In any case, however, where the subcharacteristic distribution is regular, and the manifold N exists (e.g., G_γ acts freely and properly on M_γ), the latter has the reduced Poisson structure via $(M_\gamma, TG(x))$, for the reasons quoted at the end of the proof of Theorem 7.31. In particular, this regularity hypothesis makes unnecessary to ask that γ be a regular value for all $J/_{S(x)}$, and it is enough to ask that γ be a regular value of J itself. This is the result given in [MR].

Another fact is that many applications also require level sets M_γ for non-regular values γ, and then, one should rather care for good Poisson algebras of differentiable functions, than for Poisson manifolds. We refer to [ACG] and [WA] for the corresponding theory in the symplectic case. A similar Poisson theory seems plausible.

We shall end this section by explaining why reduction is interesting for mechanics. First we show

7.32. Proposition [MR]. *Let (M_a, N_a, E_a) $(i_a : N_a \subseteq M_a)$ be Poisson reducible triples with the reduced manifolds P_a $(a = 1, 2)$. Let $\Phi : M_1 \to M_2$ be a Poisson map such that $\phi(N_1) \subset N_2$, and $\Phi_*(E_1) \subseteq E_2$. Then Φ induces a unique Poisson map $\hat{\Phi} : P_1 \to P_2$ such that $\pi_2 \circ \Phi = \hat{\Phi} \circ \pi_1$.*

Proof. We have put $\pi_a : N_a \to N_a/_{(E_a \cap TM)} = P_a$. $\hat{\Phi}$, which is to be called the *reduction of* Φ, obviously exists since $\Phi_*(E_1 \cap TM_1) \subseteq (E_2 \cap TM_2)$ and, therefore, Φ is a projectable mapping. We must only check that this projection $\hat{\Phi}$ is a Poisson mapping.

Let f, g be (local) C^∞-functions on P_2, and F, G (local) extensions of $f \circ \pi_2, g \circ \pi_2$ in M_2, with $dF = dG = 0$ on E_2. Then, $\forall v \in E_1$ we have $\phi_* v \in E_2$, hence

$$d(F \circ \phi)(v) = dF(\phi_* v) = 0,$$

and, similarly, $d(G \circ \phi)(v) = 0$. Therefore, $F \circ \phi$, $G \circ \phi$ are extensions of $f \circ \pi_2 \circ \Phi = f \circ \hat{\Phi} \circ \pi_1$ and $g \circ \hat{\Phi} \circ \pi_1$, respectively, with differentials vanishing on E_1. Hence, if we use (7.10), we get

$$\{f \circ \hat{\Phi}, g \circ \hat{\Phi}\}_{P_1} \circ \pi_1 = \{F \circ \Phi, G \circ \Phi\}_{M_1} \circ i_1 = \{F, G\}_{M_2} \circ \phi \circ i_1 =$$
$$= \{F, G\}_{M_2} \circ i_2 \circ \phi = \{f, g\} \circ \pi_2 \circ \phi = \{f, g\}_{P_2} \circ \hat{\Phi} \circ \pi_1.$$

Since π_1 is onto, this yields $\{f \circ \hat{\phi}, g \circ \hat{\phi}\}_{P_1} = \{f, g\}_{P_2} \circ \hat{\Phi}$. Q.e.d.

Now, the interest for mechanics comes from the possibility of reducing a Hamiltonian dynamical system as shown by

7.33 Theorem [MR]. *Let P be the reduced Poisson manifold of M via the pair $(N \subseteq M, E)$. Let $H \in C^\infty(M)$ be a Hamiltonian function such that the flow Φ_t of X_H preserves the submanifold N and the bundle E, and $dH = 0$ on E. Then Φ_t induces a flow $\hat{\Phi}_t$ of Poisson automorphisms of P, whose velocity field is the Hamiltonian field $X_{\hat{H}}$ of the unique function $\hat{H} \in C^\infty(P)$ determined by $\hat{H} \circ \pi = H/_M$. Moreover, X_H and $X_{\hat{H}}$ are π-related vector fields ($\pi : M \to P$).*

Proof. The existence of $\hat{\Phi}_t$ was proven in Proposition 7.32, and then, the tangent field Y of $\hat{\Phi}_t$ is π-related to X_H. On the other hand, the fact that $dH = 0$ on E shows that H is projectable to P, and provides us with the function $\hat{H}$. Let $f \in C^\infty(P)$, and let F be an extension of $f \circ \pi$ with $dF = 0$ on E. Then, $\forall x \in N$, we have

$$X_{\hat{H}}(\pi(x))(f) = \{\hat{H}, f\}_P(\pi(x)) = \{H, F\}_M(x) =$$
$$= X_H(x)(F) = X_H(x)(f \circ \pi) = (\pi_* X_H(x))(f)$$

because, by our hypotheses, $X_H \in TN$. Hence, $X_{\hat{H}}$ is π-related with X_H i.e., $Y = X_{\hat{H}}$. Q.e.d.

$\hat{H}$ is the *reduced Hamiltonian* of H, and the theorem tells us that the dynamics of X_H *projects* or *reduces* to that of $X_{\hat{H}}$ on P. In particular, such situations are encountered for dynamical systems with a group G of symmetries, if H is invariant by G. In this case we have the famous *Noether's theorem*

7.34. Proposition [Mk]. *Consider a Hamiltonian action, with momentum map J, of the group G on the Poisson manifold (M, w). If H is invariant by G, J is invariant by X_H.*

Proof. By the invariance of J we mean that J is constant along the trajectories of X_H i.e., $\forall X \in \mathcal{G}^*, X_H(J(X)) = 0$. But, using (7.15), we have

$$X_H(J(X)) = \{H, J(X)\} = -X_{J(X)}H = -X_M H = 0.$$

Q.e.d.

Consequently, under the hypotheses of Proposition 7.34, the level sets $M_\gamma = J^{-1}(\gamma)$ ($\gamma \in \mathcal{G}^*$) are preserved by X_H, and $dH = 0$ on the orbits of G. Moreover, since G acts on M by Poisson mappings, this action keeps X_H invariant hence, conversely, X_H preserves the orbits of G. Therefore, if M can be reduced via $(M_\gamma, T(\text{orbits } G))$, Theorem 7.33 can be used to reduce the dynamics of X_H accordingly.

8 Symplectic Realizations of Poisson Manifolds

8.1 Local symplectic realizations

The basic idea to be discussed in this Chapter is that of looking for symplectic structures that coinduce a given Poisson structure, in the sense of Definition 7.2, and it turns out that this idea is fruitful and very important. It can be traced back to S. Lie [Lie], and, in our era, it appears in Karasev and Maslov [Kr], [KM1,2], then made precise by Weinstein [We3].

First, following [We3] we give

8.1. Definition. Let (M, w) be a Poisson manifold. Then, a *symplectic realization* of (M, w) is a symplectic manifold V, with the symplectic 2-form σ, together with a surjective submersion $r : V \to M$ such that w is coinduced by r from the Poisson structure $\tilde{\sigma}$ defined by σ.

As a matter of fact, in [We3] this is called a *full realization*, and since this is the only case that we shall consider, we omit the adjective full.

It is important to notice that, if we start with a symplectic manifold (V, σ), and look for Poisson manifolds (M, w) such that (V, σ) is a realization of (M, w) we see that these Poisson manifolds are the quotients of V by foliations $\mathcal{F}$ that are such that the symplectic orthogonal distribution of $T\mathcal{F}$ is a foliation (i.e., *symplectically complete foliations* [Lb3] *or Libermann foliations*). This follows from Proposition 7.18 and Remark 7.20 applied to $(V, \sigma, \mathcal{F})$.

In what follows we shall prove the local existence and uniqueness of symplectic realizations.

8.2. Theorem [We3]. *Any point x of a Poisson manifold (M, w) has an open neighbourhood U such that $(U, w/_U)$ is realizable by a symplectic manifold of dimension* $2(\dim M - (1/2)\ \mathrm{rank}_x w)$.

Proof. In view of the Splitting Theorem 2.16, it is clearly enough to prove only the case $\mathrm{rank}_x w = 0$ i.e., to discuss only the transversal part of w. Hence, we may think of our problem as that of finding a symplectic structure on $\mathbb{R}^{2n} = \{(x^i, y^i)_{i=1}^n\}$ that, together with the projection $(x^i, y^i) \mapsto (x^i)$, yields a realization of $(\mathbb{R}^n, \{x^i, x^j\} = w^{ij}(x))$, where $w^{ij}(0) = 0$.

In particular, the canonical structure $\sum_i dx^i \wedge dy^i$ yields such a realization for the trivial structure $w^{ij} \equiv 0$, which suggests that we look for the realization of w by looking for *good* coordinates $\tilde{x}^i = \phi^i(x, y)$ instead of the x^i. That is, we shall look for the requested symplectic form as

$$\sigma = \sum_{i=1}^{n} d\phi^i \wedge dy^i = \sum_{i,j} \frac{\partial \phi^i}{\partial x^j} dx^j \wedge dy^i - \frac{1}{2} \sum_{i,j} \left(\frac{\partial \phi^i}{\partial y^j} - \frac{\partial \phi^j}{\partial y^i} \right) dy^i \wedge dy^j, \quad (8.1)$$

where rank $(\partial\phi^i/\partial x^j) = n$.

Suppose that the Hamiltonian vector field of x^i with respect to σ is

$$X_i = \xi_i^j \frac{\partial}{\partial x^j} + \eta_i^j \frac{\partial}{\partial y^j}. \tag{8.2}$$

Then, σ realizes w iff $\{x^i, x^j\}_\sigma = \xi_i^j = w^{ij}$, and $i(X_i)\sigma = -dx^i$ yields

$$\sum_h \eta_i^h \frac{\partial\phi^h}{\partial x^k} = -\delta_i^k, \sum_k \left\{ w^{ik}\frac{\partial\phi^h}{\partial x^k} - \left(\frac{\partial\phi^k}{\partial y^h} - \frac{\partial\phi^h}{\partial y^k}\right)\eta_i^k \right\} = 0.$$

Hence $\eta_i^j = -\frac{\partial x^i}{\partial \tilde{x}^j}$, and we remain with the system of equations

$$w^{ik}\frac{\partial\tilde{x}^h}{\partial x^k} + \sum_k \left(\frac{\partial\tilde{x}^k}{\partial y^h} - \frac{\partial\tilde{x}^h}{\partial y^k}\right)\frac{\partial x^i}{\partial\tilde{x}^k} = 0 \tag{8.3}$$

or, equivalently,

$$\frac{\partial\phi^i}{\partial y^j} - \frac{\partial\phi^j}{\partial y^i} = -w^{hk}\frac{\partial\phi^i}{\partial x^h}\frac{\partial\phi^j}{\partial x^k} = -\{\phi^i, \phi^j\}. \tag{8.3$'$}$$

Now, it is natural (see the explanations of [We3]) to look for $\tilde{x}^i$ as a deformation of x^i along a simple w-Hamiltonian field that contains the y^i as parameters. Let us take the function $f^y(x) = \sum_{i=1}^n x^i y^i$, and let $X^w_{f^y}$ be its Hamiltonian vector field, where f^y is seen as a function in x. Denote by φ_y^s the flux of $X^w_{f^y}$, and define

$$\tilde{x}^i = \phi^i(x, y) = \int_0^1 x^i \circ \varphi_y^s ds. \tag{8.4}$$

(There is no loss of generality in assuming the field $X^w_{f^y}$ to be complete, in order to ensure the existence of all the φ_y^s desired. Indeed, since our theorem is local, we may replace w by $\tilde{w}$ that equals w in a neighbourhood of 0, and has a compact support. Then, the new $X^w_{f^y}$ will have a compact support.)

For $y = 0$, $\tilde{x}^i = x^i$, and $\det(\partial\tilde{x}^i/\partial x^i) \neq 0$. Hence $\operatorname{rank}(\partial\phi^i/\partial x^j) = n$ in a neighbourhood of 0, as it is needed for σ of (8.1) to be nondegenerate.

Now, we shall prove by a straightforward computation that the functions (8.4) satisfy (8.3$'$), thereby ending the proof of the theorem. In order to do this we need a result that is less well known, and for this reason we prove it as an auxiliary lemma called the Perturbation Theorem)

8.3. Lemma [AR]. *Let ξ, η be vector fields, with compact support on a differentiable manifold M. Let $\varphi^\xi, \varphi^{\xi+\lambda\eta}$ ($\lambda \in \mathbb{R}$) denote the flows of ξ and $\xi + \lambda\eta$, respectively. Then, $\forall x \in M$, $\forall t \in \mathbb{R}$ one has*

$$\left.\frac{d}{d\lambda}\right|_{\lambda=0} (\varphi_s^{\xi+\lambda\eta}(x)) = \int_0^s (\varphi_t^\xi)_*(\eta(\varphi_{s-t}^\xi(x))dt. \tag{8.5}$$

Proof. With respect to λ, $\varphi_s^{\xi+\lambda\eta}(x)$ is a path, and (8.5) yields the tangent vector of this path at $y=\varphi_s^{\xi}(x)$. We may also see (8.5) as a calculation of the velocity of the perturbation of the flow φ^{ξ} when ξ is perturbed by η. The compact support hypothesis ensures that the vector fields used in the lemma are complete.

From the definition of the point y above, we have $x=\varphi_{-s}^{\xi}(y)$, and we shall denote the left-hand side of (8.5) by

$$\zeta_s(y)=\left.\frac{d}{d\lambda}\right|_{\lambda=0}(\varphi_s^{\xi+\lambda\eta}\circ\varphi_{-s}^{\xi}(y))\in T_yM. \tag{8.6}$$

This expression suggests that we consider the path given by

$$\zeta_\tau(y)=\left.\frac{d}{d\lambda}\right|_{\lambda=0}(\varphi_\tau^{\xi+\lambda\eta}\circ\varphi_{-\tau}^{\xi}(y))\qquad(0\le\tau\le s) \tag{8.7}$$

in the fixed linear space T_yM, and then write that

$$\zeta_s(y)=\int_0^s\left.\frac{d}{d\tau}\right|_{\tau=t}(\zeta_\tau(y))dt=\int_0^s\left.\frac{d}{d\tau}\right|_{\tau=0}(\zeta_{\tau+t}(y))dt. \tag{8.8}$$

Now,

$$\begin{aligned}\zeta_{\tau+t}(y)&=\left.\frac{d}{d\lambda}\right|_{\lambda=0}(\varphi_{\tau+t}^{\xi+\lambda\eta}\circ\varphi_{-\tau-t}^{\xi}(y))= \qquad (8.9)\\ &=\left.\frac{d}{d\lambda}\right|_{\lambda=0}\left[\varphi_t^{\xi+\lambda\eta}\left(\varphi_\tau^{\xi+\lambda\eta}(\varphi_{-\tau-t}^{\xi}(y))\right)\right]=\zeta_t(y)+(\varphi_t^{\xi})_*\left(\zeta_\tau(\varphi_{-t}^{\xi}(y))\right),\end{aligned}$$

where the two terms correspond to the two appearances of λ. As for the value of ζ at other points than y, we use (8.7) at the respective point.

If we apply $d/d\tau$ to (8.9) we get

$$\begin{aligned}\left.\frac{d}{d\tau}\right|_{\tau=0}&(\zeta_{\tau+t}(y))=(\varphi_t^{\xi})_*\left\{\left.\frac{\partial}{\partial\tau}\right|_{\tau=0}\left.\frac{\partial}{\partial\lambda}\right|_{\lambda=0}\left[(\varphi_\tau^{\xi+\lambda\eta}(\varphi_{-\tau-t}^{\xi}(y))\right]\right\}=\\ &=(\varphi_t^{\xi})_*\left\{\left.\frac{\partial}{\partial\lambda}\right|_{\lambda=0}\left.\frac{\partial}{\partial\tau}\right|_{\tau=0}\left[\varphi_\tau^{\xi+\lambda\eta}(\varphi_{-\tau-t}^{\xi}(y))\right]\right\}= \qquad (8.10)\\ &=(\varphi_t^{\xi})_*\left\{\left.\frac{d}{d\lambda}\right|_{\lambda=0}\left[(\xi+\lambda\eta)(\varphi_{-t}^{\xi}(y))+(\varphi_0^{\xi+\lambda\eta})_*\xi(\varphi_{-t}^{\xi}(y))\right]\right\}.\end{aligned}$$

Here again we derived the two appearances of τ. Furthermore, since $\varphi_0^{\xi+\lambda\eta}=$ id., we get

$$\left.\frac{d}{d\tau}\right|_{\tau=0}(\zeta_{\tau+t}(y))=(\varphi_t^{\xi})_*(\eta(\varphi_{-t}^{\xi}(y)), \tag{8.11}$$

Finally, if we put back $y = \varphi_s(x)$, and use (8.6), (8.8) and (8.11), we obtain precisely (8.5). Q.e.d.

Now, we can come back and finish the proof of Theorem 8.2. Namely, we have to compute $\partial\phi^i/\partial y^j$ for the functions ϕ^i defined by (8.4). We have

$$\frac{\partial\phi^i(y^j)}{\partial y^j}\bigg|_{y^j} = \frac{\partial\phi^i(y^j+\lambda)}{\partial(y^j+\lambda)}\bigg|_{\lambda=0} = \frac{\partial\phi^i(y^j+\lambda)}{\partial\lambda}\bigg|_{\lambda=0}, \tag{8.12}$$

where we mentioned only the chosen variable y^j, and the other y^i remain the same in the two sides of the formula. But, by (8.4), $\phi^i(y^j+\lambda)$ are the integrals of the coordinates of the points along the flow of

$$X^w_{fy^j+\lambda} = \sum_{i,k} w^{ik} y^i \frac{\partial}{\partial x^k} + \lambda \sum_k w^{jk}\frac{\partial}{\partial x^k} = X^w_{fy^j} + \lambda X^w_{x^j}. \tag{8.13}$$

Hence, the derivative required is given by a corresponding application of (8.5) (the compact support hypothesis of Lemma 8.3 may be accepted because we are only interested in a neighborhood of a given point) i.e.,

$$\frac{\partial\phi^i}{\partial y^j} = \int_0^1 \frac{\partial}{\partial\lambda}\bigg|_{\lambda=0}\left[x^i\left(\varphi_s^{X_{fy}+\lambda X_{x^j}}(x)\right)\right] ds =$$

$$= \int_0^1 (dx^i)\left[\int_0^s \left(\varphi_t^{X_{fy}}\right)_*\left(X_{x^j}(\varphi_{s-t}^{X_{fy}}(x))\right) dt\right] ds =$$

$$= \int\int_{0\le t\le s\le 1}\left\{x^j, x^i\circ\varphi_t^{X_{fy}}\right\}_w \circ \varphi_{s-t}^{X_{fy}} dt ds =$$

$$= \int\int_{0\le t\le s\le 1}\left\{x^j\circ\varphi_{s-t}^{X_{fy}}, x^i\circ\varphi_s^{X_{fy}}\right\} dt ds,$$

because $\varphi^{X_{fy}}$ is a Hamiltonian flow, and it preserves the Poisson bracket. Then, by the change of variables $(s,t)\mapsto(s,\tau=s-t)$, we have

$$\frac{\partial\phi^i}{\partial y^j} = \int\int_{0\le\tau\le s\le 1}\left\{x^j\circ\varphi_\tau^{X_{fy}}, x^i\circ\varphi_s^{X_{fy}}\right\} d\tau\ \mathrm{ds}. \tag{8.14}$$

Similarly, and then with an interchange of (τ, s), we get

$$\frac{\partial\phi^j}{\partial y^i} = -\int\int_{0\le s\le\tau\le 1}\left\{x^j\circ\varphi_\tau^{X_{fy}}, x^i\circ\varphi_s^{X_{fy}}\right\} d\tau ds. \tag{8.15}$$

Hence, by looking at the two domains of integration, we see that

$$\frac{\partial\phi^i}{\partial y^j} - \frac{\partial\phi^j}{\partial y^i} = \int_0^1\int_0^1 \{x^j\circ\varphi_\tau^{X_{fy}}, x^i\circ\varphi_s^{X_{fy}}\} d\tau ds =$$

$$\left\{\int_0^1 x^j\circ\varphi_\tau^{X_{fy}} d\tau, \int_0^1 x^i\circ\varphi_s^{X_{fy}} ds\right\} = \{\phi^j, \phi^i\}.$$

Q.e.d.

8.4. Remarks. 1) The number $2\dim M\text{-rank}_x w$ that appears in Theorem 8.2 is the smallest possible dimension of a realization $r: V \to M$ over a domain that contains $x \in M$. Indeed, r has a canonical local coordinate expression $x^i = y^i$ ($i = 1, \ldots, \dim M$), where (x^i) are coordinates of M around x, and (y^i, y^α) ($\alpha = 1, \ldots, \dim V - \dim M$) are coordinates of V. Since r is a Poisson mapping, the symplectic form σ of V has a $(\text{rank}_x w)$-dimensional part in the coordinates y^i alone, and the number of the coordinates y^α must be $\geq \dim M - \text{rank}_x w$). In other words, $\dim V - \dim M \geq \dim M - \text{rank}_x w$, which was exactly the assertion of the remark.

2) Theorem 8.2 was local but, one can prove that any Poisson manifold has also global realizations by so-called local symplectic groupoids. Such realizations will be obtained by gluing up local realizations along the given Poisson manifold [We5], [CDW], [AD1,2]. More explanations about this will be given later on in Chapter 9.

Now, we shall discuss the local essential uniqueness of the symplectic realization of a Poisson manifold.

8.5. Definition. Let $r_a : V_a \to M$ $(a = 1, 2)$ be symplectic realizations of the Poisson manifold (M, w). We shall say that r_1 is *subordinated* to r_2, and denote $r_1 \angle r_2$ if there is a symplectic embedding $i : V_1 \to V_2$ such that $r_1 = r_2 \circ i$. Two realizations r_a of (M, w) are called *equivalent* if $r_1 \angle r_2$ and $r_2 \angle r_1$ by means of symplectomorphisms i, i^{-1}. Furthermore, r_1 and r_2 are called *essentially equivalent* if there are realizations $r' \angle r_a$ $(a = 1, 2)$ where r'_1, r'_2 are equivalent.

Clearly, essential equivalence is an equivalence relation.

8.6. Theorem [We3]. *Let $r_a : V_a \to M$ be two arbitrary symplectic realizations of (M, w) with the symplectic forms σ_a $(a = 1, 2)$. Then, $\forall x \in M$, and $\forall v_a \in V_a$ with $r_a(v_a) = x$, one has neighborhoods U, W_a, respectively, such that $r_a/_{W_a}$ are essentially equivalent symplectic realizations of $(U, w/_U)$.*

Proof. Let $(p_\alpha, q^\alpha, x^\nu)$ $(\alpha = 1, \ldots, (1/2)\text{rank}_x w;\ \nu = 1, \ldots, \dim M - \text{rank}_x w)$ be canonical coordinates of M around x, in the sense of Definition 2.18. If we "forget" to write composition by r_a, the same $(p_\alpha, q^\alpha, x^\nu)$ are a part of systems of local coordinates of V_a around v_a, and the fact that r_a are symplectic realizations implies

$$\sigma_a = \sum_\alpha dp_\alpha \wedge dq^\alpha + \sigma'_a \qquad (a = 1, 2), \tag{8.16}$$

where σ'_a is a symplectic 2-form with some canonical local coordinates z^u_a $(u = 1, \ldots, \dim V_a - \text{rank}_x w)$. With respect to these new coordinates, we must have

$$dx^\nu = A^{\nu\alpha} dp_\alpha + B^\nu_\alpha dq^\alpha + C^\nu_u dz^u_a,$$

and, since $\{p_\alpha, x^\nu\} = \{q^\alpha, x^\nu\} = 0$, (8.16) yields $A^{\nu\alpha} = 0, B^\nu_\alpha = 0$. Therefore, x^ν are functionally independent functions of z^u_a only, in a neighborhood

V_a of v_a, and these functions are in involution at v_a since $\{x^{\nu_1}, x^{\nu_2}\}(v_a) = \{x^{\nu_1}, x^{\nu_2}\}(x) = 0$.

Accordingly, it is possible to find a system of functionally independent functions $\Phi_a^\nu(z_a^u)$ which are in involution (i.e., have vanishing Poisson brackets) on V_a, and $\Phi_a^\nu(v_a) = x^\nu(v_a) = 0$ (for instance), and $d\Phi_a^\nu(v_a) = dx^\nu(v_a)$. Indeed, $dx^\nu(v_a) = 0$ defines a coisotropic tangent subspace of the space of the coordinates (z_a^u), and the functions Φ_a^ν may be obtained by writing down equations of a coisotropic submanifold tangent to this subspace at v_a. Such a submanifold obviously exists. Then, a theorem of Darboux-Lie (e.g., [Lb2,3], [Va8]) provides us with the necessary local coordinates y^ν, z_a^θ such that

$$\sigma_a = \sum_\alpha dp_\alpha \wedge dq^\alpha + \sum_\nu d\Phi_a^\nu(x, y_a) \wedge dy_a^\nu + \sigma_a'', \tag{8.17}$$

where we made all the dx apparent since, by the condition imposed on Φ^ν, rank $(\partial \Phi_a^\nu / \partial x^{\nu'}) = \dim M - \mathrm{rank}_x w$, and σ_a'' are symplectic forms with canonical coordinates z_a^θ.

Clearly, the first two terms of (8.17) define local realizations of (M, w) around x that are *locally subordinated* to r_a, and, furthermore, the sets of functions ϕ_a^ν $(a = 1, 2)$ must satisfy the conditions deduced in the proof of Theorem 8.2. In particular, the Hamiltonian vector fields of x^i can be seen again as given by (8.2), where the matrix $-(\eta_i^j)$ is the inverse of $(\partial \phi_a^\nu / \partial x^{\nu'})$.

Hence, in order to finish the proof of Theorem 8.6 it suffices to show that $\sum_\nu d\phi_a^\nu(x, y) \wedge dy_a^\nu$ are two equivalent symplectic forms. Denote

$$\tilde{x}_a^\nu = \phi_a^\nu(x^{\nu'}, y_a^{\nu'}).$$

Because of the rank condition for $(\partial \phi / \partial x)$, one has local solutions $x^\nu = \psi_a^\nu(\tilde{x}_a, y)$, and we may define a mapping that has the equations

$$\tilde{x}_2^\nu = -\phi_2^\nu(\psi_1^\nu(\tilde{x}_1), y_2), \quad y_2 = y_1.$$

We claim that this is the requested symplectic equivalence. Indeed, it leads to

$$\sum_\nu d\tilde{x}_2^\nu \wedge dy_2^\nu = -\sum_\nu \frac{\partial \phi_2^\nu}{\partial x^{\nu'}} \frac{\partial x^{\nu'}}{\partial \tilde{x}_1^{\nu''}} d\tilde{x}_1^{\nu''} \wedge dy_2^\nu - \sum_{\nu'} \frac{\partial \Phi_2^\nu}{\partial y_2^{\nu'}} dy_2^{\nu'} \wedge dy_2^\nu =$$

$$= \sum_\nu (\eta^{-1})_{\nu'}^\nu \eta_{\nu''}^{\nu'} d\tilde{x}_1^\nu \wedge dy^\nu + 0 = \sum_\nu d\tilde{x}_1^\nu \wedge dy_1^\nu.$$

In this computation, the matrix η^{-1} is the inverse of the same matrix η itself because of the identification $y_2 = y_1$, and the term $dy^{\nu'} \wedge dy^\nu$ vanishes because of (8.3) and because the functions Φ^α are in involution. Q.e.d.

8.2 Dual pairs of Poisson manifolds

In order to formulate another interesting theorem about symplectic realizations we follow [We3], and define the notion of a *function group* on a symplectic manifold (V,σ) as the set $C^\infty_{\text{proj}}(\mathcal{F})$ of the functions that are constant along the leaves of a foliation $\mathcal{F}$ of V, if this set is closed by Poisson brackets. By the symplectic version of Proposition 7.18 and Remark (7.20), this happens iff the σ-orthogonal distribution $\mathcal{F}'$ of $\mathcal{F}$ is also a foliation, and, then $C^\infty_{\text{proj}}(\mathcal{F}')$ will also be a *function group* called the *polar* or *dual* group of $C^\infty_{\text{proj}}(\mathcal{F})$. The polar group obviously consists of the functions that Poisson-commute with those in the group itself. It is clear that, if $r:(V,\sigma)\to(M,w)$ is a realization of a Poisson manifold then $\{f\circ r/f\in C^\infty(M)\}$ is a function group on V. Accordingly, it makes sense to define

8.7. Definition. The Poisson manifolds (M_a,w_a) $(a=1,2)$ will be called *dual* to each other if they are realizable by the same symplectic manifold (V,σ), and the function groups of V associated with these realizations are polar function groups or, equivalently, the fibers of the realization mappings $r_a:V\to M_a$ are σ-orthogonal.

8.8. Theorem [We3]. *Let (M_a,w_a) be dual Poisson manifolds symplectically realized by $r_a:(V,\sigma)\to(M_a,w_a)$. Then, $\forall v\in V$, the transversal Poisson structures of w_a at $x_a=r_a(v)$ are Poisson anti-isomorphic (i.e., the corresponding Poisson brackets are opposite in sign).*

Proof. The result is local hence, we may restrict ourselves to neighborhoods of v and of x_a respectively. By lifting canonical coordinates $(p^a_{\alpha_a},q^{\alpha_a}_a,x^{\nu_a}_a)$ of (M_a,w_a) at x_a to V around v, as in the proof of Theorem 8.6, we get the same functions on V, and, because M_a are dual pairs, the functions with a different index a Poisson commute. Hence $(p^1_{\alpha_1},p^2_{\alpha_2},q^{\alpha_1}_1,q^{\alpha_2}_2)$ may be completed by (p_θ,q^θ) so as to get a system of symplectic-canonical coordinates on V at v. Moreover, because of Poisson-commutation, $x^{\nu_a}_a=x^{\nu_a}_a(p_\theta,q^\theta)$. Thus, if we look at the neighbourhood described by $\{(p_\theta,q^\theta)\}$, endowed with the symplectic form $\sum_\theta dp_\theta\wedge dq^\theta$, we see that it offers (locally) symplectic realizations of $tr_{x_a}w_a\overset{\text{def}}{=}$ the transversal Poisson structure of w_a at x_a, making these last structures into a dual pair.

Therefore, it will be enough to prove Theorem 8.8 in the case where $\text{rank}_{x_a}w_a=0$ only. Then, we have only the coordinates $x^{\nu_a}_a$ (that may be taken 0 at x_a), and the result means that (M_a,w_a) are Poisson anti-isomorphic. For the lifts to V, we have $\{x^\nu_a,x^{\nu'}_a\}(v)=0$, $\{x^\nu_1,x^{\nu'}_2\}=0$. If we express the Poisson brackets by means of the symplectic form of V, the previous equations show that the tangent spaces at v of the fibers of r_a through v are coisotropic symplectically orthogonal subspaces of T_vV, which cannot happen unless these tangent spaces are Lagrangian (and equal). This implies, of course,

$$\dim M_1 = \dim M_2 = (1/2) \dim \text{ V}.$$

Now, let L be a Lagrangian submanifold of V that passes through v, and is transversal here to $T_v(r^{-1}(x_1)) = T_v(r^{-1}(x_2))$. Then, $r_a/_L$ are local diffeomorphisms by means of which we may identify (locally, again) M_1 and M_2 with L or, equivalently, build a diffeomorphism φ between neighborhoods of $x_a \in M_a$ $(a = 1, 2)$.

In order to prove the Theorem, we have to show that, when transposed to L, the Poisson brackets with respect to w_1 and w_2 differ only in sign.

Obviously, $x_a^\nu \mid_L$ yield local coordinate systems on L, and a result of local symplectic geometry known as the theorem of Darboux-Lie (e.g. [Lb2,3]) tells us that V has local canonical coordinates (s^ν, t^ν) around v such that

$$s_a^\nu \mid_L = x_a^\nu \mid_L, t_a^\nu \mid_L = 0, \quad \sigma = \sum_\nu ds_a^\nu \wedge dt_a^\nu. \tag{8.18}$$

(Look at L as a leaf of a Lagrangian foliation in order to be able to apply the Darboux-Lie theorem.This is possible since everything here is local.)

Moreover, if we take the coordinates of M_a such that φ is expressed by the identity coordinate mapping, we shall have $x_1^\nu \mid_L = x_2^\nu \mid_L$, and we get a single coordinate system (s^ν, t^ν) where (8.18) holds.

With respect to these coordinates, we may write

$$x_a^\nu = s^\nu + \sum_{\nu'} \lambda_{a\nu'}^\nu t^{\nu'} \qquad (a = 1, 2), \tag{8.19}$$

where $\lambda_{a\nu'}^\nu$ are local functions on V. Accordingly, we get

$$\{x_a^\nu, x_a^\mu\} \mid_L = (\lambda_{a\nu}^\mu - \lambda_{a\mu}^\nu) \mid_L, \tag{8.20}$$

and these relations define w_a with respect to the coordinates used above. But, we also have, similarly

$$\{x_1^\nu, x_2^\mu\} \mid_L = (\lambda_{1\nu}^\mu - \lambda_{2\mu}^\nu) \mid_L = 0, \tag{8.21}$$

because of duality. From (8.20) and (8.21) it follows clearly that

$$\{x_1^\nu, x_1^\mu\} \mid_L = -\{x_2^\nu, x_2^\mu\} \mid_L .$$

Q.e.d.

8.9. Example [We3]. Let V be a symplectic manifold endowed with a Hamiltonian action of a connected Lie group G, and with an equivariant momentum map $J : V \to \mathcal{G}^*$, which is a Poisson morphism by Proposition 7.30. Suppose also that V/G is a manifold. Then, it is clear from Proposition 7.3 that $\pi : V \to V/G$ coinduces a Poisson structure on V/G. On the other hand, formula (7.27) shows that the fibers of π (i.e., the orbits of G) are symplectically orthogonal to the fibers $J^{-1}(\gamma)$ of J ($\gamma \in \mathcal{G}^*$). Hence, the Poisson manifolds V/G with the coinduced structure, and $\mathcal{G}^*$ with its Lie-Poisson structure are dual Poisson manifolds. (As a matter of fact, to be technically exact, we must either ask J to be a surjective submersion or use the definition of [We3] which doesn't ask for surjectivity instead of our Definition 8.1.)

8.3 Isotropic realizations

In this section we shall follow [Dz3], [Dz5] and [CS], and discuss symplectic realizations $r : (V, \sigma) \to (M, w)$ such that the fibers of r are isotropic submanifolds of V. These will be called *isotropic realizations*, and, among other things, they are interesting because the geometric structure that they present generalizes the famous Arnold-Liouville theorem about the existence of action-angle coordinates (e.g., [LM]). First, we notice

8.10. Proposition. *If $r : (V^{2n}, \sigma) \to (M^m, w)$ is an isotropic realization, the Poisson structure w is necessarily regular, and the codimension of its symplectic foliation $\mathcal{S}(M)$ equals the dimension of the fibers of r.*

Proof. From Proposition 7.18, we know that the σ-orthogonal distribution $T'(f^{-1}(x))$ ($x \in M$) of the fibers of f is completely integrable, and its projection by f_* onto M yields the symplectic foliation $\mathcal{S}(M)$. Since the fibers of f are isotropic, the orthogonal foliation is coisotropic, and its (sub)characteristic distribution is $T(f^{-1}(x))$. Hence, if $\dim f^{-1}(x) = k$, we get

$$\dim \mathcal{S}(M) = \dim T'(f^{-1}(x)) - \dim T(f^{-1}(x)) =$$
$$= 2n - k - k = 2(n-k) = \text{ const.}$$

On the other hand, we clearly have $m = 2n - k$, hence

$$m - \dim \mathcal{S}(M) = k.$$

Q.e.d.

8.11. Corollary. *An isotropic realization is necessarily a realization of the smallest possible dimension.*

Proof. The smallest possible dimension was indicated in Remark 8.4, and, in our case, it must be $2n$. Q.e.d.

Another basic fact is contained in

8.12. Proposition. *The fibers of an isotropic realization* $r : (V, \sigma) \to (M, w)$ *have a naturally induced locally affine structure.*

Proof. For every Casimir function φ of (M, w), the Hamiltonian vector field $X^{\sigma}_{\varphi\circ r}$ projects to $X^{w}_{\varphi} = 0$ hence, it is tangent to the fibers of r. If we declare these Hamiltonian fields parallel, we get the requested locally affine structure on $r^{-1}(x)$ $(x \in M)$ since, by Proposition 8.10, there are exactly $\dim(r^{-1}(x))$ functionally independent Casimirs at $x \in M$. Let us also notice, then, that $\forall \alpha \in \mathrm{Ann}\, \mathcal{S}_x(M)$ on M, the vector field defined on $r^{-1}(x)$ by

$$Y_\alpha \overset{\mathrm{def}}{=} \#_\sigma(r^*\alpha) \qquad (\text{i.e., } i(Y_\alpha)\sigma = -r^*\alpha) \tag{8.22}$$

is a parallel vector field, since α is a linear combination of differentials of Casimir functions. Q.e.d.

8.13. Definition. If the fibers of a symplectic realization have a certain property $\mathcal{P}$, we shall say that the realization has that property. In particular, we shall speak of *connected complete isotropic realizations (c.c.i. realizations,* or *Libermann isotropic realizations* [Dz3]) if the fibers are connected, and the locally affine structure of Proposition 8.12 is complete.

8.14. Remark. As motivated in [Dz3], it is sometimes necessary to use symplectic realizations by non-Hausdorff manifolds (V, σ). In that case, however, M, and the fibers $r^{-1}(x)$, $\forall x \in M$, will always be supposed to be Hausdorff manifolds, and this name will not be added to the realization, in spite of Definition 8.13.

In what follow, $r : (V^{2n}, \sigma) \to (M^n, w)$ will always be a c.c.i. realization of w. In this case, the following fundamental fact occurs: *the parallel fields* (8.22) *yield an action of the abelian group* $\nu_x\mathcal{S} = \mathrm{Ann}_x\mathcal{S}(M)$ = *the fiber of the conormal vector bundle* $\nu^*\mathcal{S}$ *of* $\mathcal{S}(M)$ *on the fiber* $r^{-1}(x)$ $(x \in M)$ *of the realization* r *defined by*

$$a_\alpha(z) = \exp(1Y_\alpha)(z) \quad (\alpha \in \nu_x\mathcal{S}, x = r(z)). \tag{8.23}$$

Indeed, we can see that $\{Y_\alpha\}$ $(\alpha \in \nu^*_x\mathcal{S})$ is an abelian Lie algebra of vector fields on $r^{-1}(x)$. This follows since $\alpha = \Sigma_a c_a dx^a$, where x^a are local transversal coordinates of $\mathcal{S}(M)$, and c_a are constant coefficients, and then

$$[Y_{dx^a}, Y_{dx^b}] = [X^\sigma_{x^a\circ r}, X^\sigma_{x^b\circ r}] = X^\sigma_{\{x^a,x^b\}\circ r} = 0,$$

because $\{x^a, x^b\}_w = 0$. Furthermore, since the fields Y_α for the various α commute, we get

$$a_{(\alpha+\beta)}(z) = \exp(Y_\alpha + Y_\beta)(z) = (\exp Y_\alpha) \circ (\exp Y_\beta)(z) = a_\alpha \circ a_\beta(z),$$

simply by choosing coordinates where Y_α, Y_β are partial derivatives.

The existence of the action (8.23) allows us to clarify the structure of the isotropic realizations.

First, notice that the action (8.23) is transitive on $r^{-1}(x)$, $\forall x \in M$, since any two points of $r^{-1}(x)$ can be joined by a broken geodesic of the affine flat connection defined by the parallel fields Y_α. Hence

$$r^{-1}(x) \approx (\nu_x^* \mathcal{S})/\text{Isotropy }(z) \qquad (z \in r^{-1}(x)),$$

and because $\dim(r^{-1}(x)) = \dim(\nu_x^* \mathcal{S})$, Isotropy (z) must be discrete. Moreover, since we are in the abelian case, Isotropy (z), which usually is conjugate with Isotropy (z'), is now equal to the latter, and we have a common discrete isotropy subgroup of all $z \in r^{-1}(x)$, which we shall denote by

$$\mathcal{R}_x = \{\alpha \in \nu_x^* \mathcal{S} / \forall z \in r^{-1}(x), a_\alpha(z) = z\}. \tag{8.24}$$

We shall also denote

$$\mathcal{R} = \cup_{x \in M} \mathcal{R}_x. \tag{8.25}$$

With this notation we get [Dz3], [CS]

8.15. Theorem. *i) $\mathcal{R}$ is a closed m-dimensional sub-manifold of $\nu^* \mathcal{S}$; ii) $\nu^* \mathcal{S}/\mathcal{R}$(fiberwise) has an induced structure of a (Hausdorff) differentiable manifold such that* proj: $\nu^* \mathcal{S} \to (\nu^* \mathcal{S})/\mathcal{R}$ *is a local diffeomorphism; iii) locally, the induced mapping $\pi' : (\nu^* \mathcal{S})/\mathcal{R} \to M$ is an isotropic realization of (M, w), equivalent to (V, σ).*

Proof. i) The basic starting remark of the proof is that for every local section $s : U \to V$ ($U \subseteq M$, U-open), the mapping $\alpha \mapsto a_\alpha(s(\pi(\alpha)))$ ($\pi : \nu^* \mathcal{S} \to M$ is the bundle projection) is a local diffeomorphism $\varphi_s : \pi^{-1}(U) \to r^{-1}(U)$. Indeed, in view of (8.23), this mapping is clearly differentiable. Then, if we fix $\alpha_0 \in \pi^{-1}(U)$, $\pi(\alpha_0) = x_0$, and take a tangent basis in $T_{\alpha_0}(\nu^* \mathcal{S})$ that consists of independent vectors $\xi_1, \ldots, \xi_{2n-m}$ transversal to the fiber, and independent vectors $\eta_1, \ldots, \eta_m$ tangent to the fiber, $(\varphi_s)_*(\xi_i)$ will be independent in $T_{\varphi_s(\alpha)} V$, and transversal to the fibers of V, since $\pi_*, s_*, (\exp \varphi_\alpha)_*$ preserve the linear independence, and $(\varphi_s)_*(\eta_j)$ will be tangent to the fibers of V and independent since they are computed by (8.22) and because σ is nondegenerate. Hence $\text{rank}_{\alpha_0} \varphi_s = 2n$.

Now, let us fix x_0 and α_0 as above. If $\alpha_0 \in \mathcal{R}$, $\varphi_s(\alpha_0) = s(x_0)$, and, perhaps on a smaller neighborhood U, there exists a unique local cross section $\theta : U \to \nu^* \mathcal{S}$ such that $\theta(x_0) = \alpha_0$ and $\varphi_s \circ \theta = s$ (namely, $\theta = \varphi^{-1}_{s(\text{local})} \circ s$). Moreover, we have $s(x) = \varphi_s(\theta(x)) = a_{\theta(x)}(s(x))$, and we see that $\theta(x) \in \mathcal{R}$, $\forall x \in U$. Furthermore, since $\mathcal{R}_x$ are discrete subsets of the fibers of $\nu^* \mathcal{S}$, it is clear that α_0 has a neighbourhood W, which is a neighbourhood of $\theta(U)$, and such that $W \cap \mathcal{R} = \theta(U)$. This form of the intersection $W \cap \mathcal{R}$ shows that $\mathcal{R}$ is an embedded m-dimensional submanifold of the total space of $\nu^* \mathcal{S}$.

The local cross sections θ of $\mathcal{R}$ constructed above are a very useful tool in the study of isotropic realizations. In the sequel we shall say that θ is the *s-distinguished cross section of* $\mathcal{R}$.

Similarly, if we take $\alpha_0 \notin \mathcal{R}$, we can get a cross section $\theta(x)$, $x \in U$, where $\theta(x)$ is never in $\mathcal{R}$, and a neighbourhood W of α_0 that is disjoint of $\mathcal{R}$. Therefore $\mathcal{R}$ is closed in $\nu^*\mathcal{S}$.

ii) in order to obtain the differentiable structure of $\nu^*\mathcal{S}/\mathcal{R}$, we remember the following general result (e.g., [Bk]): let M be a differentiable manifold endowed with an equivalence relation $\sim$ whose graph is $Q \subseteq M \times M$; if Q is a (closed) submanifold of $M \times M$, and $pr_1 : Q \to M$ is a submersion, then $M/\sim$ is a (Hausdorff) differentiable manifold.

In our case, the equivalence relation is defined by $\mathcal{R}$ and its graph is

$$Q = \{(\alpha, \beta) \in \nu^*\mathcal{S} \times \nu^*\mathcal{S} / \alpha - \beta \in \mathcal{R}\}.$$

If we use s-distinguished cross sections θ_1, θ_2 as in the proof of i), with the initial values α, β, and corresponding neighborhoods W_1 W_2 of α, β we shall have $Q \cap (W_1 \times W_2) = \theta_1(U) \times \theta_2(U)$, and we see that Q is a closed submanifold of $\nu^*\mathcal{S} \times \nu^*\mathcal{S}$. Moreover, the mapping

$$\gamma \mapsto (\gamma, \gamma + \theta_2(\pi(\gamma)) - \theta_1(\pi(\gamma))) \qquad (\gamma \in \nu^*\mathcal{S}/_U)$$

is a local cross section of $pr_1 : Q \mapsto \nu^*\mathcal{S}$, and the existence of such cross sections ensures that pr_1 is a submersion. Hence, the conclusions of ii) follow. (The last part of these conclusions is rather obvious.)

iii) From the definition of the mapping φ_s associated to a cross section $s : U \to V$ it is clear that φ_s induces a bundle equivalence over U

$$((\nu^*\mathcal{S})/\mathcal{R})/_U \xrightarrow{\hat{\varphi}_s} r^{-1}(U) \tag{8.26}$$

(i.e., $r \circ \hat{\varphi}^* = \pi'$), and if we take the symplectic form $\hat{\varphi}_s^*\sigma$ on $((\nu^*\mathcal{S})/\mathcal{R})/_U$, we are done. Q.e.d.

The result included in formula (8.26) shows that, locally, V consists of a union of fibers that are products of a torus by a Euclidean space. The coordinates on the torus are the *angle coordinates*, and this configuration appears as a generalization of the *action-angle coordinates* [Dt]. An important case is that of a c.c.i. realization of the Lie-Poisson structure of a Lie algebra $\mathcal{G}^*$, or of a general affine Poisson structure on $\mathcal{G}^*$. Then, under suitable hypotheses, the symplectic orthogonal foliation of the isotropic fibers consists of the orbits of a Hamiltonian group action [Dz3], [DD].

The existence of the distinguished $\mathcal{R}$-valued cross sections θ of the previous proof gives us also the interesting information that $\pi/_{\mathcal{R}}$ is a local diffeomorphism. Hence $\mathcal{R}$ can be interpreted as a subsheaf of the sheaf $\mathcal{C}^\infty(\nu^*\mathcal{S})$ of germs of sections of $\nu^*\mathcal{S}$ (e.g., [Gn]). More exactly, we have [Dz3], [CS]

8.16. Proposition. *i) If α is a cross section of $\nu^*\mathcal{S}$, then*

$$a^*_\alpha \sigma - \sigma = -r^* d\alpha. \tag{8.27}$$

ii) $\mathcal{R}$ is a subsheaf of the sheaf $\mathcal{Z}(\nu^*\mathcal{S})$ of d-closed germs of $C^\infty(\nu^*\mathcal{S})$.

Proof. i) It should be understood that if (8.27) is evaluated at $v \in V$, a^*_α means $a^*_{\alpha(r(v))}$. By (8.23), we have

$$\begin{aligned} a^*_\alpha \sigma - \sigma &= \int_0^1 \frac{d}{dt}\bigg|_{\tau=t} [\exp(\tau Y_\alpha)^* \sigma] dt = \int_0^1 [(\exp tY_\alpha)^*(L_{Y_\alpha}\sigma)] dt = \\ &= \int_0^1 [(\exp tY_\alpha)^*(\mathrm{di}(Y_\alpha)\sigma)] dt = -\int_0^1 [(\mathrm{exptY}_\alpha)^*(dr^*\alpha)] dt = \\ &= -d\int_0^1 [(r \circ \exp tY_\alpha)^*(\alpha)] dt = -d(r^*\alpha) = -r^* d\alpha \end{aligned}$$

($r \circ \exp sY_\alpha = r$ since Y_α is tangent to the fibers).
ii) If $\alpha \in \mathcal{R}$, then $a^*_\alpha \sigma = \sigma$, and, by (8.27), we get $r^* d\alpha = 0$. This implies $d\alpha = 0$ since r is a submersion. Q.e.d.

8.17. Corollary. *$d : C^\infty(\nu^*\mathcal{S}) \to \mathcal{Z}(\nu^*\mathcal{S} \otimes T^*M)$ induces an operator $\hat{d}$: $C^\infty(\nu^*\mathcal{S}/\mathcal{R}) \to \mathcal{Z}(\nu^*\mathcal{S} \otimes T^*M)$ by $d[\alpha] = d\alpha$, where, again, $\mathcal{Z}$ denotes the sheaf of germs of d-closed cross sections.*

Proof. The sections of $\nu^*\mathcal{S}/\mathcal{R}$ are to be taken with respect to the projection π', and $\hat{d}$ is well defined because of Proposition 8.16 ii). Q.e.d.

8.18. Proposition. *Put $\lambda = \iota^*\beta$, where $\iota : \nu^*\mathcal{S} \subseteq T^*M$, and β is the Liouville form of T^*M (i.e., $\beta = \theta_i dx^i$ with x^i coordinates on M, and θ_i covector coordinates e.g.,* [LM], [Va8]) *. Let Ω be the 2-form induced by $d\lambda$ on $(\nu^*\mathcal{S})/\mathcal{R}$. Then, with the notation of* (8.26) *one has*

$$\hat{\varphi}^*_s \sigma = \pi'^* s^* \sigma - \Omega. \tag{8.28}$$

Proof. Ω exists since, if $\alpha \in \mathcal{R}$, its additive action on $\nu^*\mathcal{S}$ sends λ to $\lambda + \pi^*\alpha$, and $d\pi^*\alpha = 0$ by Proposition 8.16 ii). Furthermore, if we look at α as $\alpha : M \to \nu^*\mathcal{S}$, the definition of the Liouville form implies

$$\alpha^*\lambda = \alpha, \quad \alpha^* d\lambda = d\alpha, \quad [\alpha]^*_{\mathcal{R}}\Omega = d\alpha. \tag{8.29}$$

Here, the last relation follows from the second because, if $pr : \nu^*\mathcal{S} \to \nu^*\mathcal{S}/\mathcal{R}$ is the submersion of Theorem 8.15, we have

$$[\alpha]_{\mathcal{R}} = (pr) \circ \alpha, d\lambda = (pr)^*\Omega \qquad ([\alpha]_{\mathcal{R}} : M \to \nu^*\mathcal{S}/\mathcal{R}). \tag{8.30}$$

Now, if we replace the last value of $d\alpha$ of (8.29) in (8.27) we get

$$a_\alpha^*\sigma = \sigma - ([\alpha]_\mathcal{R} \circ r)^*\Omega.$$

If this relation is pulled back to M by a local cross section $s : U \to V$, and if we notice that the local diffeomorphism φ_s used in (8.26) satisfies $\varphi_s \circ \alpha = a_\alpha \circ s$, because of its definition, and that $r \circ s = \mathrm{id.}$, we get

$$(\varphi_s \circ \alpha)^*\sigma = s^*\sigma - [\alpha]_\mathcal{R}^*\Omega \tag{8.31}$$

or, equivalently,

$$(\hat{\varphi}_s \circ [\alpha]_\mathcal{R})^*\sigma = s^*\sigma - [\alpha]_\mathcal{R}^*\Omega \tag{8.32}$$

since the left hand sides of (8.31) and (8.32) are obviously the same.

But, it is clear that $\pi' \circ [\alpha]_\mathcal{R} = \mathrm{id.}$, hence $s^*\sigma = [\alpha]_\mathcal{R}^*\pi'^*s^*\sigma$. Therefore, (8.32) becomes

$$[\alpha]_\mathcal{R}^*\{\hat{\varphi}_s^*\sigma - \pi'^*s^*\sigma + \Omega\} = 0, \tag{8.33}$$

and, since this is true for every α, the requested formula (8.28) must be true. Q.e.d.

8.4 Isotropic realizations and nets

The previous results, and particularly Theorem 8.15, show the importance of the set $\mathcal{R}$, called the *net* of the realization, for the structure of the isotropic realizations. Such a net was first studied by Duistermaat [Dt] in the Lagrangian case. $\mathcal{R}$ leads to other important results as well.

In particular, it is of course interesting to look at $\mathrm{rank}_x\mathcal{R} \overset{\mathrm{def}}{=} \mathrm{rank}\ \mathcal{R}_x$ in the sense of the theory of abelian groups. One has [Dz3], [CS]

8.19. Proposition. *$m(x) = \mathrm{rank}_x\mathcal{R}$ is a lower semi-continuous function, and it is constant along the leaves of $\mathcal{S}(M)$. If $m(x)$ is constant on $M, \mathcal{R}$ is a covering space of M.*

Proof. The rank of $\mathcal{R}_x$ is the maximal number of elements of $\mathcal{R}_x$ that are linearly independent over the integers. If we take such a *basis* of $\mathcal{R}_x$ (i.e., a basis of the free part of $\mathcal{R}_x$), and build the associated s-distinguished cross sections of $\mathcal{R}$, then, because of continuity, we get elements of the fibers of $\mathcal{R}$ that are independent over the integers along a whole neighbourhood of x. Therefore, $m(x)$ can only increase on that neighbourhood i.e., $m(x)$ is lower semicontinuous.

Furthermore, if $m(x) = \mathrm{const.}$, the s-distinguished cross sections of $\mathcal{R}$ yield local trivializations that make $\pi : \mathcal{R} \to M$ into a locally trivial fiber bundle with a discrete fiber i.e., into a covering space of M.

Finally, in order to prove that $m(x)/_S = \mathrm{const.}$ for any leaf S of $\mathcal{S}(M)$, we make the following local considerations. An isotropic realization $r : V \to M$

is of the smallest possible dimension (Corollary 8.11). Hence, it must be locally equivalent (Theorem 8.6) with the realization described in the proof of Theorem 8.2. In other words, since w is a regular Poisson structure, given $x_0 \in M$, $z_0 \in r^{-1}(x_0)$, one has coordinate neighborhoods (U, x^a, y^u) ($x^a=$ const. defines $\mathcal{S}(M)$, $y^u=$ canonical coordinates on the leaves of $\mathcal{S}(M)$) of x_0 in M, and $(\tilde{U}, x^a, y^u, z^a)$ of z_0 in V such that $r(x^a, y^u, z^a) = (x^a, y^u)$, and

$$\sigma = \sum_a dx^a \wedge dz^a + \text{ (a canonical symplectic form in } y).$$

The corresponding local coordinates of $\alpha \in \pi^{-1}(U) \subseteq \nu^*\mathcal{S}$ are (x^a, y^u, α_a), where $\alpha = \alpha_a dx^a$, and, accordingly,

$$Y_\alpha = -\sum_a \alpha_a X^\sigma_{x^a \circ r} = -\sum_a \alpha_a \frac{\partial}{\partial z^a}. \tag{8.34}$$

Now, if we have, locally, $\alpha_a = \alpha_a(x^a)$, the fields Y_α also do not depend on y^u, and the same holds for the corresponding flows. Hence, the action $a_\alpha(x^a, y^u, z^a)$ has equations that are locally independent on the y^u. Therefore, if we start at $x_0 \in M$ with some α_0, and build a local projectable cross section α of $\nu^*\mathcal{S}$ with $\alpha(x_0) = \alpha_0$, this section will either be entirely inside $\mathcal{R}$, or entirely outside $\mathcal{R}$. In particular, if we start with $\alpha_0 \in \mathcal{R}_{x_0}$, the values of α can be seen as images of α_0 by isomorphisms $\mathcal{R}_{x_0} \to \mathcal{R}_x$, and these groups must have the same rank on a small enough neighborhood of x_0. The conclusion $m(x)/_S =$ const. is now obvious. Q.e.d.

But, what we would really like to do is to go in the opposite direction. Namely, according to the previous theory, if we have a regular Poisson manifold (M, w), and if we want to find c.c.i. realizations of it, we should start by fixing a corresponding net as defined by

8.20. Definition. Let $\mathcal{S}(M)$ be the symplectic foliation of (M, w). A subsheaf $\mathcal{R}$ of $\mathcal{Z}(\nu^*\mathcal{S})$ is called a *net* if a) $\forall x \in M$, the stalk $\mathcal{R}_x$ of $\mathcal{R}$ is a discrete subgroup of $\nu^*_x\mathcal{S}$; b) $\mathcal{R}$ is a closed submanifold of $\nu^*\mathcal{S}$.

Then, we may look for c.c.i. realizations such that the given net $\mathcal{R}$ is the net of the realization. The interesting point is that this second step is of a cohomological nature [DD], [Dz3], [CS], and we shall describe the corresponding solution of this problem.

The basic observation is that a c.c.i. realization $r : V \to M$ of net $\mathcal{R}$ has a naturally associated *characteristic class* (*Chern class* in the literature quoted above). Namely, let $\{U_i\}$ be an open covering of M, endowed with local cross sections $s_i : U_i \to V$ of r. Then, there are well defined cross sections

$$\mu_{ij} : U_i \cap U_j \to (\nu^*\mathcal{S})/\mathcal{R}$$

such that, $\forall x \in U_i \cap U_j$ we have

$$s_j(x) = a_{\mu_{ij}}(x)s_i(x). \tag{8.35}$$

Moreover, if we look at the corresponding mappings (8.26), we see from the definition of $\hat{\varphi}_s$ there that, $\forall \alpha \in \nu^*\mathcal{S}$, one has

$$\hat{\varphi}_{s_i}^{-1}(\hat{\varphi}_{s_j}([\alpha]_{\mathcal{R}})) = [\alpha]_{\mathcal{R}} + \mu_{ij}(x) \qquad (x = \pi(\alpha)), \tag{8.36}$$

and it is an easy consequence of this relation that $\mu_{ij} + \mu_{jk} = \mu_{ik}, \forall i, j, k$, i.e., $\{\mu_{ij}\}$ is a 1-cocycle of $\{U_i\}$ with values in the sheaf $\mathcal{C}^\infty((\nu^*\mathcal{S})/\mathcal{R})$ of germs of sections of π', defined in the obvious way. A corresponding cohomology class $\mu \in H^1(M, \mathcal{C}^\infty(\nu^*\mathcal{S}/\mathcal{R}))$ is thereby defined, and, by an argument similar to the one that gave (8.36), it follows that μ does not depend on the choice of the cross sections s_i.

Furthermore, since it is clear that $\mathcal{C}^\infty(\nu^*\mathcal{S}/\mathcal{R})$ is isomorphic to the quotient sheaf $[\mathcal{C}^\infty(\nu^*\mathcal{S})]/\mathcal{R}$, we have the exact sequence of sheaves over M

$$0 \to \mathcal{R} \to \mathcal{C}^\infty(\nu^*\mathcal{S}) \to \mathcal{C}^\infty(\nu^*\mathcal{S}/\mathcal{R}) \to 0, \tag{8.37}$$

and, since $\mathcal{C}^\infty(\nu^*\mathcal{S})$ is a fine sheaf, we get isomorphisms

$$\delta : H^k(M, \mathcal{C}^\infty(\nu^*\mathcal{S}/\mathcal{R})) \to H^{k+1}(M, \mathcal{R}), \tag{8.38}$$

for $k \geq 1$. Using δ, we define

8.21. Definition. The cohomology class $\nu(r) = \delta\mu \in H^2(M, \mathcal{R})$ is called the *characteristic (Chern) class* of the isotropic realization r.

Now, let us remember that a foliated manifold $(M, \mathcal{F})$, where $\mathcal{F}$ is a regular foliation has a *relative de Rham cohomology* with respect to $\mathcal{F}$, defined by

$$H^k_{\mathrm{rel}}(M, \mathcal{F}) = \frac{\ker(d : \Lambda^k_0 M \to \Lambda^{k+1}_0 M)}{\operatorname{im}(d : \Lambda^{k-1}_0 M \to \Lambda^k_0 M)}, \tag{8.39}$$

where $\Lambda^k_0 M$ is the space of those k-forms on M that have a vanishing pull-back to the leaves of $\mathcal{F}$.

This cohomology also has a sheaf-theoretic interpretation as shown by

8.22. Theorem [Va2], [DD]. *Let $\mathcal{Z}(\Lambda^h_0 M)$ be the sheaf of germs of d-closed elements of $\Lambda^h_0 M$. Then*

$$H^k_{\mathrm{rel}}(M, \mathcal{F}) = H^{k-h}(M, \mathcal{Z}(\Lambda^h_0 M)) \qquad (h \geq 1). \tag{8.40}$$

Proof. The sequence of sheaves of germs

$$0 \to \mathcal{Z}(\Lambda_0^h M) \xrightarrow{\subseteq} C^\infty(\Lambda_0^h M) \xrightarrow{d} C^\infty(\Lambda_0^{h+1} M) \xrightarrow{d} C^\infty(\Lambda_0^{h+2} M) \to \dots \tag{8.41}$$

is obviously well defined, and all its terms, except $\mathcal{Z}(\Lambda_0^h M)$ are fine sheaves.

Furthermore, fix a complementary distribution of $T\mathcal{F}$ as $\nu\mathcal{F}$, and use the type decomposition of forms and of d as described in Chapter 4, formula (4.34). Then, we shall have

$$\Lambda_0^k M = \oplus_{i+j=k, i\geq 1} \Lambda^{ij} M.$$

If $\lambda = \lambda_{1,k-1} + \lambda_{2,k-2} + \ldots + \lambda_{k,0} \in \Lambda_0^k M$, and $d\lambda = 0$, the d-Poincaré lemma gives locally

$$\lambda = d(\theta_{0,k-1} + \theta_{1,k-2} + \ldots + \theta_{k-1,0}),$$

whence $d''\theta_{0,k-1} = 0$. Now, the d''-Poincaré lemma yields $\theta_{0,k-1} = d''\alpha_{0,k-2} = d\alpha_{0,k-2} - d'\alpha_{0,k-2} - \partial\alpha_{0,k-2}$, and we get (always locally)

$$\lambda = d(-d'\alpha_{0,k-2} - \partial\alpha_{0,k-2} + \theta_{1,k-2} + \ldots + \theta_{k-1,0}),$$

i.e., the sequence (8.41) is exact.

Hence, (8.41) is a fine resolution of $\mathcal{Z}(\Lambda_0^h M)$, and we are done. Q.e.d.

In particular, if (M, w) is a regular Poisson manifold, there exist 2-forms ω that pull back to the symplectic forms of the leaves of the symplectic foliation $\mathcal{S}(M)$ (the forms ω such that $w = \#_w\omega$). Hence $d\omega$ vanishes on these leaves, and it is a closed element of $\Lambda_0^3 M$. Of course, ω is defined only up to the addition of an arbitrary form in $\Lambda_0^2 M$, but such an additional term will not change the cohomology class defined by $d\omega$ in $H^3_{\mathrm{rel}}(M, \mathcal{S}(M))$. This justifies

8.23. Definition. The cohomology class $\xi(w) \stackrel{\mathrm{def}}{=} [d\omega] \in H^3_{\mathrm{rel}}(M, \mathcal{S}(M))$ will be called the *characteristic form-class* of the Poisson structure w.

The characteristic classes $\nu(r)$ and $\xi(w)$ are related by

8.24. Proposition. *$\xi(w) = \hat{d}_*(\nu(r))$, where ν is seen as its isomorphic image (8.38) in $H^1(M, C^\infty(\nu^*\mathcal{S}/\mathcal{R}))$, ξ is seen as its isomorphic image (8.40) in $H^1(M, \mathcal{Z}(\Lambda_0^2 M))$, and $\hat{d}_*$ is induced in theČech cohomology of the corresponding sheaves by the mapping $\hat{d}$ of Corollary* 8.17.

Proof. If we represent $\nu(r)$ by a cocycle $\{\mu_{ij}\}$, as we did previously, we have from (8.27),

$$a^*_{\mu_{ij}}\sigma - \sigma = -r^* d\mu_{ij}.$$

Then, if we apply here s_i^*, and use (8.35) and $r \circ s_i = \mathrm{id.}$, we get

$$s_j^*\sigma - s_i^*\sigma = d\mu_{ij}. \tag{8.42}$$

Therefore, $\{s_j^*\sigma - s_i^*\sigma\}$ defines a 1-cocycle for the sheaf $\mathcal{C}^\infty(\Lambda_0^2 M)$, and, because this sheaf is fine, the cocycle is a coboundary i.e.,

$$s_j^*\sigma - s_i^*\sigma = \lambda_j - \lambda_i, \text{ where } \lambda_i \in \Lambda_0^2 U_i. \tag{8.43}$$

Accordingly, $\{s_i^*\sigma - \lambda_i\}$ glue up to a 2-form ω whose pull back to the leaves of $\mathcal{S}(M)$ is the symplectic structure of the leaves. Then, $\xi(w)$ is represented in $H^3_{\text{rel}}(M,\mathcal{F})$ by $d(s_i^*\sigma - \lambda_i) = -d\lambda_i$, and if we express the isomorphism (8.40) in agreement with sheaf theory (e.g. [Va2]), the corresponding 1-cocycle with values in $\mathcal{Z}(\Lambda_0^2 M)$ is

$$\lambda_j - \lambda_i = \hat{d}\mu_{ij} \tag{8.44}$$

(because of (8.42), (8.43)). Q.e.d.

In what follows we shall denote by

$$\tilde{d} : H^2(M,\mathcal{R}) \to H^3_{\text{rel}}(M,\mathcal{S}(M)) \tag{8.45}$$

the isomorphism obtained by composing $\hat{d}_*$ of Proposition 8.24 with the isomorphisms (8.38), (8.40), and, from Definition 8.20, it follows that $\hat{d}$ and $\tilde{d}$ exist for abstract nets as well.

In view of Proposition 8.24, in order to find c.c.i. realizations of a fixed regular Poisson manifold (M,w) with a fixed net $\mathcal{R}$, we have to consider the cohomology classes of $\tilde{d}^{-1}(\xi(w))$, and see whether these classes are characteristic classes of realizations. The answer is affirmative since one has [Dt], [DD], [Dz3], [CS]

8.25. Theorem. *If $\nu \in H^2(M,\mathcal{R})$, and $\tilde{d}\nu = \xi(w)$, there exist c.c. isotropic realizations (V,σ) of (M,w) whose net is $\mathcal{R}$, and whose characteristic class is ν.*

Proof. Let $\{U_i\}$ be a covering of M by canonical coordinate neighborhoods in the sense of the Splitting Theorem 2.16 (see Definition 2.18). Then, on U_i, the symplectic form of the leaves of $\mathcal{S}(M)$ has the canonical coordinate expression, which can be seen as a closed 2-form σ_i on U_i. In other words, σ_i is a local closed extension of the symplectic structure of the leaves.

On the other hand, since $\tilde{d}\nu = \xi(w)$, we shall have a representative cocycle of ν in the covering $\{U_i\}$ that consists of cross sections μ_{ij} of $\nu^*\mathcal{S}/\mathcal{R}$ satisfying

$$\sigma_j - \sigma_i = \hat{d}\mu_{ij}. \tag{8.46}$$

Of course, we should notice that, by the conditions imposed on the notion of a net, $\nu^*\mathcal{S}/\mathcal{R}$ is a Hausdorff differentiable manifold, just as in Theorem 8.15.

Now, it is clear that

$$V = \{\cup_i[(\nu^*\mathcal{S})/\mathcal{R}\,|_{U_i}\}/\!\sim\,, \tag{8.47}$$

where two points α_i, α_j over U_i, U_j, respectively, are equivalent ($\alpha_i \sim \alpha_j$) iff $\alpha_j = \alpha_i + \mu_{ij}$, is a Hausdorff differentiable manifold endowed with a natural submersion $r : V \to M$. Moreover, for every U_i one has a corresponding mapping (8.26) for some bundle equivalences $\hat{\varphi}_i$, that are connected by the equation

$$\hat{\varphi}_i \circ \hat{\varphi}_j^{-1}(y) = y + \mu_{ij}(r(y)) \qquad (y \in V/_{U_j}). \tag{8.48}$$

Denote $V_i = r^{-1}(U_i) = V/_{U_i}$. Every V_i is endowed with the symplectic form

$$\Theta_i = r^*\sigma_i - (\varphi_i^{-1})^*\Omega, \tag{8.49}$$

where Ω was defined in Proposition 8.18, and, since

$$\Theta_j - \Theta_i = r^*(\sigma_j - \sigma_i - \hat{d}\mu_{ij}),$$

because of (8.48) and of the fact that Ω is a global form, (8.46) shows that $\Theta_j - \Theta_i = 0$. Hence, V has a global symplectic form Θ given by (8.49), which also shows that r is a Poisson mapping, and it has isotropic fibers. By (8.47) the fibers are also connected and complete, and the net of r is $\mathcal{R}$. Obviously, the characteristic class of r is the given class ν. Q.e.d.

In fact, a more complete result can be found in the literature quoted above. Namely, look at the exact sequence of sheaves

$$0 \to \mathcal{R} \to \mathcal{Z}(\nu^*\mathcal{S}) \to \mathcal{Z}(\nu^*\mathcal{S})/\mathcal{R} \to 0, \tag{8.50}$$

and at the corresponding long exact sequence

$$\begin{aligned} H^1(M,\mathcal{R}) \to H^1(M,\mathcal{Z}(\nu^*\mathcal{S})) \to H^1(M,\mathcal{Z}(\nu^*\mathcal{S})/\mathcal{R}) \to \\ \to H^2(M,\mathcal{R}) \to H^2(\mathcal{Z}(\nu^*\mathcal{S})). \end{aligned} \tag{8.51}$$

Here, we may replace the second term by $H^2_{\rm rel}(M,\mathcal{S}(M))$, and the last term by $H^3_{\rm rel}(M,\mathcal{S}(M))$, because of (8.40), and we get the exact sequence

$$\begin{aligned} H^1(M,\mathcal{R}) \xrightarrow{\tilde{d}} H^2_{\rm rel}(M,\mathcal{S}(M)) \to H^1(M,\mathcal{Z}(\nu^*\mathcal{S})/\mathcal{R}) \to \\ \longrightarrow H^2(M,\mathcal{R}) \xrightarrow{\tilde{d}} H^3_{\rm rel}(M,\mathcal{S}(M)), \end{aligned} \tag{8.52}$$

where the last $\tilde{d}$ is (8.45), and the first one is just a new notation. Then, a more thorough analysis shows that the equivalence classes of the c.c.i. realizations of a given net $\mathcal{R}$ are in a 1-1 correspondence with the elements of the space $H^2_{\rm rel}(M,\mathcal{S}(M))/\tilde{d}H^1(M,\mathcal{R})$ with $\tilde{d}$ of (8.52). We refer the reader to [DD], [Dz3], [CS] for the corresponding proof.

We also refer to [Dz3,5] for other results on isotropic realizations, in particular for the analysis of the non-Hausdorff case.

9 Realizations of Poisson Manifolds by Symplectic Groupoids

9.1 Realizations of Lie-Poisson structures

Realizations of a Poisson manifold by *symplectic groupoids*, if they exist, are a very important tool in both the geometry and the quantization of the Poisson manifolds. This theory can be seen as starting with the papers of Karasev and Maslov [KM1,2], and it was developed by A. Weinstein, and then by P. Dazord, G. Hector and others [We5-10], [CDW], [DH], [MW], [AD1,2], [AC2], etc. The problem of finding such realizations can be seen as a generalization of the famous Lie's third theorem in the theory of the Lie groups, and we shall motivate it precisely from this viewpoint.

Namely, let $\mathcal{G}$ be a Lie algebra, and $\mathcal{G}^*$ its dual space endowed with the Lie-Poisson structure (see Chapter 3). Lie's third theorem tells us that there exist connected Lie groups G such that the given Lie algebra $\mathcal{G}$ is exactly the Lie algebra of G, and the point to be made here is that such a group G leads to an interesting symplectic realization of $\mathcal{G}^*$ by a structure that is well known in symplectic geometry: the cotangent bundle T^*G with the symplectic form $\sigma = d\lambda$, where λ is the Liouville form of T^*G [AM], [LM].

Indeed, there are two left actions of G on G namely, by left translations L_g and by right translations $R_{g^{-1}}$ $(g \in G)$, and we shall denote by Φ_g, Ψ_g the lift of these actions to T^*G defined by $\Phi_g = L^*_{g^{-1}}$, $\Psi_g = R^*_g$. It turns out that Φ and Ψ are Hamiltonian actions that have equivariant momentum maps $J^\Phi, J^\Psi : T^*G \to \mathcal{G}^*$, defined by the natural mappings

$$J^\Phi(\xi) = -R^*_g\xi, J^\Psi(\xi) = L^*_g\xi, \quad (\xi \in T^*_gG). \tag{9.1}$$

(Remember that if e is the unit of G then $\mathcal{G} = T_eG$ and $\mathcal{G}^* = T^*_e$G.)

We shall prove these results as in [LM]. First, we notice that we have a natural identification $T^*G = G \times \mathcal{G}^*$ given by the projections $\pi : T^*G \to G$ (natural), and $p : T^*G \to \mathcal{G}^*$ defined by $p(\xi) = L^*_{\pi(\xi)}\xi = J^\Psi(\xi)$ (the *left trivialization* [LM]). This also induces an identification $T_\xi T^*G = \mathcal{G} \times \mathcal{G}^*$ given by the projections $\pi' = L^{-1}_{g*} \circ \pi_*$, and $p' = p_*$ $(\xi \in T^*_gG)$. Now, the actions Φ, Ψ are seen to be expressible by

$$\pi \circ \Phi_g = L_g \circ \pi, p \circ \Phi_g = p, \tag{9.2}$$

$$\pi \circ \Psi_g = R_{g^{-1}} \circ \pi, p \circ \Psi_g = (\mathrm{Coad}\ g) \circ p, \tag{9.3}$$

and we may compute accordingly the corresponding infinitesimal actions defined as in (7.13), and denoted by $X^\Phi_{T^*G}, X^\Psi_{T^*G}$ $(X \in \mathcal{G})$. Namely, for $\xi \in T^*_gG$

one has

$$\pi'[X^{\Phi}_{T^*G}(\xi)] = L^{-1}_{g*} \left.\frac{d}{dt}\right|_{t=0} [\exp(-tX)(g)] = \tag{9.4}$$
$$= -L_{g^{-1}*}R_{g*}(X) = -(\mathrm{Ad}\ g^{-1})(X),$$

$$p'[X^{\Phi}_{T^*G}(\xi)] = 0, \tag{9.5}$$

$$\pi'[X^{\Psi}_{T^*G}(\xi)] = L^{-1}_{g*} \left.\frac{d}{dt}\right|_{t=0} [g\exp(tX)] = X \tag{9.6}$$

$$p'[X^{\Psi}_{T^*G}(\xi)] = -\left.\frac{d}{dt}\right|_{t=0} [\mathrm{Coad}(\exp tX)L^*_g\xi] = -(\mathrm{coad}\ X)(p(\xi)), \tag{9.7}$$

where $\forall X \in \mathcal{G}$, ad X and coad X are the infinitesimal actions of $\mathcal{G}$ on $\mathcal{G}$ and $\mathcal{G}^*$ associated with Ad and Coad on the same spaces. (Remember the classical formula $(\mathrm{ad}\ X)(Y) = [X, Y]$ e.g., [LM] Appendix 5.)

Now, let λ be the Liouville form, and $\Omega = d\lambda$ be the canonical symplectic form of T^*G. (We have already encountered these forms on a general cotangent bundle T^*M.) Let $\xi \in T^*_gG$, $\Xi, \Xi_1, \Xi_2 \in T_\xi T^*G$, and $\tilde{\Xi}_1, \tilde{\Xi}_2 =$ the values at $\tilde{\xi} \in T^*G$ of the vector fields given by the cross sections of $TT^*G = G \times \mathcal{G} \times \mathcal{G}^*$ over G that have the same (constant) projections on $\mathcal{G}$ and $\mathcal{G}^*$ as Ξ_1, Ξ_2. Then the definition of λ (e.g., [LM]) yields

$$\lambda_\xi(\Xi) = \xi(\pi_*\Xi) = p(\xi)(\pi'(\Xi)), \tag{9.8}$$

$$(d\lambda)_\xi(\Xi_1, \Xi_2) = \Xi_1(\lambda(\tilde{\Xi}_2)) - \Xi_2(\lambda(\tilde{\Xi}_1)) - \lambda_\xi([\tilde{\Xi}_1, \tilde{\Xi}_2]) = \tag{9.9}$$
$$= \Xi_1(p(\tilde{\xi})(\pi'(\tilde{\Xi}_2)) - \Xi_2(p(\tilde{\xi})(\pi'(\tilde{\Xi}_1)) - p(\xi)([\pi'\tilde{\Xi}_1, \pi'\tilde{\Xi}_2]) =$$
$$= p'(\Xi_1)(\pi'(\Xi_2)) - p'(\Xi_2)(\pi'(\Xi_1)) - p(\xi)([\pi'\tilde{\Xi}_1, \pi'\tilde{\Xi}_2]),$$

where the last equality follows since, by the definition of $\tilde{\Xi}_a$, $\pi'(\tilde{\Xi}_a) =$ const. $(a = 1, 2)$.

Furthermore, in order to obtain the desired result, we have to compute $i(X^{\Phi,\Psi}_{T^*G})d\lambda$, for $X \in \mathcal{G}$, and at a point $\xi \in T^*_gG$. Let $Y \in T_\xi T^*G$ be extended by a field $\tilde{Y}$ as we did for Ξ_a above. Then, using (9.4)–(9.9), we get

$$[i(X^{\Phi}_{T*G})d\lambda](Y) = (d\lambda)_\xi(X^{\Phi}_{T*G}, Y) = p'(Y)((\mathrm{Adg}^{-1})(X)) + \tag{9.10}$$
$$+ p(\xi)([(\mathrm{Adg}^{-1})(X), \pi'Y]),$$

$$[i(X^{\Psi}_{T*G})d\lambda](Y) = (d\lambda)_\xi(X^{\Psi}_{T*G}, Y) = -[(\mathrm{coad}\ X)(p(\xi))](\pi'Y) - \tag{9.11}$$
$$- (p'Y)(X) - p(\xi)([X, \pi'Y]) = -(p'(Y))(X).$$

On the other hand, for $X \in \mathcal{G}$, take J^{Φ}_X, J^{Ψ}_X as defined by 7.14 for the mappings (9.1), and compute their derivatives in the direction of the field $\tilde{Y}$ used above, at $\xi \in T^*_g$G. We have

$$J^{\Phi}_X(\xi) = -(R^*_g\xi)(X) = -(R^*_gL^*_{g^{-1}}p(\xi)(X) = -(\mathrm{Coad}\ g)(p(\xi))(X), \tag{9.12}$$

whence

$$(dJ^{\Phi}_X)_\xi(\tilde{Y}) = YJ^{\Phi}_X = \frac{d}{dt}\bigg|_{t=0} \{J^{\Phi}_X[\exp t\tilde{Y}(\xi)]\} = \tag{9.13}$$
$$= - \frac{d}{dt}\bigg|_{t=0} \{\mathrm{Coad}[(\exp t\pi_*\tilde{Y})(g)][(\exp tp'\tilde{Y})(p(\xi))]\}(X) =$$
$$= - \frac{d}{dt}\bigg|_{t=0} \{\mathrm{Coad}[g(\exp t\pi'Y)][p(\xi) + tp'Y]\}(X) =$$
$$= - \mathrm{Coad}\ g([\mathrm{coad}(\pi'Y)(p(\xi))])(X) - [(\mathrm{Coad}\ g)(p'Y)](X) =$$
$$= -p(\xi)([(\mathrm{Ad}\ g^{-1})X, \pi'Y]) - (p'Y)(\mathrm{Ad}\ g^{-1}(X)).$$

Then, similarly,

$$J^{\Psi}_X(\xi) = L^*_g\xi(X) = p(\xi)(X),$$
$$(dJ^{\Psi}_X)_\xi(\tilde{Y}) = YJ^{\Psi}_X = \frac{d}{dt}\bigg|_{t=0} \{J^{\Psi}_X[\exp t\tilde{Y}(\xi)]\} = \tag{9.14}$$
$$= \frac{d}{dt}\bigg|_{t=0} [\exp t(p'\tilde{Y})(p(\xi))](X) = (p'(Y))(X).$$

Now, by a comparison of (9.13), (9.14) with (9.10), (9.11), respectively, we see that $X^{\Phi}_{T*G}, X^{\Psi}_{T*G}$ are precisely the Hamiltonian vector fields of J^{Φ}_X, J^{Ψ}_X hence, J^{Φ}, J^{Ψ} are momentum mappings, and the actions Φ, Ψ are Hamiltonian actions (Proposition 7.25). Moreover, from (9.1) we get by a straightforward computation that $\forall\gamma \in G$

$$J^{\Phi}(\Phi_\gamma\xi) = (\mathrm{Coad}\ \gamma)(J^{\Phi}\xi), \tag{9.15}$$

$$J^{\Psi}(\Psi_\gamma\xi) = (\mathrm{Coad}\ \gamma)(J^{\Psi}_\gamma\xi), \tag{9.16}$$

hence, the momentum mappings J^{Φ}, J^{Ψ} are equivariant.

Therefore, according to Proposition 7.30, J^{Φ} and J^{Ψ}(which, obviously, are surjective submersions) are symplectic realizations of $\mathcal{G}^*$ on the symplectic manifold T^*G, as claimed.

9.2 The Lie groupoid and symplectic structure of $\mathbf{T^*G}$

Now, what is special about the symplectic realization T^*G is that the manifold T^*G has the interesting algebraic structure of a *groupoid*, a structure that we recall as follows:

9.1. Definition. On a set Γ, a partially defined binary operation, denoted multiplicatively by xy or $x.y$, makes Γ into a *groupoid*, if it satisfies the following axioms: a) *if one of the products $x(yz), (xy)z$ exists, so does the other one, and they are equal*; b) $\forall x \in \Gamma$, *well defined elements $l(x), r(x) \in \Gamma$ exist such that $l(x)x = x$, and $xr(x) = x$*; c) *there exists a well defined mapping $i : \Gamma \to \Gamma$ such that $\forall x \in \Gamma$, $x.i(x) = l(x), i(x).x = r(x)$*; d) *$xy$ is defined iff $l(y) = r(x)$.*

The groupoid product is not defined for all the pairs (x, y), unless Γ is a group, and we shall denote by $\Gamma^{(2)}$ the subset of $\Gamma^2 = \Gamma \times \Gamma$ that consists of pairs (x, y) where $x.y$ is defined. $\Gamma^{(2)}$ is characterized by axiom d). Whenever, we write down a product we also imply that it exists. Property a) is associativity. $l(x), r(x)$ are called the *left unit* or *target unit* of x and the *right unit* or *source unit* of x, and $i(x)$ is the *inverse* of x, that will be also denoted by x^{-1}. The names of source and target come from the other possible definition of a groupoid as the set of morphisms of a small category with objects $A, B, \ldots$, where every morphism $x \in \mathrm{Hom}(A, B)$ is an isomorphism, therefore, it has the inverse $x^{-1} \in \mathrm{Hom}(B, A)$, and where $xy = x \circ y$, $l(x) = 1_B$, $r(x) = 1_A$ (e.g., [Va2]).

The simplest example is the so-called *banal groupoid* $B(M)$ associated with an arbitrary set M. Namely, one defines $B(M) = M \times M$, $l(x, y) = (x, x)$, $r(x, y) = (y, y)$, $i(x, y) = (y, x)$, and $(x, y).(y, z) = (x, z)$. Then, the axioms of a groupoid are easily checked. The set of units of $B(M)$ is the diagonal of $M \times M$, and we may identify it with M.

Let us also indicate a few other elementary algebraic facts:

9.2. Proposition. *In any groupoid Γ, one has:*

i) $\forall x \in \Gamma, \quad l(x).l(x) = l(x), \quad r(x).r(x) = r(x);$

ii) $\forall x \in \Gamma, \quad r(l(x)) = l(x), \quad l(r(x)) = r(x),$
$l(l(x)) = l(x), \quad r(r(x)) = r(x);$

iii) $\forall x, y \in \Gamma, \quad l(x.y) = l(x), \quad r(x.y) = r(y);$

iv) $\forall x \in \Gamma, \quad r(x^{-1}) = l(x), \quad l(x^{-1}) = r(x);$

v)
$$x.y_1 = x.y_2 \Rightarrow y_1 = y_2;$$

vi)
$$x_1.y = x_2.y \Rightarrow x_1 = x_2;$$

vii)
$$(x^{-1})^{-1} = x;$$

viii)
$$(xy)^{-1} = y^{-1}x^{-1};$$

ix) the inverse mapping $i : \Gamma \to \Gamma$ is a bijection.

Proof. i) Multiplying $l(x).x = x$ by x^{-1}, and using associativity, we get exactly $l(x).l(x) = l(x)$; the second relation follows similarly.

ii) This is only a different way of reading i).

iii) By axioms b) and a) of Definition 9.1 we have $l(x).x.y = x.y$ and $x.y.r(y) = x.y$ as requested.

iv) If iii) and i) are applied to $l(x) = x.x^{-1}$ we get $rl(x) = l(x) = r(x.x^{-1}) = r(x^{-1})$. Similarly, starting with $r(x) = x^{-1}.x$ we have $lr(x) = r(x) = l(x^{-1}.x) = l(x^{-1})$.

v) Multiplying the first equality by x^{-1}, we get $r(x)y_1 = r(x)y_2$. But, the existence of the equal products also yields $r(x) = l(y_1) = l(y_2)$ (axiom d)). If these values of $r(x)$ are replaced in the previous result, we get exactly $y_1 = y_2$.

vi) has a proof similar to v).

vii) follows from vi) and iv) because of

$$(x^{-1})^{-1}x^{-1} = r(x^{-1}) = l(x) = xx^{-1}.$$

viii) The existence of xy and iv) ensures the existence of $y^{-1}x^{-1}$. Then, straightforward computations show that $y^{-1}x^{-1}$ satisfies the properties of axiom c) for the element xy.

ix) Because of vii), i is a surjection; on the other hand, if $\gamma_1^{-1} = \gamma_2^{-1}$ $(\gamma_1, \gamma_2 \in \Gamma)$ we get

$$l(\gamma_2) = r(\gamma_2^{-1}) = r(\gamma_1^{-1}) = l(\gamma_1) = \gamma_1\gamma_1^{-1} = \gamma_1\gamma_2^{-1}$$

hence, $(\gamma_1\gamma_2^{-1})\gamma_2 = \gamma_2$ i.e., $\gamma_1 = \gamma_2$, and we see that i is also injective. Q.e.d.

9.3. Remarks. 1) Property ii) above shows that $\operatorname{im}(l : \Gamma \to \Gamma) = \operatorname{im}(r : \Gamma \to \Gamma)$, and we shall denote by Γ_0 this common image. It is called the *set of units* of the groupoid Γ, and we also see (from ii), again) that Γ_0 is the fixed point set of the mappings l and r. It is obvious that Γ_0 reduces to one element iff Γ is a group, and $\Gamma_0 = \Gamma$ iff $l = r = \mathrm{id}$. i.e., the only existing products are those of the form $x.x$. In this last case Γ is called the *zero groupoid.*

2) An interesting case of a groupoid appears if we require that $l = r$. Then, $\forall x \in \Gamma_0, l^{-1}(x)$ is a group. In the general case, $\forall x \in \Gamma_0$, we also have an associated group, $\Gamma_x = r^{-1}(x) \cap l^{-1}(x)$, called the *isotropy group* of x. In the categorical interpretation Γ_x is the group of the (iso)morphisms of one object.

3) The categorical interpretation also leads to other interesting notions. Namely, $\forall x \in \Gamma_0, l(r^{-1}(x))$ is the set of targets that can be reached from a given source x. For this reason, it is called the *orbit* of x in Γ. Obviously, this set of targets reachable from x is simultaneously the set of sources for which x can be a target i.e., $r(l^{-1}(x))$. If all the orbits are equal to Γ_0 (i.e., $(r,l) : \Gamma \to \Gamma_0 \times \Gamma_0$ is a surjection) Γ is called a *transitive groupoid.*

Now, since we are interested in differentiable manifolds, it is natural to consider the following definition (where we follow the conventions of [CDW])

9.4. Definition. A groupoid Γ which is also a differentiable manifold (possibly non-Hausdorff) is called a *Lie groupoid* if the following properties are satisfied:

1) Γ_0 *is a Hausdorff submanifold of* Γ *called the unit submanifold;*

2) l *and* r *are differentiable submersions;*

3) *the multiplication is a differentiable mapping of* $\Gamma^{(2)}$ *into* Γ (*notice that* $\Gamma^{(2)} = (r \times l)^{-1}$ (diagonal of $\Gamma_0 \times \Gamma_0$), *and by* 2) *above* $r \times l : \Gamma \times \Gamma \to \Gamma_0 \times \Gamma_0$ *is transversal to this diagonal hence,* $\Gamma^{(2)}$ *is a submanifold of* $\Gamma \times \Gamma$, *and condition* 3) *makes sense*);

4) i (*i.e., the mapping* $x \mapsto x^{-1}$) *is a diffeomorphism of* Γ *onto itself.*

For instance, if M is a differentiable manifold, the banal groupoid $B(M)$ is a Lie groupoid. Moreover, this groupoid structure lifts obviously to the *fundamental groupoid* i.e., the groupoid of homotopy classes with fixed ends of paths in M, $\Pi(M) \approx (\tilde{M} \times \tilde{M})/\pi_1(M)$ (here, M is connected, $\tilde{M}$ is its universal covering manifold, and $\pi_1(M)$ is the Poincaré group of M), by means of the covering map $p : \Pi(M) \to M \times M$ that sends the homotopy class of a path to its ends [Sp]. Therefore, $\Pi(M)$ is also a Lie groupoid.

We also indicate two other basic examples of Lie groupoids, while leaving the details to the reader.

a) Let $\pi : E \to M$ be a G-principal differentiable fiber bundle (G is a Lie group), and put $\mathcal{E} = (E \times E)/\sim$, where $(z_1, z_2) \sim (z'_1, z'_2)$ iff $z'_1 = z_1 g$, $z'_2 = z_2 g$ for the right translation by some $g \in G$. Then, if we define

$$\begin{aligned} l([z_1, z_2]) &= [z_1, z_1], \\ r([z_1, z_2]) &= [z_2, z_2], \\ i([(z_1, z_2]) &= [z_2, z_1], \\ [z_1, z_2].[z_2, z_3] &= [z_1, z_3], \end{aligned}$$

we get a Lie groupoid structure on $\mathcal{E}$, called the *gauge groupoid* E. The unit manifold is the quotient of the diagonal of $E \times E$.

b) Let Γ be a Lie groupoid, and M a manifold endowed with a submersion $\varphi : M \to \Gamma_0 =$ the unit manifold of Γ. Put $H = \{(g, x) \in \Gamma \times M / r(g) = \varphi(x)\}$. A *left action* α of Γ on M is $\alpha : H \to M$ with the properties

$$\varphi(\alpha(g, x)) = l(g); \quad \alpha(\varphi(x), x) = x; \quad \alpha(g_1, \alpha(g_2, x)) = \alpha(g_1.g_2, x),$$

if $g_1.g_2$ exists. Then, H becomes a Lie groupoid with the unit manifold equivalent to M if we put

$$r(g, x) = (r(g), x), l(g, x) = (l(g), \alpha(g, x)), i(g, x) = (g^{-1}, \alpha(g, x)),$$
$$(g_1, x_1)(g_2, x_2) = (g_1.g_2, x_2),$$

if $g_1.g_2$ exists, and $x_1 = \alpha(g_2, x_2)$. This is the *action groupoid* of Γ on M. The identification of M to the unit manifold of this groupoid is by $x \leftrightarrow (\varphi(x), x)$. The categorical interpretation is obtained by looking at the points of M as objects, and at the action of α as defining morphisms $(g, x) \in \mathrm{Hom}(x, \alpha(g, x))$. The simplest case is that of a usual action of a Lie group on a manifold.

Now, we can prove

9.5. Proposition. *For any Lie group G, T^*G has a natural structure of a Lie groupoid with the unit submanifold $\mathcal{G}^*$, the source projection $r = J^\Psi$, and the target projection $l = -J^\Phi$. Furthermore, the graph of the multiplication $T^{*(2)}G \to T^*G$ is a Lagrangian submanifold of $T^*G \times T^*G \times T^*G$, endowed with the symplectic structure $d\lambda \oplus d\lambda \oplus (-d\lambda)$, where λ is the Liouville form.*

Proof. The differential of the multiplication of G defines an operation on the tangent bundle TG given by

$$X \vee Y = R_{g_2^*} X + L_{g_1^*} Y \qquad (X \in T_{g_1}G, Y \in T_{g_2}G), \tag{9.17}$$

and it is very easy to see that $\vee : T_{g_1}G \times T_{g_2}G \to T_{g_1 g_2}G$ is an epimorphism with $\ker \vee = \{(X, Y)/R_{g_2^*}X = -L_{g_1^*}Y\}$.

Accordingly, the formula

$$(\xi_1 \vee \xi_2)(X \vee Y) = \xi_1(X) + \xi_2(Y) \tag{9.18}$$

yields a well defined operation between $\xi_a \in T^*_{g_a}G$ $(a = 1, 2)$ with the result in $T^*_{g_1 g_2}G$ iff (9.18) vanishes on $\ker \vee$.

But $(X, Y) \in \ker \vee$ is equivalent to $X = -R_{g_2^{-1}*} L_{g_1 *} Y$, and if we insert this in (9.18) we get

$$L^*_{g_1} R^*_{g_2^{-1}} \xi_1 = \xi_2. \tag{9.19}$$

Since left and right translations commute, and in view of (9.1), (9.19) means precisely

$$J^\Psi(\xi_1) = -J^\Phi(\xi_2). \tag{9.20}$$

This result shows that, if $\vee$ satisfies axioms a), b), c) of Definition 9.1, it will also satisfy axiom d) for r and l indicated in Proposition 9.5.

Furthermore, using (9.17), (9.18) and (9.19) it is easy to check that if $\xi_1 \vee \xi_2$ exists it is given exactly by

$$\xi_1 \vee \xi_2 = \frac{1}{2}\left(R^*_{g_2^{-1}} \xi_1 + L^*_{g_1^{-1}} \xi_2\right), \tag{9.21}$$

and this formula allows for a straightforward verification of axioms a) and b). It also allows us to see that

$$\xi^{-1} = L^*_g R^*_g \xi \quad (\xi \in T^*_g G) \tag{9.22}$$

defines the inverse mapping needed for axiom c).

Hence, the groupoid structure required exists and, moreover, since (9.21) and (9.22) define differentiable operations, and the mappings (9.1) are submersions, we clearly have here the case of a Lie groupoid with the unit submanifold $\mathcal{G}^* = T^*_e G$ ($e =$ the unit of G).

For the last assertion of Proposition 9.5, let us recall a classical result of symplectic geometry (e.g., [Va8]) namely, that for every submanifold N of a differentiable manifold M, the conormal bundle $\nu^* N = \mathrm{Ann(TN)} \subseteq T^*M/_N$ is a Lagrangian submanifold of $(T^*M, d\lambda)$, where λ is the Liouville form. Indeed, using local coordinates (x^α, x^σ) of M such that N is $x^\alpha = 0$, and denoting by (ξ_α, ξ_σ) the corresponding covector coordinates, we see that $\nu^* N$ has the local equations $x^\alpha = 0$, $\xi_\sigma = 0$, hence $\dim \nu^* N = \dim M = (1/2)(\dim T^* M)$, and $d\lambda = d\xi_\alpha \wedge dx^\alpha + d\xi_\sigma \wedge dx^\sigma$ vanishes on $\nu^* N$.

In our case, let $\mathfrak{G}$ be the submanifold of $G \times G \times G$ defined by the equation $g_3 = g_1 g_2$. Then, (9.18) shows that the corresponding conormal bundle is $\nu^* \mathfrak{G} = \{(\xi_1, \xi_2, -\xi_1 \vee \xi_2)\}$. Therefore, this is a Lagrangian submanifold of $(T^*G \times T^*G \times T^*G, d\lambda \oplus d\lambda \oplus d\lambda)$. But, the graph of the multiplication mapping $T^*G^{(2)} \to T^*G$ differs from $\nu^*\mathfrak{G}$ only by the sign of the third component, and, since this change of sign is a symplectomorphism of $(T^*G, d\lambda)$ to $(T^*G, -d\lambda)$ we are done. Q.e.d.

The last property in Proposition 9.5 is to be understood as a condition of compatibility between the multiplication and the symplectic structure, which is modeled on the well known fact that a mapping $\varphi : M \to N$ of two symplectic manifolds is a symplectomorphism iff Graph φ is a Lagrangian submanifold of $M \times (-N)$.

9.6. Remark. Proposition 9.5 extends to the case of the cotangent bundle $T^*\Gamma$ of a Lie groupoid Γ, with the multiplication $\mu : \Gamma^{(2)} \to \Gamma$, and the unit manifold Γ_0, instead of a Lie group G. Namely $T^*\Gamma$ is a Lie groupoid with a multiplication whose graph is diffeomorphic to the conormal bundle of the graph of μ in a way that makes it be a Lagrangian submanifold of $T^*\Gamma \times T^*\Gamma \times (-T^*\Gamma)$ ($-T^*\Gamma$ is $T^*\Gamma$ with the opposite of the Liouville form), and whose unit manifold is the conormal bundle of Γ_0 in $T^*\Gamma$.

This can be seen along the same lines as in the proof of Proposition 9.5. The main technical difficulty is that L_g, R_g are not always defined in a groupoid, but one can overcome it. Indeed, if $\xi \in T^*_g\Gamma$, $g \in \Gamma$, and if $\gamma \in \Gamma$ is such that $\gamma^{-1}g$ exists (i.e., $r(\gamma^{-1}) = l(\gamma) = l(g)$, we may choose a local differentiable isomorphism

$$T_{\gamma^{-1}g}\Gamma \approx T_{l(\gamma^{-1}g)}\Gamma_0 + \ker l_*(\gamma^{-1}g) \tag{9.23}$$

(since l is a submersion), and then define $L^*_\gamma \xi \in T^*_{\gamma^{-1}g}\Gamma$ to be zero on $T_{l(\gamma^{-1}g)}\Gamma_0$ and to be given on

$$X = \frac{d}{dt}\bigg|_{t=0} \sigma(t) \qquad (\sigma(0)) = \gamma^{-1}g, l(\sigma(t)) = l(\gamma^{-1}g) = \text{ const.})$$

by $(L^*_\gamma\xi)(X) = \xi(L_{\gamma*}X)$, where

$$L_{\gamma*}X \stackrel{\text{def}}{=} \left.\frac{d}{dt}\right|_{t=0} (\gamma\sigma(t))$$

($\gamma\sigma(t)$ exists since $l(\sigma(t)) = l(\gamma^{-1}g) = l(\gamma^{-1}) = r(\gamma)$, in view of Proposition 9.2). $R^*_\gamma\xi$ will be obtained similarly.

The definition of $L^*_\gamma\xi$ and $R^*_\gamma\xi$ depends, generally, on the choice of the decomposition (9.23). But, we have a natural choice at the points of Γ_0, and this allows us to define $J^\Phi, J^\Psi : T^*\Gamma \to \text{Ann}\,(T\Gamma_0)$ as in the case of a group, and by applying the previous construction for $\gamma = g$. Then, we can use some fixed choices of (9.23) to define an operation similar to (9.21), and this operation will be invariant because its graph is invariant namely, it is Ann T(Graph μ) with the sign of the third component changed.

Therefore, we can extend Proposition 9.5 to Lie groupoids, as claimed. (See [CDW] for a detailed proof.)

9.3 General symplectic groupoids

The geometric situation that exists in Proposition 9.5 and in Remark 9.6 justifies

9.7. Definition. Let Γ be a Lie groupoid. A *symplectic groupoid structure* on Γ is a symplectic 2-form σ on the manifold Γ such that the graph γ_m of the multiplication m of Γ is a Lagrangian submanifold of $\Gamma \times \Gamma \times (-\Gamma)$ ($-\Gamma$ is Γ endowed with $-\sigma$). Then, the pair (Γ, σ) is called a *symplectic groupoid.*

It is worthwhile to notice that the notion of a symplectic groupoid can be naturally extended to that of a *Poisson groupoid* by asking only for a Poisson structure on Γ, with the condition that the graph γ_m is coisotropic in $\Gamma \times \Gamma \times (-\Gamma)$. This notion was studied in [We6].

Usually we shall just talk of the symplectic groupoid Γ instead of (Γ, σ), and we can immediately give the following examples:

a) The example of Remark 9.6 i.e., $T^*\Gamma$ where Γ is any Lie groupoid.

b) For any differentiable manifold M, the cotangent bundle T^*M has the natural symplectic structure $d\lambda$, where λ is the Liouville form, and it has the structure of a Lie groupoid with the multiplication m equal to the addition of covectors along the fibers. Clearly, the graph γ_m consists of triples $(\xi, \eta, \xi + \eta) \in T^*M \times T^*M \times T^*M$, and the conormal bundle of the diagonal $\Delta = \{(x, x, x)/x \in M)$ consists of triples $(\xi, \eta, -(\xi + \eta))$. Since the latter is a Lagrangian submanifold of $(T^*M)^3$ with its Liouville form, γ_m is Lagrangian for the form $d\lambda \oplus d\lambda \oplus (-d\lambda)$, and we are in the case of a symplectic groupoid.

Accordingly, notice that the cotangent bundle $T^*\Gamma$ of a Lie groupoid Γ has two different structures of a symplectic groupoid. See [CDW] for details about the relations that exist between these two structures, and a corresponding notion of a *double symplectic groupoid.*

c) Let (M,σ) be a symplectic manifold, and let $-M$ be $(M,-\sigma)$. Let $\Pi(M)$ be the fundamental groupoid of M, with the covering map $p:\Pi(M)\to B(M)=(-M)\times M=$ the banal groupoid. We know that $\Pi(M)$ and $B(M)$ are Lie groupoids. Now, $\tilde{\sigma}=p^*[(-\sigma)\oplus\sigma]$ is a symplectic form on $\Pi(M)$, and p is a *homomorphism of groupoids* (i.e., it preserves multiplication), and a local symplectic equivalence. Hence, the graphs of the multiplications in $\Pi(M)$ and $B(M)$ are Lagrangian submanifolds, simultaneously. Therefore, $(\Pi(M),\tilde{\sigma})$ is a symplectic groupoid.

The notion of a symplectic groupoid has been derived from a certain symplectic realization. In what follows we shall see that this is always the case, namely, we shall see that the unit manifold of a symplectic groupoid is a Poisson manifold, and the groupoid is a symplectic realization of the latter.

9.8. Proposition. *Let (Γ,σ) be a symplectic groupoid. Then its unit manifold Γ_0 is a Lagrangian submanifold of Γ and the inverse mapping $i(\xi)=\xi^{-1}$ $(\xi\in\Gamma)$ is an antisymplectomorphism (i.e., it sends σ to $-\sigma$).*

Proof. In order to relate Γ_0 to the graph γ_m of the multiplication of Γ, we shall notice that

$$\Gamma_0=\{x\in\Gamma/\exists u\in\Gamma, xu=u\} \tag{9.24}$$

(i.e., x is the left unit of some $u\in\Gamma$). Equivalently, we can write

$$\Gamma_0=pr_1(\gamma_m\cap\Delta(\Gamma\times\Gamma)), \tag{9.25}$$

where $\Delta:\Gamma\times\Gamma\to\Gamma\times(\Gamma\times\Gamma)$ is defined by $\Delta(x,u)=(x,u,u)$, and $pr_1:\Gamma\times\Gamma\times\Gamma\to\Gamma$ is the projection on the first factor.

Now, a tangent vector of $\Delta(\Gamma\times\Gamma)$ is obviously of the form (X,Y,Y), where X,Y are tangent vectors of Γ, and (A,B,C) is symplectic orthogonal to it in $\Gamma\times\Gamma\times(-\Gamma)$ iff $A=0$, $B=C$. This shows that $\Delta(\Gamma\times\Gamma)$ is a coisotropic submanifold, and pr_1 is the projection along its characteristic foliation, given by the symplectic orthogonal spaces (Chapter 7).

It is a well known result of symplectic reduction theory (e.g., [Va8] Proposition 2.2.11 and Theorem 3.3.6) that such a projection sends the intersection of $T\Delta(\Gamma\times\Gamma)$ with Lagrangian subspaces, particularly $T\gamma_m$, to Lagrangian subspaces of the quotient space, $T\Gamma$ in our case. Moreover, the mappings $u\mapsto(l(u),u)\mapsto(l(u),u,u))$ embed Γ onto $\gamma_m\cap\Delta(\Gamma\times\Gamma)$ in $\Gamma\times\Gamma\times\Gamma$ hence, the intersection is a submanifold with tangent spaces equal to the image of $T\Gamma$ by the embeddings. This image is $\{(X,Y,Y)/X,Y\in T\Gamma\}$ where, if Y is tangent to a path $\sigma(t)$, X is tangent to $l(\sigma(t))$ i.e., we have exactly $T(\gamma_m)\cap T(\Delta(\Gamma\times\Gamma))$. Hence $\gamma_m\cap\Delta(\Gamma\times\Gamma)$ is a clean intersection, and $T\Gamma_0$ is exactly the projection of $T(\gamma_m)\cap T(\Delta(\Gamma\times\Gamma))$ by pr_1.

It follows that $T\Gamma_0$ are Lagrangian subspaces of $T\Gamma$, which is the first assertion of Proposition 9.8.

The proof of the second part is similar. Namely, it is clear that i is an antisymplectomorphism iff Graph i is a Lagrangian submanifold of $\Gamma \times \Gamma$ with the form $\sigma \oplus \sigma$. But,

$$\begin{aligned} \text{Graph } i &= \{(x,y)/x,y \in \Gamma \text{ and } xy = l(x)\} = \\ &= \tilde{l}^{-1}[\gamma_m \cap (\Gamma \times \Gamma \times \Gamma_0)], \end{aligned} \tag{9.26}$$

where $\tilde{l} : \Gamma \times \Gamma \to \Gamma \times \Gamma \times \Gamma$ is defined by

$$\tilde{l}(x,y) = (x,y,l(x)).$$

Here again, $\Gamma \times \Gamma \times \Gamma_0$ is coisotropic in $\Gamma \times \Gamma \times (-\Gamma)$ since Γ_0 is Lagrangian in Γ by the first part of the present proof. Then, γ_m is Lagrangian, and the intersection that appears in (9.26) is clean. Since $\tilde{l}^{-1}$ is the projection on $\Gamma \times \Gamma$ we have by the results of symplectic reduction theory mentioned above that Graph i is Lagrangian in $\Gamma \times \Gamma$. Q.e.d.

Now, we can obtain

9.9. Theorem. *For any symplectic groupoid (Γ, σ), the unit manifold Γ_0 has a well defined Poisson structure w coinduced by σ via the source projection r, and by $-\sigma$ via the target projection l, and (Γ, σ) is a symplectic realization of (Γ_0, w). The symplectic leaves of w are the connected components of the orbits of the groupoid Γ.*

Proof. We shall prove that the fibers of the projection $l : \Gamma \to \Gamma_0$ constitute the σ-orthogonal foliation of the fibers of $r : \Gamma \to \Gamma_0$. Since l, r are submersions, and $\dim \Gamma_0 = (1/2)\dim \Gamma$, by Proposition 9.8, we have dim(fiber l) + dim(fiber r) = $(1/2)\dim\Gamma + (1/2)\dim\Gamma = \dim\Gamma$, and it will be enough to show that $\forall z \in \Gamma$, $T_z l^{-1}(l(z))$ and $T_z(r^{-1}(r(z)))$ are σ-orthogonal subspaces.

If $X \in T_z l^{-1}(l(z))$, we have

$$X = \left.\frac{d}{dt}\right|_{t=0} \tau(t)$$

where $\tau(0) = z, l(\tau(t)) = l(z)$ for all t. Then, $(l(z), \tau(t), l(z)\tau(t) = \tau(t))$ is a path in the multiplication graph γ_m with the tangent vector $(0, X, X) \in T_{(l(z),z,z)}\gamma_m$. Similarly, $\forall Y \in T_z r^{-1}(r(z))$, we have $(Y, 0, Y) \in T_{(z,r(z),z)}\gamma_m$. If $z \in \Gamma_0$, we have $l(z) = r(z) = z$, and we have two vectors in a tangent space of a Lagrangian submanifold of $\Gamma \times \Gamma \times (-\Gamma)$ hence,

$$\sigma(0,Y) + \sigma(X,0) - \sigma(X,Y) = -\sigma(X,Y) = 0,$$

as required.

If $z \notin \Gamma_0$, we need an instrument that allows us to transfer everything to a point of Γ_0, and this is given by the notion of *a bi-cross section* of a groupoid. We shall study it before going on with the proof of Theorem 9.9.

9.10. Definition. A subset Σ of the groupoid Γ is a *(local) bi-cross section* if it is the common image of two (local) cross sections s_l, s_r of the surjections l, r respectively. ($s_l(\Gamma_0) = s_r(\Gamma_0)$ but, not pointwise!)

Clearly, if Σ is given $s_l = (l/_\Sigma)^{-1}$, $s_r = (r/_\Sigma)^{-1}$, and these mappings are well defined. The bi-cross sections are important because they act, on the left and on the right on Γ (or a part of it, if they are local) by means of the formulas

$$\Sigma_l(z) = s_r(l(z)).z, \quad \Sigma_r(z) = z.s_l(r(z)) \quad (z \in \Gamma). \tag{9.27}$$

In the case of a Lie groupoid we shall use differentiable bi-cross sections i.e., where s_l and s_r are differentiable mappings. For a symplectic groupoid the important case is that where Σ is a Lagrangian submanifold of (Γ, σ), and, then, we shall say that Σ is a *Lagrangian bi-cross section.*

Through every point $z \in \Gamma$ there exists a local Lagrangian bi-cross section. Indeed, we may identify a neighbourhood of z with $T_z\Gamma$ by means of a Darboux chart. Then the requested section will be defined by a Lagrangian plane of $T_z\Gamma$ that is transversal to both the l-fiber and the r-fiber at z, and such a plane can be obtained by a slight perturbation (if necessary) of an arbitrary Lagrangian plane of $T_z\Gamma$. In the sequel, we shall denote by $\mathcal{L}(\Gamma)$ the set of the local Lagrangian bi-cross sections of Γ.

9.11. Lemma. *If $\Sigma \in \mathcal{L}(\Gamma)$, the mappings Σ_l, Σ_r of* (9.27) *preserve the symplectic form σ of Γ.*

Proof. It follows from the formulas (9.27) that we can look at a couple $(x, \Sigma_l(x)) \in \text{Graph } \Sigma_l \subseteq \Gamma \times (-\Gamma)$ as being the projection of a triple $(u, x, \Sigma_l(x)) \in \gamma_m \cap (\Gamma \times \Gamma \times (-\Gamma))$ where $u = s_r(l(x)) \in \Sigma$. Conversely, if $(u, x, v) \in \gamma_m \cap (\Sigma \times \Gamma \times (-\Gamma))$ we have $ux = v$, and $u = s_r(l(y))$ for some $y \in \Gamma$. Hence, $l(x) = r(s_r(l(y))) = l(y)$, and $(x, v) \in \text{Graph } \Sigma_l$. Therefore, Graph Σ_l is the projection of $\gamma_m \cap (\Sigma \times \Gamma \times (-\Gamma))$ onto $\Gamma \times (-\Gamma)$. Since γ_m is Lagrangian and $\Sigma \times \Gamma \times (-\Gamma)$ is coisotropic, because Σ is Lagrangian, we conclude, as in the proof of Proposition 9.8, that Graph Σ_l is Lagrangian in $\Gamma \times (-\Gamma)$ hence, Σ_l is a symplectomorphism. The same result for Σ_r follows similarly using triples $(x, u, \Sigma_r(x))$ with $u = s_l(r(x)) \in \Sigma$. Q.e.d.

9.12. Lemma. *$\mathcal{L}(\Gamma)$ acts transitively by Σ_l on the fibers of r, and by Σ_r on the fibers of l.*

Proof. That Σ_l preserves the fibers of r is clear from (9.27) and Proposition 9.2. iii). Now, let $u \in \Gamma_0$ and $x, y \in r^{-1}(u)$. Since also $u = l(y^{-1})$, $z = xy^{-1}$ exists. Now, let Σ be a local Lagrangian bi-cross section through z. Then the equality $x = zy$ may be interpreted as $x = \Sigma_l(y)$, since $r(z) = r(y^{-1}) = l(y)$. This shows the transitivity of $\mathcal{L}(\Gamma)$ on the fibers of r. The result for the fibers of l and the right action of $\mathcal{L}(\Gamma)$ is similar. Q.e.d.

Now, we may come back to the proof of Theorem 9.9. Remember that we had to prove that the r-fibers and the l-fibers are symplectically orthogonal in (Γ, σ), and that we proved that this result holds at the points of the unit submanifold Γ_0. Take $z \in \Gamma$, and $\Sigma \in \mathcal{L}(\Gamma)$ such that $r(z) = \Sigma_l(z)$. Σ exists because of Lemma 9.12, and Σ_l sends the tangent space to $r^{-1}(r(z))$ at z to the tangent space of the same fiber at $r(z)$, and the tangent space to $l^{-1}(l(z))$ at z to that of $l^{-1}(r(z))$ at $r(z)$. This latter assertion follows since (9.27) shows that Σ_l sends l-fibers to l-fibers (use Proposition 9.2 iii)). Since the spaces obtained at $r(z)$ are σ-orthogonal, and Σ_l is a symplectomorphism (Lemma 9.11), the fibers are σ-orthogonal at z as well.

Now, the existence of the r-coinduced Poisson structure w follows from P. Libermann's symplectic version of Proposition 7.18 (see also Lemma 7.19 and Remark 7.20). It also follows that the symplectic foliation of w is tangent to the r-projection of the tangent spaces of the l-fibers hence, the symplectic leaves are the connected components of $r(l^{-1}(x)) = l(r^{-1}(x)) =$ the orbit of x in Γ_0.

For the same reason, Γ_0 also has an l-coinduced Poisson structure w'. But, since $l = r \circ i$ (Proposition 9.2 iv)), and i is an antisymplectomorphism (Proposition 9.8) we have

$$\{\varphi, \psi\}_{w'} \circ l = \{\varphi \circ l, \psi \circ l\}_\sigma = \{\varphi \circ r \circ i, \psi \circ r \circ i\}_\sigma =$$
$$= -\{\varphi \circ r, \psi \circ r\}_\sigma = -\{\varphi, \psi\}_w \circ r,$$

which means that $w' = -w$.

Thereby, all the assertions of Theorem 9.9 were proven. Q.e.d.

9.13. Remark [CDW]. The bi-cross sections of a groupoid Γ are very interesting objects. From the algebraic viewpoint, if $A, B \in \mathcal{P}(\Gamma) =$ the set of the subsets of Γ, one can define a product $A.B = \{\text{x.y} \ /(x, y) \in A \times B$ and $l(y) = r(x)\}$, and it makes $\mathcal{P}(\Gamma)$ into a semigroup with unit element Γ_0, and it is easy to check that the set $G(\Gamma)$ of all the global bi-cross sections is exactly the group of the invertible elements of this semigroup. Moreover, $G(\Gamma)$ acts effectively on Γ by (9.27). In the case of a symplectic groupoid one can consider a group $L(\Gamma)$ of global Lagrangian bi-cross sections of Γ.

9.4 Lie algebroids and the integrability of Poisson manifolds

Theorem 9.9 motivates the search for symplectic realizations of a given Poisson manifold by symplectic groupoids for which the given manifold is the unit manifold, and the fact that Lie's third theorem ensures the existence of such realizations for the Lie-Poisson structures of duals $\mathcal{G}^*$ of Lie algebras shows that the problem mentioned above can be seen as that of generalizing Lie's third theorem. But, as a matter of fact, this interpretation has an even stronger motivation that comes from the fact that Lie groupoids have associated *Lie algebroids*, instead of Lie algebras, and also, Poisson manifolds have associated Lie algebroids.

9.14. Definition [Pr1]. A *Lie algebroid structure* on a differentiable vector bundle $E \to M$ is a pair that consists of a Lie algebra structure $\{\ ,\ \}$ on the space $\Gamma(E)$ of the global cross sections of E, and a vector bundle morphism $\rho : E \to TM$ (the *anchor map* [Mz]) such that: i) the induced map $\rho : \Gamma(E) \to \Gamma(TM)$ is a Lie algebra homomorphism of $\{\ ,\ \}$ in $[\ ,\ ]$, the Lie bracket of vector fields; ii) $\forall f \in C^\infty(M)$, and $\forall \sigma, \tau \in \Gamma(E)$ one has

$$\{\sigma, f\tau\} = f\{\sigma, \tau\} + (\rho(\sigma)(f))\tau. \tag{9.28}$$

A triple $(E, \{\ ,\ \}, \rho)$ is called a *Lie algebroid over* M. Furthermore, if $\forall x \in M$, $\forall v \in T_xM$, $\exists \sigma \in \Gamma(E)$ such that $v = \rho(\sigma)(x)$, we say that we have a *transitive* Lie algebroid.

If $\mathcal{G}$ is a Lie algebra, then ($\mathcal{G} \to$ point, $\rho = 0$) is a Lie algebroid, but the reason for studying Lie algebroids is more profound, of course, and it consists in the fact that this notion leads to generalizations of Lie's third theorem [Pr1,2], [Mz], etc. First, we have

9.15. Theorem. *Every Lie groupoid* Γ *has two associated antiisomorphic Lie algebroid structures canonically defined on the normal bundle* $\nu\Gamma_0$ *of its unit manifold* Γ_0.

Proof. As in the case of Lie groups, the idea is to use left and right invariant vector fields, but we must take into account that, now, the left and right translations are not always defined (see Remark 9.6). Namely $\forall x \in \Gamma$, $L_x(y) = xy$ is defined only on the fiber $l^{-1}(r(x))$ and sends it diffeomorphically onto $l^{-1}(l(x))$, while $R_x(y) = yx$ is defined on $r^{-1}(l(x))$ and sends it onto $r^{-1}(r(x))$.

Accordingly, a vector field X on Γ will be *left-invariant* if it is tangent to the l-fibers, and invariant by all possible L_x, and we shall denote the set of these vector fields by $\mathfrak{L}(\Gamma)$. And, X will be *right-invariant* if it is tangent to the r-fibers and invariant by all possible R_x; the corresponding set is $\mathfrak{R}(\Gamma)$. If $X, Y \in \mathfrak{L}(\Gamma)$, the bracket $[X, Y]$ computes along the l-fibers hence, this bracket operation is compatible with the differential of the left translations, and $[X, Y] \in \mathfrak{L}(\Gamma)$. Therefore, $\mathfrak{L}(\Gamma)$ is a Lie algebra, and so is $\mathfrak{R}(\Gamma)$ too. Moreover, if $X \in \mathfrak{L}(\Gamma)$ and $Y \in \mathfrak{R}(\Gamma)$ we must have $[X, Y] = 0$ because, as in the case of a Lie group, the flows of X, Y consist of right and left translations, respectively, and these flows commute.

Now, consider the vector bundle $E_r = (\ker l_*)/_{\Gamma_0}$ whose fibers are the tangent spaces of the l-fibers at the points of Γ_0. It is clear that $\mathfrak{L}(\Gamma)$ is isomorphic as a vector space with the space of the cross sections of $E_r, \Gamma(E_r)$. Hence, the bracket in $\mathfrak{L}(\Gamma)$ makes $\Gamma(E_r)$ into a Lie algebra whose operation may be denoted by $\{\ ,\ \}$. Moreover, since the flow of a left invariant vector field consists of right translations, it is easy to see that the fields $X \in \mathfrak{L}(\Gamma)$ are r-projectable. Hence the differential $r_*/_{\Gamma(E_r)}$ is a Lie algebra homomorphism into $\Gamma(T\Gamma_0)$, and it can play the role of an anchor map. (9.28) for $f \in C^\infty(\Gamma_0)$, and $\sigma, \tau \in \Gamma(E_r)$ is rather obvious. Therefore, the general conclusion is that $(E_r, \{\ ,\ \}, r_*)$ is a Lie algebroid over Γ_0.

Since l is a submersion, the l-fibers are transversal to Γ_0, and E_r is isomorphic to the normal bundle $\nu\Gamma_0 = (T\Gamma/_{\Gamma_0})/T\Gamma_0$. The bracket $\{\ ,\ \}$ and the projection r_* may, therefore, be transferred to a bracket $\{\ ,\ \}$ and an anchor map ρ of $\nu\Gamma_0$, and we may see the latter as the Lie algebroid, required by Theorem 9.15.

In order to end the proof of the theorem, it remains to notice that the vector bundle $E_l = \ker r_*/_{\Gamma_0}$ of the tangent spaces to the r-fibers at the points of Γ_0 inherits the structure of a Lie algebroid from the bracket of the right invariant vector fields, and with anchor map l_*. By the canonical isomorphism $E_l \approx \nu\Gamma_0$, this yields a second Lie algebroid structure on $\nu\Gamma_0$. Take two fields $[X], [Y] \in \nu\Gamma_0$, and let $(X_r, Y_r), (X_l, Y_l)$ be the corresponding vectors in E_r, E_l respectively. Then $X_r - X_l, Y_r - Y_l \in T\Gamma_0$, and $[X_r - X_l, Y_r - Y_l]$ projects to $0 \in \nu\Gamma_0$. Furthermore, since left invariant and right invariant fields commute, what we get from the previous bracket is $pr_{\nu\Gamma_0}(\{X_r, Y_r\} + \{X_l, Y_l\}) = 0$. This means that the brackets of the two algebroid structures of $\nu\Gamma_0$ differ only in sign. Q.e.d.

Now, we recall that, on any Poisson manifold (M, w), the bracket $\{\ ,\ \}$ of 1-forms, studied in Chapter 4, defines, in fact, a Lie algebroid structure on T^*M, with the anchor map $\#_w : T^*M \to TM$ (Remark 4.2). The interesting fact is that the following holds

9.16. Proposition. *If Γ is a symplectic groupoid with the unit manifold Γ_0, and if w is the induced Poisson structure of Γ_0* (Theorem 9.9), *then the Lie algebroid $(T^*\Gamma_0, \{\ ,\ \}, \#_w)$ is isomorphic with the Lie algebroid $\nu\Gamma_0$ of Γ.*

Proof. Let $\tilde{\sigma}$ be the Poisson structure associated with the symplectic structure σ of Γ i.e., $\forall \theta \in T^*\Gamma_0$ we have $\#_{\tilde{\sigma}}\theta = -\#_\sigma\theta$. If $X \in T_u\Gamma$ for $u \in \Gamma_0$, it has an associated form $\varphi(X) \in T^*_u\Gamma_0$ defined as the pullback to Γ_0 of $\#_\sigma^{-1}X$ i.e., such that

$$\varphi(X)(Y) = \sigma(Y, X) \qquad (Y \in T_u\Gamma_0). \tag{9.29}$$

$\varphi : T_u\Gamma \to T^*_u\Gamma_0$ is obviously an epimorphism, and, since Γ_0 is a Lagrangian submanifold of Γ, $\ker\varphi = T_u\Gamma_0$. Hence, φ induces an isomorphism of $\nu_u\Gamma_0$ onto $T^*_u\Gamma_0$, that will again be denoted by φ, and (9.29) allows us to check that, if $\theta \in T^*_u\Gamma_0$, then

$$\varphi^{-1}(\theta) = pr_{\nu\Gamma_0}(\#_{\tilde{\sigma}} r^*\theta). \tag{9.30}$$

Since the Poisson structure w is coinduced by $\tilde{\sigma}$, the corresponding operators $\#$ are related by formula (7.3) i.e., $\#_w = r_* \circ \#_{\tilde{\sigma}} \circ r^*$ (where all these operators are at u), and we deduce that

$$\begin{array}{ccc} T^*_u\Gamma_0 & \overset{\varphi^{-1}}{\longrightarrow} & \nu_u\Gamma_0 \\ & {\scriptstyle \#_w}\searrow\ T_u\Gamma_0 \ \swarrow{\scriptstyle r_*} & \end{array}$$

is a commutative diagram.

Obviously, φ also acts on cross sections and, in order to complete our proof, we must show that the Lie brackets are preserved.

If $[X],[Y] \in \Gamma(\nu\Gamma_0)$, and $X = \#_{\tilde{\sigma}} r^*\theta, Y = \#_{\tilde{\sigma}} r^*\theta'$ are their representative vector fields, defined as in (9.30), we have defined

$$\{[X],[Y]\} = pr_{\nu\Gamma_0}([\tilde{X}_l, \tilde{Y}_l]/_{\Gamma_0}), \tag{9.31}$$

where $\tilde{X}_l, \tilde{Y}_l$ are the left-invariant vector fields generated by X and Y. We shall analyse such a field $\tilde{X}_l$ over a coordinate neighbourhood (U, x^i) of Γ_0 i.e., on $r^{-1}(U)$. Then, $\theta = \theta_i(x^j)dx^i$, and $X = \theta_i(x^j)(\#_{\tilde{\sigma}} r^* dx^i)/_{\Gamma_0}$. Hence, if we can prove that $\#_{\tilde{\sigma}} r^* dx^i$ is a left invariant vector field, it will follow that the relation $\tilde{X}_l = \#_{\tilde{\sigma}} r^*\theta$ holds over the whole manifold Γ.

Let us take any $f \in C^\infty(\Gamma_0)$, and put $Z = \#_{\tilde{\sigma}} r^* df = -\#_\sigma r^* df$ i.e., Z is the Hamiltonian vector field $X^\sigma_{f \circ r}$. Since $f \circ r$ is constant along the r-fibers, Z is σ-orthogonal to the r-fibers and, therefore, it is tangent to the l-fibers (see the proof of Theorem 9.9.) Furthermore, take $x \in \Gamma$, and look at the left translation L_x. If Σ is a local Lagrangian bi-cross section through x, then, along the l-fiber of x, L_x acts like Σ_l of (9.27) i.e., by a symplectomorphism. Then, since $f \circ r$ is invariant by left translations (Proposition 9.2 ii)), its Hamiltonian Z will be invariant by Σ_l, and by L_x.

Hence, we are done, and we may write $\tilde{X}_l = \#_{\tilde{\sigma}} r^*\theta, \tilde{Y}_l = \#_{\tilde{\sigma}} r^*\theta'$ over Γ. (In fact, what was proven above is that a vector field on Γ is left invariant iff it can be put into this form for some $\theta \in \Gamma(T^*\Gamma_0)$.) Now, we can go on and compute

$$[\tilde{X}_l, \tilde{Y}_l] = [\#_{\tilde{\sigma}} r^*\theta, \#_{\tilde{\sigma}} r^*\theta'] = \#_{\tilde{\sigma}}(\{r^*\theta, r^*\theta'\}_{\tilde{\sigma}}) = \#_{\tilde{\sigma}} r^*\{\theta, \theta'\}_w,$$

where the last equality follows from the fact that w is coinduced by r. (This is straightforward if θ, θ' are differentials, and it extends to Pfaff forms by (4.2).) Q.e.d.

Now, it is clear that we have the following well defined questions in the vein of Lie's third theorem:

1) Is any Lie algebroid E associated to a certain Lie groupoid Γ?

2) Is any Poisson manifold (Γ_0, w) the unit manifold of a symplectic groupoid with the induced Poisson structure?

In this second question, of course, we shall have to find a symplectic Lie groupoid Γ with unit manifold Γ_0, whose Lie algebroid is isomorphic to $T^*\Gamma_0$, and such that Γ induces the Poisson structure w of Γ_0.

9.17. Definition. If question 1) has a positive answer for the Lie algebroid E then E is said to be *integrable*. If question 2) has a positive answer for (Γ_0, w), then the latter is an *integrable Poisson manifold*.

Unfortunately, the fact is that the general answer to these questions is negative (see the references quoted in [Mz] for question 1)). However, these questions have a certain local solution which is based on the notion of a *local Lie groupoid* defined as follows [vE], [CDW].

9.18. Definition. Let $r, l : \Gamma \to \Gamma_0$ be two submersions of a differentiable (possibly non-Hausdorff) manifold Γ onto its Hausdorff submanifold Γ_0, and let $i : \Gamma \to \Gamma$ be an involutive diffeomorphism of Γ ($i^2 = \mathrm{id}$). Then a structure of a *local Lie groupoid* on Γ consists of a differentiable multiplication $(x, y) \mapsto x.y$ defined on an open neighbourhood U of $\Gamma_0 \approx \{(u, u)/u \in \Gamma_0\}$ in $\Gamma^{(2)} = \{(x, y)//r(x) = l(y)\}$, and which satisfies the axioms:

1) $\forall x \in \Gamma$, $(l(x), x) \in U$, $(x, r(x)) \in U$ and $l(x).x = x, x.r(x) = x$;

2) $\forall x \in \Gamma$, $(x, x^{-1} \stackrel{\text{def}}{=} i(x)) \in U$, $(x^{-1}, x) \in U$ and $x.x^{-1} = l(x)$, $x^{-1}.x = r(x)$;

3) if $(x, y) \in U$ then $(y^{-1}, x^{-1}) \in U$ and $i(x.y) = y^{-1}.x^{-1}$;

4) if $x.y$, $y.z$ and $x.(y.z)$ exist, so does $(x.y).z$, and $x.(y.z) = (x.y).z$.

If, moreover, Γ is a symplectic manifold and the graph of the multiplication is a Lagrangian submanifold of $\Gamma \times \Gamma \times (-\Gamma)$, then Γ is a *local symplectic groupoid.*

It was proven by Pradines [Pr2] that any Lie algebroid is associated with a local Lie groupoid. Furthermore, in the theory of the Poisson manifolds, it was proven by Weinstein [We5], [CDW] [AD1,2] that any Poisson manifold (Γ_0, w) is integrable by a local symplectic groupoid Γ. The latter can be obtained by gluing up carefully local symplectic realizations in the sense of Theorem 8.2., and defining the multiplication by means of an adequate use of the flows of the Hamiltonian vector fields of functions $f \circ r$, where $f \in C^\infty(\Gamma_0)$ and support f is compact, and r is the source projection of Γ. These Hamiltonian vector fields are complete, iff the local groupoid Γ turns out to be a global symplectic groupoid.

We shall not develop this local theory here, and we refer the reader to [CDW] and [AD1,2] for exact results and proofs. Instead, since until now our only example was that of the Lie-Poisson structure, we shall give two more examples of global integrability.

First, let us notice that the Poisson structure of a symplectic manifold is integrable. Indeed, it suffices to consider the banal groupoid $M \times M$ with the symplectic form $-\sigma \oplus \sigma = r^*\sigma - l^*\sigma$, that was introduced in Example c) of a symplectic groupoid. Then, it is clear that r is a Poisson mapping, and we are done. Moreover, from the same Example c) we see that the fundamental groupoid of M is also a solution of the problem of integrating the Poisson structure of M.

Second, let us look at the simple case of the Poisson manifold $(\mathbb{R}^2, w = (x^2 + y^2)(\partial/\partial x) \wedge (\partial/\partial y))$, that we also discussed in Example 5.8. This manifold is integrable, and the corresponding symplectic groupoid was constructed in [ADH]. By Theorem 9.9, if we want to *integrate* a Poisson manifold, we have to look for a symplectic groupoid whose orbits are given by the symplectic leaves of the former. In the case under discussion, there are just two symplectic leaves: the origin 0, and $\mathbb{R}^2 \backslash \{0\}$, and they are preserved by rotations and homotheties. An analysis of this situation led the authors of [ADH] to the following construction, which is a modified version of an action groupoid (Lie groupoids, Example b)).

Let us identify $\mathbb{R}^2$ with the plane of the complex variable $z = x + iy$. Then,

$$w = -2iz\bar{z}\frac{\partial}{\partial z} \wedge \frac{\partial}{\partial \bar{z}},$$

the leaf $\mathbb{R}^2 \backslash \{0\} = \mathbb{C}^* = \{z \in \mathbb{C}/z \neq 0\}$, and the transformations $\zeta(z) = e^{\zeta\bar{z}}z$ $(\zeta \in \mathbb{C})$ preserve the symplectic leaves of w. Hence, if we take $\Gamma = \mathbb{C} \times \mathbb{C}$ with the operation

$$(\zeta_1, z_1) \cdot (\zeta_2, z_2) = (\zeta_1 e^{\bar{\zeta}_2 z_2} + \zeta_2, z_2), \tag{9.32}$$

which is defined iff $z_1 = e^{\zeta_2 \bar{z}_2} z_2$ and expresses the composition of transformations if we see a pair (ζ, z) as a mapping $z \mapsto e^{\zeta\bar{z}}z$, it is easy to check that a Lie groupoid structure has been obtained on Γ. Indeed, if we take into account the existence condition of the product, (9.32) becomes

$$(\zeta_1, z_1)(\zeta_2, z_2) = (\zeta_1(\bar{z}_1/\bar{z}_2) + \zeta_2, z_2), \tag{9.33}$$

and, from (9.33), associativity follows easily. It is also easy to check that the left and right units are given by

$$l(\zeta, z) = (0, e^{\zeta\bar{z}}z), \quad r(\zeta, z) = (0, z), \tag{9.34}$$

and the inverse map is

$$i(\zeta, z) = (-\zeta e^{-\bar{\zeta}z}, e^{\zeta\bar{z}}z). \tag{9.35}$$

The unit manifold is $\Gamma_0 = \{(0, z)/z \in \mathbb{C}) \approx \mathbb{C}$, and the orbits of the groupoid structure are exactly the orbits of the Poisson structure w.

Now, the symplectic form induced by w in its leaf $\mathbb{C}^*$ is

$$\sigma = \frac{i}{2z\bar{z}}dz \wedge d\bar{z} = \frac{i}{2}d(\ln z) \wedge d(\ln \bar{z}) \tag{9.36}$$

and, if we restrict ourselves to $l^{-1}(\mathbb{C}^*) \cap r^{-1}(\mathbb{C}^*) = \mathbb{C} \times \mathbb{C}^*$, we get there the 2-form

$$\tilde{\sigma} = r^*\sigma - l^*\sigma. \tag{9.37}$$

But, in view of (9.34), we have

$$\begin{aligned}\tilde{\sigma} = \frac{i}{2}d(\ln z) \wedge d(\ln \bar{z}) - \frac{i}{2}(d(\zeta\bar{z}) + d(\ln z))\wedge \\ \wedge(d(\bar{\zeta}z) + d(\ln \bar{z})) = -\frac{i}{2}(dz \wedge d\bar{\zeta} + d\zeta \wedge d\bar{z} - \zeta\bar{\zeta}dz \wedge d\bar{z} + \\ + z\bar{z}d\zeta \wedge d\bar{\zeta} + z\zeta d\bar{z} \wedge d\bar{\zeta} + \bar{z}\bar{\zeta}d\zeta \wedge dz),\end{aligned} \tag{9.38}$$

and this is a global symplectic 2-form on the whole groupoid Γ. (This was the reason for looking at the transformation $z \mapsto e^{\zeta\bar{z}}z$ rather than at the simpler one $z \mapsto e^{\zeta}z$. In the latter case, (9.37) does not extend to a global form on Γ [ADH].)

Moreover, we can see that $(\Gamma, \tilde{\sigma})$ is a symplectic groupoid that integrates $(\mathbb{R}^2, w)$. This can be seen either by a lengthy computation that, using (9.33) and (9.38), will show that the graph of the multiplication is a Lagrangian submanifold of $\Gamma \times \Gamma \times (-\Gamma)$, or by the following indirect argument: (9.37) compared with the example of a banal symplectic groupoid, shows that $\mathbb{C} \times \mathbb{C}^*$ is a symplectic groupoid that integrates the symplectic manifold $(\mathbb{C}^*, \sigma)$ (everything necessary is of a local character and it holds because it holds in the locally equivalent banal case); hence, for $\mathbb{C} \times \mathbb{C}^*$ the graph is Lagrangian as requested, and it will remain Lagrangian for Γ because of continuity. The same argument shows that $r : (\Gamma, \tilde{\sigma}) \to (\mathbb{C} = \mathbb{R}^2, w)$ is a Poisson mapping.

The conclusion is that $(\mathbb{R}^2, w)$ is integrable, and $(\Gamma, \tilde{\sigma})$ is a corresponding symplectic groupoid.

9.5 Further integrability results

In the remaining part of this chapter, we review briefly some of the main results that were obtained in the study of the global integrability of Poisson manifolds. There will be (almost) no proofs, and the aim is to guide the interested reader through the corresponding literature.

We begin with the special interesting case of the Heisenberg-Poisson manifolds, defined and studied by Weinstein [We8], and in this case we shall explain the geometric part of the proofs.

Let (M, w) be a Poisson manifold, and define

$$M' = M \times \mathbb{R}, \quad w'(df, dg) = tw(d_M f, d_M g), \tag{9.39}$$

where $t \in \mathbb{R}$; $f, g \in C^\infty(M')$, and d_M denotes the differential along M only. Then, (M', w') is, obviously, a new Poisson manifold such that

$$\mathrm{rank}/_{(x \in M, t \neq 0)} w' = \mathrm{rank}_x w; \ \mathrm{rank}_{(x,0)} w' = 0, \tag{9.40}$$

and the symplectic leaves along $M \times \{t \neq 0\}$ are those of M, while those through $(x, 0)$ are just points. If $M = \mathbb{R}^{2n}$ with the canonical symplectic structure, M' is exactly the dual of a well known Lie algebra called the *Heisenberg algebra*, which leads to [We8]

9.19. Definition. The manifold (M', w') defined above is called the *Heisenberg-Poisson manifold* or, shortly, the *HP manifold* of (M, w).

With this construction, one has the following main result

9.20. Theorem [We8]. *Let (M, ω) be a symplectic manifold such that the de Rham cohomology class $[\omega]$ lifts to a multiple of an integral cohomology class of the universal covering manifold $\tilde{M}$ of M. Then, the corresponding Heisenberg-Poisson manifold M' is integrable by a symplectic groupoid.*

The proof of this theorem demands rather lengthy preparations. We only indicate the main ideas and steps, and we refer the reader to [We8] for details.

One starts with a construction which is valid for a symplectic manifold (M, ω) endowed with a prequantization complex line bundle K as defined in Chapter 6. Particularly, K is endowed with a Hermitian connection ∇ of curvature form $\Omega = -2\pi i\omega$. Let $\mathcal{K}$ be the total space of the principal circle bundle to which K is associated (i.e., the set of the unit vectors of the fibers of K), and let θ be the real valued global connection 1-form of ∇ on $\mathcal{K}$. It is known that θ is a contact form on $\mathcal{K}$ (the Boothby-Wang construction e.g., [Bl]), and $(\mathcal{K}, \theta)$ is a contact manifold.

Furthermore, if (θ, λ^i) is a local basis of $T^*\mathcal{K}$, the Liouville form of $T^*\mathcal{K}$ is of the form $\lambda = \alpha\theta + \zeta_i\lambda^i$, and it follows that the submanifold $L = \{\alpha\theta\} \subseteq T^*\mathcal{K}$ given by $\zeta_i = 0$ has a closed 2-form induced by $d\lambda$:

$$\Theta = d\alpha \wedge \theta + \alpha d\theta = d\alpha \wedge \theta - 2\pi\alpha\omega \tag{9.41}$$

(since connection theory says that $d(i\theta)$ is the curvature Ω) that is symplectic on $L_0 \subseteq L$ where $\alpha \neq 0$. Moreover, let E be the vector field defined on $\mathcal{K}$ by $\theta(E) = 1$, $i(E)d\theta = 0$, and look at the manifold $L^* = \{\xi E\} \subseteq T\mathcal{K}$, and at the mapping $j : L^* \to L$ defined by $j(\xi E) = \frac{1}{\xi}\theta$ ($\xi \neq 0$). Then,

$$-\frac{1}{2\pi} j^*\Theta = \frac{1}{2\pi\xi^2} d\xi \wedge \theta + \frac{1}{\xi}\omega \tag{9.42}$$

is a symplectic 2-form on $L^* \backslash \{0\}$, and the corresponding Poisson bivector

$$w_{L^*} = -2\pi\xi^2 \frac{\partial}{\partial\xi} \wedge E + \xi w_\omega, \tag{9.43}$$

where w_ω is the Poisson structure of ω, is a globally defined Poisson structure on L^* of rank 0 at the points of the zero section. (This is a general construction on contact manifolds, due to C. LeBrun [We8].) From (9.43), it follows straightforwardly that the mapping $\rho : L^* \to M'$ given by $\xi E \mapsto (pr_M E, \xi)$ relates w_{L^*} with the Poisson structure w' of M' = the HP-manifold of M, and therefore, ρ is a Poisson submersion. (The result is seen in [We8] by using canonical coordinates.) As a matter of fact, this submersion is a symplectic realization outside the zero section of L^*. The nice property that occurs in this situation is *completeness* in the sense that $\forall f \in C^\infty(M)$ such that the Hamiltonian field X_f is complete on M, $X_{f\circ\rho}$ is complete on L^*. This follows because $X_{f\circ\rho} = 0$ along $\xi = 0$, and for $\xi \neq 0$ ρ is proper.

The second step in the proof of Theorem 9.20 consists in proving that, under the hypotheses, the fundamental groupoid $\Pi(M) = \tilde{M} \times \tilde{M}/_{\pi_1(M)}$ ($\pi_1(M)$ = the Poincaré group of M), which is a symplectic manifold, has a prequantization bundle that yields a Lie groupoid with the unit manifold M.

First, by the symplectic version of Theorem 6.8 [Ur], there exists $k \in \mathbb{R}$ such that the symplectic manifold $(\tilde{M}, k(\text{lift } \omega))$ admits a prequantization bundle K, and we shall consider for this K the elements $\mathcal{K}, \theta$, etc. that were used for the bundle K of the previous step.

Let $\{U_i\}$ be a covering of $\tilde{M}$ by trivializing neighborhoods of K, endowed with the local unitary bases b_i. Let K^* be the dual bundle of K with the local dual bases b_i^*. Let

$$b_j = \varphi_{ij} b_i, \quad b_j^* = \varphi_{ij}^{-1} b_i^* = \varphi_{ji} b_i^*$$

be the corresponding transition relations. Then, there is a well defined Hermitian line bundle $Q = K \cdot K^*$ over $\tilde{M} \times \tilde{M}$ defined by local unitary bases β_{ij} over $U_i \times U_j$ which satisfy the transition relations

$$\beta_{hk} = \varphi_{ih} \varphi_{kj} \beta_{ij} \tag{9.44}$$

over $(U_i \times U_j) \cap (U_h \times U_k)$. There also exists a natural surjection $\varphi : K \times K^* \to Q$ defined by $\varphi(b_i, b_j^*) = \beta_{ij}$, and φ does not depend on the choice of bases.

Furthermore, Q is associated with a principal circle bundle $\tilde{Q}$ on which $\theta \oplus (-\theta)$ becomes the 1-form associated with a Hermitian connection that makes Q into a prequantization bundle of $\tilde{M} \times (-\tilde{M})$. We also have the well defined restriction mapping $\varphi : \mathcal{K} \times \mathcal{K}^* \to \tilde{Q}$, and $\mathcal{K}^*$ and $\mathcal{K}$ can be identified by means of the Hermitian metric of K. Hence we may speak of the diagonal Δ of $\mathcal{K} \times \mathcal{K}^*$, and $\Lambda = \varphi(\Delta)$ is a $2n$-dimensional submanifold of $\tilde{Q}$ where $\theta \oplus (-\theta)$ pulls back to 0 i.e., Λ is what one calls a *Legendrian submanifold* in the theory of contact manifolds. Λ will be called the *distinguished Legendrian submanifold* of $\tilde{Q}$.

Finally, we shall see that $\tilde{Q}$ has the structure of a Lie groupoid with the unit manifold $\tilde{M}$. This structure comes by φ from the gauge groupoid of the principal circle bundle $\mathcal{K} \to \tilde{M}$, but we shall describe it here straightforwardly. Namely, let us denote by $pr_{1,2} : \tilde{Q} \to \tilde{M} \times \tilde{M}$ the two natural projections, and take $z_{1,2} \in \tilde{Q}$ such that $pr_1 z_1 = x \in U_i$, $pr_2 z_1 = pr_1 z_2 = y \in U_j$, $pr_2 z_2 = u \in U_k$. Put

$$z_1 = \zeta_1 \beta_{ij}(x, y), \quad z_2 = \zeta_2 \beta_{jk}(y, u). \tag{9.45}$$

Then, we shall define the multiplication

$$z_1 * z_2 = \zeta_1 \zeta_2 \beta_{ik}(x, u), \tag{9.46}$$

and (9.44) shows that the result does not depend on the choice of the local bases. The Lie groupoid axioms are easily verified. Particularly, the right and left units are defined by

$$r(z_1) = \beta_{jj}(y, y), \quad l(z_2) = \beta_{jj}(y, y), \quad i(z_1) = \frac{1}{\zeta_1} \beta_{ji}(y, x), \tag{9.47}$$

and, again, (9.44) shows that all the results are invariant. The unit manifold is the *diagonal section* $\{\beta_{ii}(x,x)/x \in M\}$ i.e., just the submanifold Λ, and we see that we may identify it with $\tilde{M}$.

The next point is that the action of the Poincaré group $\pi_1(M)$ on $\tilde{M} \times (-\tilde{M})$ can be lifted to $\tilde{Q}$ by acting on Λ as on $\tilde{M}$, and then by transferring this action to $\tilde{Q}$ using parallel translation. Moreover, the whole previous structure goes over to the quotient space $\Gamma(M) \overset{\text{def}}{=} \tilde{Q}/\pi_1(M)$ [We8] such that the latter becomes the total space of the principal circle bundle associated with a prequantization bundle of the fundamental symplectic groupoid $\Pi(M) = \tilde{M} \times (-\tilde{M})/_{\pi_1(M)}$, and a Lie groupoid with the unit manifold $\tilde{M}/\pi_1(M) = M$.

In the third step of the proof of Theorem 9.20, the construction of step one is applied to the symplectic manifold $\Pi(M)$ endowed with the contact manifold $\Gamma(M)$ of the prequantization bundle described above. As a result, one obtains a Poisson manifold $(L^*(\Gamma(M)), w_{L*})$ endowed with a Poisson submersion ρ over something that turns out to be equivalent to $M' \times (-M'), M'$ being the *HP*-manifold of M. More exactly, the groupoid structure of $\Gamma(M)$ yields a groupoid structure of $L^*(\Gamma(M))$ with the unit manifold M', and the two projections on the factors of $M' \times (-M')$ are the target-source projections. As in the first step, $L^*\backslash\{0\}$ is a symplectic groupoid over the regular part of M'.

As seen during step one, $L^*(\Gamma(M))$ is in fact $\Gamma(M) \times \mathbb{R}$, and the next idea is that of replacing $\Gamma(M) \times \{0\}$ by "something better" in order to get a symplectic groupoid for the whole manifold M'. Accordingly, one uses the differential topology technique of *explosion* [We8]. Briefly, let X^n be a differentiable manifold, and $Y^k a$ submanifold. Then a new manifold $E(X,Y)$ the *explosion* of X along Y is defined by

$$E(X,Y) = [X \times (\mathbb{R}\backslash\{0\})] \cup (T_Y X/TY),$$

with the usual differentiable structure on the first term, and local coordinates at the points of the second term obtained as follows. Let $(y^1,\ldots,y^k,z^{k+1},\ldots,z^n)$ be coordinates of X at the points of Y such that $z^{k+1}=0,\ldots,z^n=0$ on Y. Then, introduce a new coordinate t, and $n+1$ coordinates defined by $(y;tz;t)$ for $t \neq 0$ and by (y; coordinates of the projection of the tangent of the path $(0,tz)$ at $t=0$ to $T_Y X/TY; 0)$ for $t=0$.

The proofs given in [We8] show that if the contact manifold $\Gamma(M)$ is exploded along its distinguished Legendrian submanifold, and then the result is exploded along the conormal bundle of the image of the contact distribution of $\Gamma(M)$ (i.e., the distribution defined by the annulation of the contact form) one gets $L^*(\Gamma(M))$ with its zero section conveniently replaced such as to provide the symplectic groupoid required by Theorem 9.20.

Now, let us turn towards the regular Poisson manifolds. The study of their integrability by a symplectic groupoid has been done by Dazord [Dz4], Dazord and Hector [DH], Hector [Hc], Alcalde Cuesta [AC1, 2], etc. To some extent the problem is simpler now, since no explosion techniques are needed.

Let (M, w) be a regular Poisson manifold with the symplectic foliation $\mathcal{S}$. Then, one can construct a fundamental groupoid $\Pi(\mathcal{S})$ of the foliation $\mathcal{S}$, which consists of the leafwise fixed-end-points homotopy classes of paths of the leaves of $\mathcal{S}$, and if (Γ, σ) is a symplectic groupoid that has connected and simply connected r-fibers (hence l-fibers too) and integrates (M, w), one proves in [Dz4] that there exists a groupoid homomorphism $\varphi : (\Gamma, \sigma) \to (\Pi(\mathcal{S}),$ lift $w)$ which is an isotropic realization (see Chapter 8). Therefore, one has to look for such realizations first and, then, determine if one of them is the symplectic groupoid required. In [Dz4] this study is completed for the case where $\mathcal{S}$ is a locally trivial fibration.

Furthermore, in the general case of a regular foliation $\mathcal{S}$, we may use complementary distributions $\nu\mathcal{S}$, and represent w by forms ω of the type (0,2) such that $d''\omega = 0$ (see Chapters 3, 4 and also [HMS]). In particular, if ω extends to a closed (exact) 2-form, we shall say that w is a *covariantly closed (covariantly exact)* regular Poisson structure. (In [DH] the terms *presymplectic* and *exact* are used instead.)

In [DH] one proves that w can be lifted to $\Pi(\mathcal{S})$ making the latter into a *Poisson groupoid* that *integrates* (M, w). If $\tilde{w} =$ lift of w is covariantly closed on $\Pi(\mathcal{S})$ for the lifted foliation $\tilde{\mathcal{S}}$, it turns out that the canonical bundle $\nu^*\tilde{\mathcal{S}}$ is an isotropic realization of $\tilde{w}$, and a symplectic groupoid that integrates (M, w). Therefore, the covariant closedness of $\tilde{w}$ is a sufficient condition for the integrability of (M, w).

For instance, let (M, ω) be a symplectic manifold, and $\mathcal{S}$ be a foliation with symplectic leaves on M. Then, the Dirac bracket of $(M, \omega, \mathcal{S})$ (Definition 3.7) is *minimal* in the sense of [DH]. Indeed, if we use the ω-orthogonal distribution of $\mathcal{S}$, ω has a type decomposition $\omega = \omega_{(2,0)} + \omega_{(0,2)}$, and $d\omega = 0$ implies $d'\omega_{(0,2)} = 0$, which is just the definition of minimality in [DH]. (See Chapter 4 for the definition of bigrades and of d'.) Then, Proposition 5.7 of [DH] shows that $\tilde{w}$ is covariantly closed. Hence, *any Dirac bracket of a symplectic manifold is integrable by a symplectic groupoid.*

Next, in [DH], again, one gives sufficient conditions that are of a more concrete character, and ensure the previous sufficient conditions of integrability. Namely, one proves

9.21. Theorem [DH]. *If the symplectic foliation $\mathcal{S}$ of a regular Poisson manifold (M, w) is totally aspherical, then $\tilde{w}$ is covariantly closed, and (M, w) is integrable.*

In this theorem, totally aspherical means that: a) the second homotopy group $\pi_2(S) = 0$ for every leaf S of $\mathcal{S}$, and b) $\mathcal{S}$ has no nontrivial *vanishing cycles*. A *vanishing cycle* is a mapping $A : S^1 \times [0,1] \to M$ such that, $\forall t \in [0,1]$, A_t is a loop of a leaf S_t of $\mathcal{S}$, and the loops A_t are null-homotopic in S_t for $t > 0$; triviality occurs if A_0 is also null-homotopic in S_0. A different use of vanishing cycles to get an integrability obstruction for regular Poisson manifolds can be found in [Hc].

The fundamental groupoid of a foliation was also used by F. Alcalde Cuesta [AC1] for the more particular case of a compact Poisson manifold (M, w) whose symplectic foliation $\mathcal{S}$ is Riemannian (i.e., it has a bundle-like metric). He proves that $\Pi(\mathcal{S})$ is locally trivial in this case, and deduces that, if $\dim \mathcal{S} = 2$, and one of the following two conditions holds: i) $\mathcal{S}$ is aspherical, ii) $\mathcal{S}$ is a sphere bundle and the leaves have a constant symplectic volume, then (M, w) is integrable by a symplectic groupoid.

More recently, important general results on the integrability of regular Poisson manifolds with no vanishing cycles were given in [AC2].

On the other hand, P. Xu [Xu1] uses Dazord's results [Dz4] in order to construct a symplectic groupoid that integrates a Poisson manifold of the form $M = S \times P$, where S is a symplectic manifold, and the symplectic leaves of M are $S \times \{p\}$ $(p \in P)$ with a symplectic structure that varies in a certain specific way with the point p.

We end this list of results on the global integration of Poisson manifolds by quoting the important results of Xu [Xu2,3,4] about the construction of the symplectic groupoid of reductions of Poisson manifolds, and the construction due to Lu and Weinstein [LW1] of a double symplectic groupoid associated with a *Poisson-Lie group*. The notions of *Poisson-Lie group* and *Poisson action* will be defined and studied in the next chapter. (Notice also the study of the *PL*-cohomology of such Poisson-Lie groups in [Lu1] and that of the star product in [MV].)

We hope that the previous theory has already convinced the reader that symplectic groupoids are important geometric structures. But, their original and basic motivation was the idea of using them for quantization theory (see Chapter 6). Indeed, if a Poisson manifold (M, w) is seen as the phase space of a classical mechanical system, the symplectic groupoid (Γ, σ) that integrates (M, w) (if it exists) can also be interpreted as a phase space of the system, which has more coordinates than really needed but, instead it has some kind of *symmetries*, the left and right translations of the groupoid. Hence, it is natural to look at the quantization of (Γ, σ) as the corresponding quantum system, and study whether it also has similar symmetries. This program was carried out to some extent in [We9], [We10], [WX1], [Zk], [Pu], etc.

Particularly, in [WX1] one looks at the principal circle bundle E of a prequantization of the symplectic groupoid (Γ, σ) of (M, w), in conjunction with a conveniently defined so-called identity smooth groupoid cohomology of Γ. It turns out that, under certain natural hypotheses, E has a nice groupoid structure itself, and the Poisson cohomology class $[w]$ lifts to a groupoid 2-cocycle of Γ. In the same paper, the symplectic groupoid and its cohomology are used to discuss the existence of momentum mappings for Poisson actions of Lie groups.

The cohomology considerations of [WX1] also led to the result that the LP-cohomology spaces $H^*_{LP}(M, w)$ (Chapter 5) are isomorphic to the cohomology spaces of the cochain of left invariant forms of the corresponding symplectic groupoid (Γ, σ). This result was used in [Xu1] in order to compute the Lichnerowicz-Poisson cohomology of products $S \times P$ with the symplectic leaves $S \times \{p\}$ $(p \in P)$ endowed with a nicely varying symplectic structure.

10 Poisson-Lie Groups

10.1 Poisson-Lie and biinvariant structures on Lie groups

One of the important reasons for studying Poisson manifolds is the potential usefulness of such manifolds in quantization via objects called *Poisson-Lie groups*, and, then, *quantum groups* [Dr1], [Dr2]. The latter are not really groups, but noncommutative algebras obtained by a deformation quantization (Chapter 6) of Poisson-Lie groups. From the purely geometric viewpoint it is also completely natural to define and study Poisson-Lie groups.

The present chapter is an introduction to this study. We shall give various characterizations of the Poisson-Lie groups, and discuss the correspondence with an associated class of Lie algebras, called Poisson-Lie algebras or Lie bialgebras. We shall study a method of finding Lie bialgebras, and Poisson-Lie groups known as the Yang-Baxter equation method, as well as another method based on the identification of Lie bialgebras with Manin triples of Lie algebras. In particular, this latter method yields a Poisson-Lie group structure on $SU(n)$. Finally, we shall introduce briefly the important notions of Poisson group actions and dressing transformations. Indications about various generalizations will be given at the end.

10.1. Definition. Let G be a Lie group, endowed with a Poisson structure γ. Then, (G,γ) is a *Poisson-Lie group* if the multiplication

$$\mu : G \times G \to G \tag{10.1}$$

is a Poisson mapping that relates $\gamma \oplus \gamma$ with γ.

In order to write down various properties of γ, notice the easy to establish formula

$$\mu_{*(g_1,g_2)}(X_{g_1} \oplus Y_{g_2}) = L_{g_1^*}Y + R_{g_2^*}X, \tag{10.2}$$

and the induced formula for 1-forms

$$(\mu^*\lambda)_{(g_1,g_2)}(X_{g_1} \oplus Y_{g_2}) = (L^*_{g_1}\lambda)(Y_{g_2}) + (R^*_{g_2}\lambda)(X_{g_1}). \tag{10.3}$$

Using them, we can express μ-relatedness of tensors, particularly if t_{g_1}, t_{g_2} and $t_{g_1g_2}$ are either totally contravariant or totally covariant tensors, it makes sense to check whether the relations

$$\mu_*(t_{g_1} \oplus t_{g_2}) = t_{g_1g_2} \quad \text{or} \quad \mu^* t_{g_1g_2} = t_{g_1} \oplus t_{g_2}, \tag{10.4}$$

hold good, correspondingly, which leads to

10.2. Definition. Let t be either a $(p,0)$-tensor field or a $(0,q)$-tensor field on G. Then t is said to be a *multiplicative field* if, $\forall g_1, g_2 \in G$, the corresponding part of (10.4) holds for the values of t at these points.

Definitions 10.1 and 10.2 imply

10.3. Proposition. *The Poisson structure γ makes the Lie group G into a Poisson-Lie group iff γ is a multiplicative tensor field i.e., $\forall g_1, g_2 \in G$ one has*

$$\gamma_{g_1 g_2} = L_{g_1^*}\gamma_{g_2} + R_{g_2^*}\gamma_{g_1}. \tag{10.5}$$

In particular, we must have $\gamma_e = 0$ ($e =$ the unit of G).

Proof. Express (10.4) for $t = \gamma$, and use (10.3). For the last assertion, take $g_1 = g_2 = e$. Q.e.d.

Before continuing with the study of the Poisson-Lie groups, we should like to refer to another interesting situation too. In analogy to Definition 10.1, we might also ask for Poisson structures γ on a Lie group G, such that the multiplication μ of (10.1) is a Poisson mapping between $c(\gamma \oplus \gamma)$ and γ, where $c = \text{const.} \neq 0$. Then, (10.5) must by replaced by

$$\gamma_{g_1 g_2} = c(L_{g_1^*}\gamma_{g_2} + R_{g_2^*}\gamma_{g_1}),$$

and looking at $g_1 = g_2 = e$, then at $g_1 = e$, $g_2 = e$ separately we see that the only other possibility than the Poisson-Lie groups is $\gamma_e \neq 0$, $c = 1/2$, and γ invariant by both the left and the right translations.

The structure of (G, γ) where γ is a bi-invariant Poisson structure on the Lie group G is rather simple, and it is given by

10.4. Theorem. *Let G be a connected Lie group with the Lie algebra $\mathcal{G}$. Then a bi-invariant Poisson structure γ on G, if any, is obtained by translating along G an* ad$\mathcal{G}$*-invariant bivector $\gamma_0 \in \Lambda^2\mathcal{G}$. Furthermore, $\mathcal{S}_0 = \mathrm{im}\#_{\gamma_0}$ is an abelian ideal of $\mathcal{G}$ endowed with an* ad $\mathcal{G}$*-invariant structure of a symplectic vector space, the symplectic leaves of γ are all symplectomorphic, and symplectically covered by some $\mathbb{R}^{2k}$ with the canonical symplectic structure, and the Poisson structure γ is transversally constant.*

Proof. Let G have the bi-invariant Poisson structure γ, and put $\gamma_0 = \gamma_e$. Then, the bi-invariance of γ implies the AdG-invariance of γ_0 and, since G is connected, this is equivalent to the ad$\mathcal{G}$-invariance of γ_0. This latter property means that $\forall X \in \mathcal{G}$, $\forall a, b \in \mathcal{G}^*$ one has

$$\gamma_0(\mathrm{coad}\ X(a), b) + \gamma_0(a,\ \mathrm{coad}\ X(b)) = 0.$$

Of course, $\gamma_g = L_{g*}\gamma_0 = R_{g*}\gamma_0$. Conversely, if the ad-invariant γ_0 is given, the previous formula yields a left and right invariant bivector field γ, of the form

$$\gamma = \frac{1}{2}\gamma^{ij}\bar{X}_i \wedge \bar{X}_j = \frac{1}{2}\gamma^{ij}\tilde{X}_i \wedge \tilde{X}_j,$$

where $\gamma_0 = \frac{1}{2}\gamma^{ij} X_i \wedge X_j$, and $\gamma^{ij} = \text{const.}$, $X_i, X_j \in \mathcal{G}$, $\bar{X}_i, \tilde{X}_i$ = the corresponding left and right invariant vector fields, respectively. Then, since $[\bar{X}_i, \tilde{X}_j] = 0$, (1.12) shows that $[\gamma, \gamma] = 0$, and γ is the requested Poisson bivector of G.

Now, let us denote by $\mathcal{S}_0 = \text{im}\#_{\gamma_0}$ the tangent space of the symplectic leaf of γ at e. Then, from the ad-invariance condition we get $\forall a, b \in \mathcal{G}^*$ and $\forall X \in \mathcal{G}$

$$b(\text{ad } X(a^\#)) = -(\text{coad } X(b))(a^\#) =$$
$$= -\gamma_0(a, \text{ coad } X(b)) = \gamma_0(\text{coad } X(a), b) = b([\text{coad } X(a)]^\#)$$

i.e., ad $X(a^\#) = (\text{coad } X(a))^\#$, and this proves that $\mathcal{S}_0$ is an ideal of $\mathcal{G}$.

On the other hand, because of the translation invariance of γ we have

$$[a^\#, b^\#] = [\bar{a}^\#, \bar{b}^\#]_e = \{\bar{a}, \bar{b}\}_e^\#,$$

where the bar denotes the corresponding left invariant 1-forms. But,

$$\{\bar{a}, \bar{b}\} = L_{\bar{a}^\#}\bar{b} - L_{\bar{b}^\#}\bar{a} - d(\gamma(\bar{a}, \bar{b})).$$

Since $\gamma(\bar{a}, \bar{b}) = \text{const.}$, if we evaluate on the left invariant vector field $\bar{X}$ generated by $X \in \mathcal{G}$, and then at e, we get

$$\{a, b\}_e(X) = (\text{ad } X(\gamma_0))\,(a, b) = 0.$$

Therefore, $[a^\#, b^\#] = 0$, and $\mathcal{S}_0$ is abelian.

If we put $\sigma(a^\#, b^\#) = \gamma_0(a, b)$, we clearly have the structure of a symplectic vector space required by Theorem 10.4.

Now, it is clear that all the symplectic leaves of γ are symplectomorphic by translations. Therefore, it is enough to discuss the leaf through e, which is the connected abelian normal subgroup S_0 of G associated with the ideal $\mathcal{S}_0$. The symplectic structure of this leaf is the bi-invariant structure obtained by translating σ along S_0. Then, it is clear that the universal covering group of S_0 is some $\mathbb{R}^{2k}$ with the canonical symplectic structure.

Finally, let $\mathcal{S}'_0$ be a complementary subspace of $\mathcal{S}_0$ in $\mathcal{G}$, and $\mathcal{S}'$ be the distribution obtained by left translations of $\mathcal{S}'_0$ along G. In view of Definition 4.8, we want to prove that every vector field $A \in \mathcal{S}'$, such that $[V, A] \in \mathcal{S}(G)$ = translated of $\mathcal{S}_0$, if $V \in \mathcal{S}(G)$, satisfies the condition $L_A\gamma = 0$.

Let Z_i be a basis of $\mathcal{S}'_0$, and $\bar{Z}_i$ be the associated left invariant vector fields. Then $A = \varphi^i \bar{Z}_i$, and $[V, A] = (V\varphi^i)\bar{Z}_i + \varphi^i[V, \bar{Z}_i]$. Since the flow of $\bar{Z}_i$ consists of right translations in G, and $\mathcal{S}(G)$ is bi-invariant, $[V, \bar{Z}_i] \in \mathcal{S}(G)$. Therefore $[V, A] \in \mathcal{S}(G)$ iff $V\varphi^i = 0$, and φ^i are Casimir functions. But then, an easy computation with Lie derivative formulas, and for left invariant 1-forms α, β, gives

$$(L_A\gamma)(\alpha, \beta) = \varphi^i(L_{\bar{Z}^i}\gamma)(\alpha, \beta) - \alpha(\bar{Z}_i)\gamma(d\varphi^i, \beta) - \beta(\bar{Z}_i)\gamma(\alpha, d\varphi^i) = 0,$$

because γ is bi-invariant, and $(d\varphi^i)^\# = 0$. Q.e.d.

Theorem 10.4 tells us how to look for bi-invariant Poisson structures on G. Namely, we have to find an abelian ideal $\mathcal{S}_0$ in the Lie algebra $\mathcal{G}$ of G (if any), and find an ad $\mathcal{G}$-invariant symplectic structure σ on $\mathcal{S}_0$. Then, take the natural projection of the dual spaces $p : \mathcal{G}^* \longrightarrow \mathcal{S}_0^*$ (restriction of forms), which, necessarily, commutes with coad, and put $\#_{\gamma_0} = \#_\sigma \circ p : \mathcal{G}^* \longrightarrow \mathcal{S}_0$. The corresponding bivector γ_0 is ad$\mathcal{G}$-invariant, and the translation of γ_0 along G yields the requested Poisson structure. For instance, if $\mathcal{S}_0$ is an even-dimensional ideal contained in the center of $\mathcal{G}$, any symplectic form σ on $\mathcal{S}_0$ is clearly ad$\mathcal{G}$-invariant, and we can get a corresponding bi-invariant Poisson structure. Such situations can be obtained by constructing *central extensions* of given Lie algebras (e.g., [TW]).

10.2 Characteristic properties of Poisson-Lie groups

Now we come back to the Poisson-Lie groups. First, we shall notice various other properties of multiplicative tensor fields and of the Poisson bivector γ of (10.5), in particular.

In order to do so, for any tensor t at $g \in G$ we shall denote by t_l, t_r its images at the unit e of G by left and right translations, respectively. Then, we have

10.5. Proposition [LW2]. A $(p, 0)$-*tensor field* t *on the Lie group* G *is multiplicative iff one of the following equivalent properties holds:*

1°. $t_l(g_1 g_2) = t_l(g_2) + (\text{Ad } g_2^{-1}) t_l(g_1)$.

2°. $t_r(g_1 g_2) = t_r(g_1) + (\text{Ad } g_1) t_r(g_2)$.

3°. *(If G is connected)* $t(e) = 0$, and $L_X t$ *is left invariant whenever* X *is a left invariant vector field.*

4°. *(If G is connected)* $t(e) = 0$, *and* $L_X t$ *is right invariant whenever* X *is a right invariant vector field. Similar results hold for covariant fields.*

Proof. By definition,

$$t_l(g_1 g_2) = L_{(g_1 g_2)^{-1}*} t(g_1 g_2), \quad t_r(g_1 g_2) = R_{(g_1 g_2)^{-1}*} t(g_1 g_2),$$

and a simple computation shows that 1°, 2° are equivalent to (10.5) for t.

For 3°, if t is multiplicative, $t(e) = 0$ follows as in Proposition 10.3, and, since the flow of the left invariant field X is $R_{\exp(sX)}$, we have

$$(L_X t)(g) = \frac{d}{ds}\bigg|_{s=0} R_{\exp(-sX)*} t(g.\exp(sX)) \overset{(10.5)}{=}$$

$$= \frac{d}{ds}\bigg|_{s=0} R_{\exp(-sX)*} [L_{g*} t(\exp sX) + R_{\exp(sX)*} t(g)] =$$

$$= \frac{d}{ds}\bigg|_{s=0} R_{\exp(-sX)*} [L_{g*} (t(\exp sX)].$$

If we replace here g by hg ($h \in G$) we get the required invariance.

$$(L_X t)(hg) = L_{h*}[(L_X t)(g)]. \tag{10.6}$$

Conversely, if the Lie derivatives of (10.6) are expressed as derivatives along the flow of X, one gets

$$\left.\frac{d}{ds}\right|_{s=0} \{R_{\exp(-sX)*}[t(hg\exp(sX)) - L_{h*}(t(g\exp(sX)))]\} = 0. \tag{10.7}$$

By the usual trick of translating along the flow, (10.7) holds for all s, and it follows that

$$\begin{aligned} R_{\exp(-sX)*}[t(hg\exp(sX)) - L_{h*}(t(g\exp(sX))) &= \text{const.} = \\ &= t(hg) - L_{h*}(t(g)), \end{aligned} \tag{10.8}$$

where the last expression is obtained for $s = 0$. Then, if we take $g = e$, $s = 1$ in (10.8), we have

$$t(\text{hexp } X) = L_{h*}(t(\exp X)) + R_{(\exp X)*}[t(h) - L_{h*}t(e)], \tag{10.9}$$

and, since $t(e) = 0$, (10.9) expresses the multiplicativity of t in a neighbourhood of e. G being connected, we are done with 3°. 4° follows similarly. Q.e.d.

10.6. Remark. Again, if G is connected, it is clear that the left invariance of $L_X t$ (X left invariant) means $L_Y L_X t = 0$ for the right invariant vector fields Y, and $L_Y t$ (Y right invariant) is right invariant iff $L_X L_Y t = 0$ for X left invariant. For instance, using this we see immediately that the Schouten bracket $[P, Q]$ ($P \in \mathcal{V}^p G$, $Q \in \mathcal{V}^q G$) of two multiplicative fields P, Q is multiplicative [LW2]. Indeed, $P(e) = Q(e) = 0$ implies $[P, Q](e) = 0$ by (1.18). Then (1.11) gives

$$L_X[P, Q] = [L_X P, Q] + [P, L_X Q] \tag{10.10}$$

hence, for a left invariant X and a right invariant Y we shall have

$$L_Y L_X[P, Q] = [L_X P, L_Y Q] + [L_Y P, L_X Q], \tag{10.11}$$

where $L_X P, L_X Q$ are left invariant, and $L_Y P, L_Y Q$ are right invariant. By (1.12), and since the bracket of a left and a right invariant vector field is zero, the right-hand side of (10.11) vanishes, and we are done.

As a consequence of Proposition 10.5, 3°, we can also prove the following nice characterization of the Poisson-Lie groups

10.7. Proposition [We7], [DS2]. *Let G be a connected Lie group, and w a Poisson bivector on G. Then (G, w) is a Poisson-Lie group iff $w(e) = 0$, and for any two left invariant (right invariant) 1-forms α, β, the bracket $\{\alpha, \beta\}$ is also left-invariant (right invariant).*

Proof. For $\{\alpha, \beta\}$ we shall use the expression (4.6), and we shall evaluate $(L_Y\{\alpha,\beta\})(X)$, where Y is right invariant, and X is left invariant. Since $[Y, X] = 0$, we get

$$\begin{aligned}(L_Y\{\alpha,\beta\})(X) &= Y(\{\alpha,\beta\}(X)) = \\ &= Y[X(w(\alpha,\beta)) + (d\beta)(\alpha^{\#}, X) - (d\alpha)(\beta^{\#}, X)] = \\ &= Y[(L_X w)(\alpha,\beta) + w(L_X\alpha, \beta) + w(\alpha, L_X\beta) + \\ &+ (d\beta)(\alpha^{\#}, X) - (d\alpha)(\beta^{\#}, X)] = Y[(L_X w)(\alpha,\beta)],\end{aligned} \tag{10.12}$$

where we get the last form by developing $d\alpha, d\beta$, making cancellations, and using $\beta(X) = \text{const.}$, $\alpha(X) = \text{const.}$ which follows from the left invariance. Now, it is clear that $L_Y\{\alpha,\beta\} = 0$ iff $L_X w$ is left invariant, which, clearly, implies the conclusion required. In the right invariant case, we change the role of X and Y. Q.e.d.

This result led Dazord and Sondaz to define and study the more general class of *affine Poisson groups*, where the condition $w(e) = 0$ is dropped [DS2].

In order to obtain another important characterization of Poisson-Lie groups, we need to remember the general definition of Lie algebra cohomology that we gave in Chapter 6 (see formula (6.18)), and to make the following preparations. A k-vector field $Q \in \mathcal{V}^k(G)$ such that $Q(e) = 0$ defines a natural mapping $d_eQ : \mathcal{G} \to \Lambda^k\mathcal{G}$ ($\mathcal{G}$ is the Lie algebra of G), called the *intrinsic derivative* of Q at e, and given by

$$(d_eQ)(X) = (L_{\bar{X}}Q)(e), \tag{10.13}$$

where $X \in \mathcal{G}$, and $\bar{X}$ is an *arbitrary* vector field on G such that $\bar{X}(e) = X$. The result is independent on the choice of $\bar{X}$ since $L_{\bar{X}}Q$ is the Schouten bracket $[\bar{X}, Q]$ i.e., we may use (1.18), and $Q(e) = 0$. The notation d_e is justified since if we take $\bar{X}$ left invariant (right invariant), and look at Q_r as a function $Q_r : G \to \Lambda^k\mathcal{G}$ ($Q_l : G \to \Lambda^k\mathcal{G}$) we see that d_eQ is the usual differential of the function Q_r at e, i.e., $d_eQ = d_eQ_r = d_eQ_l$.

10.8. Remark [LW2]. If G is connected and Q is multiplicative, then $Q \equiv 0$ iff $d_eQ = 0$. Of course, $Q \equiv 0$ implies $d_eQ = 0$. Conversely, $d_eQ = 0$ implies $L_{\bar{X}}Q(e) = 0$ for any left invariant field $\bar{X}$, and, since $L_{\bar{X}}Q$ is left invariant because of multiplicativity, $L_{\bar{X}}Q \equiv 0$ on G. But, in turn, this relation shows that Q is right invariant, and $Q(e) = 0$ implies $Q \equiv 0$. Since the sum (difference) of two multiplicative fields $Q_{1,2}$ is obviously multiplicative, we also see that $d_eQ_1 = d_eQ_2$ is equivalent to $Q_1 \equiv Q_2$.

Furthermore, if $u \in \mathcal{V}^2(G)$ and $u(e) = 0$, we have $d_eu : \mathcal{G} \to \Lambda^2\mathcal{G}$, and this has a dualization that is denoted by

$$[\ ,\ ]_u : \mathcal{G}^* \wedge \mathcal{G}^* = \Lambda^2\mathcal{G}^* \to \mathcal{G}^*, \tag{10.14}$$

and is given by

$$[\alpha, \beta]_u(X) = (L_{\bar{X}}u)_e(\alpha, \beta) = \bar{X}_e(u(\bar{\alpha}, \bar{\beta})) = X(u(\bar{\alpha}, \bar{\beta})),$$

where $\bar{X}$ is as in (10.13), $\bar{\alpha}, \bar{\beta}$ are 1-forms on G such that $\bar{\alpha}(e) = \alpha$, $\bar{\beta}(e) = \beta$, and we used the fact that $u(e) = 0$. In other words,

$$[\alpha, \beta]_u = d_e(u(\bar{\alpha}, \bar{\beta})). \tag{10.15}$$

Now, we can formulate [LW2]

10.9. Theorem. *Let G be a Lie group, and w a bivector on G such that $w(e) = 0$. Then: 1) If w is multiplicative $d_e w : \mathcal{G} \to \Lambda^2\mathcal{G}$ is a $\Lambda^2\mathcal{G}$-valued 1-cocycle with respect to the adjoint action of $\mathcal{G}$ on $\Lambda^2\mathcal{G}$. Conversely, if G is connected and simply connected, for any ad-1-cocycle $\epsilon : \mathcal{G} \to \Lambda^2\mathcal{G}$ there exists a unique multiplicative bivector w on G such that $\epsilon = d_e w$. 2) If w is a Poisson bivector, $[\ ,\ ]_w : \Lambda^2\mathcal{G}^* \to \mathcal{G}^*$ satisfies the Jacobi identity, and it is a Lie bracket on $\mathcal{G}^*$. Furthermore, if G is connected, the multiplicative bivector w is Poisson iff $[\ ,\ ]_w$ is a Lie bracket on $\mathcal{G}^*$.*

Proof. By (6.18), we have $\forall X, Y \in \mathcal{G}$

$$\partial(d_e w)(X, Y) = \text{ ad } X(d_e w(Y)) - \text{ ad } Y(d_e w(X)) - d_e w([X, Y]). \tag{10.16}$$

Let us denote by $\bar{X}, \bar{Y}$ the left invariant fields equal to X, Y at e. If w is multiplicative, $L_{\bar{Y}}w$ is left invariant, and we get

$$\begin{aligned}\text{ad } X(d_e w(Y)) = \text{ ad } X(L_{\bar{Y}}w(e)) &= \left.\frac{d}{ds}\right|_{s=0} \text{Ad } \exp(sX)(L_{\bar{Y}}w(e)) = \\ &= \left.\frac{d}{ds}\right|_{s=0} R_{\exp(-sX)*}L_{\exp(sX)*}(L_{\bar{Y}}w(e)) = \\ &= \left.\frac{d}{ds}\right|_{s=0} R_{\exp(-sX)*}\left(L_{\bar{Y}}w(\exp(sX))\right) = \left(L_{\bar{X}}L_{\bar{Y}}w\right)(e).\end{aligned}$$

Accordingly, (10.16) becomes

$$\partial(d_e w)(X, Y) = L_{\bar{X}}L_{\bar{Y}}w(e) - L_{\bar{Y}}L_{\bar{X}}w(e) - L_{[\bar{X}, \bar{Y}]}w(e) = 0$$

i.e., $d_e w$ is a 1-cocycle.

For the second assertion of 1), we particularize the general results of §3 of [KSM1]. For the given 1-cocycle ϵ, we define the 1-form $\tilde{\epsilon}$ on G, with values in $\Lambda^2\mathcal{G}$, as follows

$$\tilde{\epsilon}_g(X_g) = \text{ Ad } g(\epsilon(L_{g^{-1}*}X_g)) \qquad (g \in G, X_g \in T_gG). \tag{10.17}$$

It is easy to check that this form satisfies the equivariance condition

$$L_g^*\tilde{\epsilon} = (\mathrm{Ad}\, g) \circ \tilde{\epsilon} \qquad (g \in G), \tag{10.18}$$

and the same holds for its exterior differential $d\tilde{\epsilon}$. Hence, if we shall prove that $(d\tilde{\epsilon})(e) = 0$, we shall be able to conclude that $\tilde{\epsilon}$ is a closed 1-form. To check this, let $X, Y \in \mathcal{G}$, and $\bar{X}, \bar{Y}$ be the corresponding left invariant vector fields. Then

$$(d_e\tilde{\epsilon})(X,Y) = X(\tilde{\epsilon}(\bar{Y})) - Y(\tilde{\epsilon}(\bar{X})) - \epsilon([X,Y]).$$

But

$$X(\tilde{\epsilon}(\bar{Y})) = \left.\frac{d}{ds}\right|_{s=0} (\tilde{\epsilon}_{\exp sX}(\bar{Y}_{\exp sX})) =$$

$$\left.\frac{d}{ds}\right|_{s=0} (\mathrm{Ad}(\exp sX)\epsilon(Y)) = (\mathrm{ad}X)(\epsilon(Y)),$$

and a similar result holds for $Y(\tilde{\epsilon}(\bar{X}))$. Hence $d_e\tilde{\epsilon} = \partial\epsilon = 0$, and we see that $d\tilde{\epsilon} = 0$ is indeed satisfied.

Now, if G is simply connected the 1-form $\tilde{\epsilon}$ must be exact i.e., $\tilde{\epsilon} = dw_r$ for a certain function $w_r : G \to \Lambda^2\mathcal{G}$. Moreover, since G is also connected, the function w_r will be uniquely defined by the initial condition $w_r(e) = 0$. Then, if we define the bivector w by $w(g) = R_{g*}(w_r(g))$, we have $d_e w = d_e w_r = \epsilon$. Finally, to prove that this w is multiplicative, it suffices to check for w_r the equality 2° of Proposition 10.5. If $\tilde{\epsilon} = dw_r$ is replaced in (10.17), we obtain

$$X_g w_r = (\mathrm{Ad}\, g)[X_g(w_r \circ L_{g^{-1}})], \tag{10.19}$$

and if, again, we extend X_g to the left invariant vector field $\bar{X}$, (10.19) means

$$\left.\frac{d}{ds}\right|_{s=0} [w_r(g \exp sX) - (\mathrm{Ad}\, g)(w_r(\exp sX))] = 0. \tag{10.20}$$

By translation along the orbit of $\exp sX$, the same holds for all s, and the values of the bracket of (10.20) at $s = 1$ is the same as the value at $s = 0$ i.e.,

$$w_r(g \exp X) - (\mathrm{Ad}\, g)(w_r(\exp X)) = w_r(g). \tag{10.21}$$

G being connected, (10.21) implies 2° of Proposition 10.5.

Notice that, in fact, what we showed by the above calculation is that if w is a bivector on the connected Lie group G, such that $w(e) = 0$, and if dw_r is equivariant, then w is multiplicative. The converse of this assertion is also true since, if w is multiplicative, we have (10.21), and we can go back from it to (10.19) and to corresponding equalities (10.17) and (10.18).

Now, we go over to part 2) of the theorem. Since $w(e) = 0$, a Taylor development (3.12) of w at e exists, i.e., we have

$$w^{ij}(x) = c_k^{ij} x^k + o(2),$$

where c_k^{ij} = constant (but not the structural constants of $\mathcal{G}$!), and x^k are coordinates of G around e. Accordingly, and with the notation of (10.15) and for $\bar{\alpha}(e) = \alpha_i dx^i$, $\bar{\beta}(e) = \beta_i dx^i$, we get

$$d_e(w(\bar{\alpha}, \bar{\beta})) = (c_k^{ij} \alpha_i \beta_j) dx^k,$$

and this shows that the bracket $[\ ,\ ]_w$ is precisely the linear approximation of the Poisson structure w at $e \in G$, as defined in Chapter 3. This justifies the first assertion of 2).

Furthermore, let G be connected, w multiplicative and $[\ ,\ ]_w$ satisfy the Jacobi identity. Then, Remark 10.6 shows that the Schouten bracket $[w, w]$ is multiplicative, and the use of (10.15) by (3.12) as above yields (see also (1.5))

$$(d_e[w,w])(\alpha, \beta, \gamma) = -2([\alpha, [\beta, \gamma]_w]_w + [\beta, [\gamma, \alpha]_w]_w + [\gamma, [\alpha, \beta]_w]_w) = 0.$$

Now, finally, Remark 10.8 shows that $[w, w] \equiv 0$. Q.e.d.

If w is as in part 2) of the previous theorem, we may also notice that $\forall \alpha, \beta \in \mathcal{G}^*$ one has

$$[\alpha, \beta]_w = \{\bar{\alpha}, \bar{\beta}\}(e) = \{\tilde{\alpha}, \tilde{\beta}\}(e), \tag{10.15'}$$

where $\bar{\alpha}, \bar{\beta}$ are the left invariant forms defined by α, β, and $\tilde{\alpha}, \tilde{\beta}$ are the right invariant forms, while $\{\ ,\ \}$ is the bracket of the Poisson calculus (Chapter 4). Indeed, with (4.6), and if $\bar{X}$ is left invariant with $X = \bar{X}(e)$, we get

$$\{\bar{\alpha}, \bar{\beta}\}_e(X) = -X(w(\bar{\alpha}, \bar{\beta})) + \beta(L_{\bar{X}} \bar{\alpha}^{\#}) - \alpha(L_{\bar{X}} \bar{\beta}^{\#}) = -X(w(\bar{\alpha}, \bar{\beta})) +$$

$$+ X(\bar{\beta}(\bar{\alpha}^{\#})) - (L_{\bar{X}} \bar{\beta})_e(\alpha^{\#}) + (L_{\bar{X}} \bar{\alpha})_e(\beta^{\#}) - X(\bar{\alpha}(\bar{\beta}^{\#})) = X(w(\bar{\alpha}, \bar{\beta})) -$$

$$- w_e(\bar{\alpha}, L_{\bar{X}} \bar{\beta}) - w_e(L_{\bar{X}} \bar{\alpha}, \bar{\beta}) = [\alpha, \beta]_w(X),$$

in view of (10.15) and since $w_e = 0$.

A similar computation will give the second part of (10.15').

10.3 The Lie algebra of a Poisson-Lie group

Theorem 10.9 is important because it clarifies the extra-structure that exists on the Lie algebra $\mathcal{G}$ of a Poisson-Lie group (G, w). Namely, let us give

10.10. Definition. A *Poisson-Lie algebra* (or *Lie-Drinfeld algebra* in [KS2]) $(\mathcal{G}, \epsilon)$ is a Lie algebra $\mathcal{G}$ endowed with a $\Lambda^2 \mathcal{G}$-valued 1-cocycle ϵ relative to the adjoint representation of $\mathcal{G}$ on $\Lambda^2 \mathcal{G}$, such that its dual, denoted by $[\ ,\ ]^*$, satisfies the Jacobi identity. Such an ϵ is also called a *Jacobian cocycle*.

10.11. Proposition. *A Poisson-Lie structure on the Lie algebra $\mathcal{G}$ is equivalent with a Lie bracket $[\ ,\ ]^*$ on the dual space $\mathcal{G}^*$ of $\mathcal{G}$ such that*

$$[\xi,\eta]^*([X,Y]) = -[(\text{coad } X)\xi,\eta]^*(Y)- \\ -[\xi,(\text{coad } X)\eta]^*(Y)+[(\text{coad } Y)\xi,\eta]^*(X)+[\xi,(\text{coad } Y)\eta]^*(X). \tag{10.22}$$

Proof. (10.22) is exactly the translation of the condition that the dual mapping $\epsilon : \mathcal{G} \to \mathcal{G}\wedge\mathcal{G}$ of $[\ ,\ ]^* : \mathcal{G}^*\wedge\mathcal{G}^* \to \mathcal{G}^*$ is a 1-cocycle for the adjoint representation i.e.,

$$(\text{ad } X)(\epsilon(Y)) - (\text{ad } Y)(\epsilon(X)) - \epsilon([X,Y]) = 0.$$

Of course, the ad-representation extends to $\mathcal{G}\wedge\mathcal{G}$ by

$$(\text{ad } X)(Y\wedge Z) = (\text{ad } X)(Y)\wedge Z + Y\wedge(\text{ad } X)(Z) = \\ = [X,Y]\wedge Z + Y\wedge[X,Z],$$

whence, a similar formula holds for coad on $\mathcal{G}^*\wedge\mathcal{G}^*$, and this is used in the translation of the cocycle condition. Q.e.d.

Accordingly, the Poisson-Lie algebra $(\mathcal{G},\epsilon)$ can be seen as $(\mathcal{G},\mathcal{G}^*,[\ ,\ ], [\ ,\ ]^*)$, and, for this reason it is called usually a *Lie bialgebra* (or a *Lie bigebra* [KSM1]). (Lie bialgebras are generalized to *Lie bialgebroids* in [MX].)

Now, Theorem 10.9 yields straightforwardly

10.12. Theorem [Dr1,2]. *If (G,w) is a Poisson-Lie group, then $(\mathcal{G},d_e w)$ is a Poisson-Lie algebra (or equivalently, $(\mathcal{G},\mathcal{G}^*)$ is a Lie bialgebra). Conversely, if $(\mathcal{G},\epsilon)$ (or $(\mathcal{G},\mathcal{G}^*)$) is a Poisson-Lie algebra, there exists a unique connected and simply connected Poisson-Lie group G whose associated Poisson-Lie algebra is the given one.*

Of course, uniqueness is up to an isomorphism, and the existence of G follows from Lie's third theorem.

In order to make an even sharper assertion, we define the notion of a *(homo)morphism* $\varphi : (G_1,w_1) \to (G_2,w_2)$ of Poisson-Lie groups by the natural conditions: i) φ is a homomorphism of Lie groups, and ii) φ is a Poisson mapping. Then, we define a *(homo)morphism* $f : (\mathcal{G}_1,\mathcal{G}_1^*) \to (\mathcal{G}_2,\mathcal{G}_2^*)$, or $f : (\mathcal{G}_1,\epsilon_1) \to (\mathcal{G}_2,\epsilon_2)$, of Poisson-Lie algebras as a Lie algebra homomorphism $f : \mathcal{G}_1 \to \mathcal{G}_2$ whose transposed map $f^* : \mathcal{G}_2^* \to \mathcal{G}_1^*$ is also a Lie algebra homomorphism or, equivalently, $(f\wedge f)\circ\epsilon_1 = \epsilon_2\circ f$. Every $\varphi : (G_1,w_1) \to (G_2,w_2)$ as above induces $\varphi_*(e) : \mathcal{G}_1 \to \mathcal{G}_2$ which is a Poisson-Lie algebra homomorphism since one checks easily that $w_{2,r}\circ\varphi = \varphi_*(e)\circ w_{1,r}$ whence, $d_e w_{2,r}\circ\varphi_*(e) = (\varphi_*(e)\wedge\varphi_*(e))\circ d_e w_{1,r}$ i.e., exactly what we need. Conversely, a morphism $f : \mathcal{G}_1 \to \mathcal{G}_2$ lifts to a corresponding $\varphi : G_1 \to G_2$ of the associated connected and simply connected Lie groups, and if f commutes correctly with the 1-cocycles ϵ_1,ϵ_2, φ will be a Poisson map of the corresponding Poisson structures constructed in the proof of Theorem 10.9.

The conclusion is that the correspondence of Theorem 10.12 is an equivalence between the category of the connected, simply connected Poisson-Lie groups and morphisms, and the category of the Poisson-Lie algebras and their morphisms.

The simplest example is that of the abelian Lie group $\mathbb{R}^n$. Using Proposition 10.5, 3°, it follows immediately that a Poisson structure w on $\mathbb{R}^n$ makes the latter into a Poisson-Lie group iff w is linear (see Chapter 3) [KS4], [CG-]. In other words, if $\mathcal{G}$ is a Lie algebra, its dual space $\mathcal{G}^*$ is a Poisson-Lie group with the Lie structure given by the (commutative) addition of vectors of $\mathcal{G}^*$, and the Poisson structure given by what we called the Lie-Poisson structure in Chapter 3. The Lie bialgebra of the Poisson-Lie group $\mathcal{G}^*$ is $(\mathcal{G}^*, \mathcal{G})$, where the bracket of $\mathcal{G}^*$ is zero. Furthermore, since the components of the linear Poisson structure w of $\mathbb{R}^n$ are not periodic, unless they are zero, the Lie group T^n = torus has no nontrivial Poisson structure that makes it into a Poisson-Lie group.

We also mention a few other simple examples following [Am] and [KS4]:

1) Let (e_1, e_2) be a basis of the non-commutative 2-dimensional Lie algebra $\mathcal{G}_2$ such that $[e_1, e_2] = e_2$. Let (ϵ^1, ϵ^2) be the dual cobasis in $\mathcal{G}_2^*$, and define $[\epsilon^1, \epsilon^2]^* = \epsilon^2$. Then, it is easy to check (10.22), and see that $\mathcal{G}_2$ is a Poisson-Lie algebra, leading, therefore, to a Poisson-Lie group.

2) Take $\mathcal{G} = gl(2, \mathbb{C})$, and its complex basis

$$e_1 = \begin{pmatrix} 1 & 0 \\ 0 & 0 \end{pmatrix}, \quad e_2 = \begin{pmatrix} 0 & 1 \\ 0 & 0 \end{pmatrix}, \quad e_3 = \begin{pmatrix} 0 & 0 \\ 1 & 0 \end{pmatrix}, \quad e_4 = \begin{pmatrix} 0 & 0 \\ 0 & 1 \end{pmatrix}.$$

Take the dual cobasis ϵ^i of $\mathcal{G}^*$, and put

$$[\epsilon^1, \epsilon^2]^* = \epsilon^1 - \epsilon^4, [\epsilon^1, \epsilon^3]^* = 0, [\epsilon^1, \epsilon^4]^* = \epsilon^3,$$
$$[\epsilon^2, \epsilon^3]^* = \epsilon^3, [\epsilon^2, \epsilon^4]^* = 0, [\epsilon^3, \epsilon^4]^* = 0.$$

Then, again, technical computations show that (10.22) holds, and we have the structure of a Lie bialgebra, leading to a Poisson-Lie group structure of $Gl(2, \mathbb{C})$.

3) Take $\mathcal{G} = sl(2, \mathbb{C})$ with the complex basis

$$e_1 = \begin{pmatrix} 0 & 1 \\ 0 & 0 \end{pmatrix}, \quad e_2 = \begin{pmatrix} 0 & 0 \\ 1 & 0 \end{pmatrix}, \quad e_3 = \begin{pmatrix} 1 & 0 \\ 0 & -1 \end{pmatrix},$$

that satisfies

$$[e_1, e_2] = e_3, \quad [e_1, e_3] = -2e_1, \quad [e_2, e_3] = 2e_2,$$

and the dual cobasis ϵ^i $(i = 1, 2, 3)$. Define

$$[\epsilon^1, \epsilon^2]^* = 2\epsilon^2, \quad [\epsilon^1, \epsilon^3]^* = 2\epsilon^3, \quad [\epsilon^2, \epsilon^3]^* = 0.$$

Then (10.22) holds, and one gets a Poisson-Lie algebra structure on $sl(2,\mathbb{C})$ and, correspondingly, a Poisson-Lie group structure on $Sl(2,\mathbb{C})$.

As a matter of fact, the dual brackets of the previous examples were not just lucky guesses, and how they were obtained will be explained at the end of the next section.

10.4 The Yang-Baxter equations

In what follows, we shall discuss some important general equations on Lie algebras, whose solutions lead to Poisson-Lie structures.

We saw that whenever $w \in \mathcal{V}^2(G)$ is multiplicative, $d_e w$ is a 1-cocycle (Theorem 10.9, 1)), and we shall say that w is ad-*exact* if this cocycle is a coboundary i.e., there exists $\mathbf{r} \in \mathbf{\Lambda}^2\mathcal{G}$ such that

$$d_e w(X) = \partial\mathbf{r}(X) \stackrel{\text{def}}{=} (\text{ad } X)(\mathbf{r}) \qquad (\forall X \in \mathcal{G}). \tag{10.23}$$

It is easy to see that $\forall \mathbf{r} \in \mathbf{\Lambda}^2\mathcal{G}$ there exists w satisfying (10.23) namely,

$$w(g) = L_{g*}\mathbf{r} - R_{g*}\mathbf{r}. \tag{10.24}$$

Indeed, w of (10.24) satisfies (10.5), and

$$(d_e w)(X) = (L_{\bar{X}} w)(e) = \left.\frac{d}{ds}\right|_{s=0} R_{\exp(-sX)*} w(\exp(sX)) =$$

$$= \left.\frac{d}{ds}\right|_{s=0} [R_{\exp(-sX)*} L_{\exp(sX)*}\mathbf{r} - \mathbf{r}] = (\text{ad } X)\mathbf{r},$$

where $\bar{X}$ is the left invariant vector field defined by X. Finally, Remark 10.8 ensures that, if G is connected, w of (10.24) is the unique bivector that satisfies (10.23).

Now, the question is for which $\mathbf{r} \in \mathcal{G}$, is w a Poisson bivector i.e., we have an ad-*exact Poisson-Lie group structure* on G?

First, notice that we may see formula (1.12) as relating to elements of an arbitrary Lie algebra $\mathcal{G}$, and use it to get a well defined *algebraic Schouten bracket* $[P, Q]$ of any $P \in \Lambda^h\mathcal{G}$, $Q \in \Lambda^k\mathcal{G}$ (exercise). Moreover, if $\forall P \in \Lambda^h\mathcal{G}$ we agree to denote by $\bar{P}$ the corresponding left-invariant h-vector field on the Lie group G of Lie algebra $\mathcal{G}$, and by $\tilde{P}$ the right invariant field, and if we notice the simple relations

$$[X, Y] = (\text{ad } X)(Y) = (L_{\bar{X}}\bar{Y})(e) = -(L_{\tilde{X}}\tilde{Y})(e) \tag{10.25}$$

for $X, Y \in \mathcal{G}$, then (1.12) yields the relation between the algebraic and the differential Schouten-Nijenhuis brackets

$$[P, Q] = [\bar{P}, \bar{Q}](e) = -[\tilde{P}, \tilde{Q}](e). \tag{10.26}$$

As a matter of fact, if $\mathcal{G}$ is identified with the algebra of the left invariant vector fields on the Lie group G, the algebraic Schouten bracket is just the restriction of the Schouten-Nijenhuis bracket to the left invariant multivector fields of G. That this set of fields is closed under Schouten-Nijenhuis brackets follows from Remark 1.3, 2).

With these preparations, we can prove the following result of Drinfeld

10.13. Proposition [Dr1,2], [CG-]. *If* $\mathbf{r} \in \Lambda^2\mathcal{G}$, *then* $w = \bar{\mathbf{r}} - \tilde{\mathbf{r}}$ *is a Poisson bivector that makes* G *into a Poisson-Lie group iff the algebraic Schouten bracket* $[\mathbf{r},\mathbf{r}]$ *is* AdG*-invariant.*

Proof. Of course, w is the same as that of (10.24), and we have

$$[w,w] = [\bar{\mathbf{r}} - \tilde{\mathbf{r}}, \bar{\mathbf{r}} - \tilde{\mathbf{r}}] = [\bar{\mathbf{r}},\bar{\mathbf{r}}] + [\tilde{\mathbf{r}},\tilde{\mathbf{r}}] - 2[\bar{\mathbf{r}},\tilde{\mathbf{r}}]. \tag{10.27}$$

Since any bracket $[\bar{X},\tilde{Y}] = 0$, formula (1.12) shows that we must have $[\bar{\mathbf{r}},\tilde{\mathbf{r}}] = 0$. On the other hand, since the Schouten-Nijenhuis brackets of φ-related fields are φ-related for any differentiable mapping φ (Remark 1.3, 2)) we see that $[\bar{r},\bar{r}]$ is left invariant, and $[\tilde{r},\tilde{r}]$ is right invariant, and this implies $[\bar{\mathbf{r}},\bar{\mathbf{r}}] = \overline{[\mathbf{r},\mathbf{r}]}$, and $[\tilde{\mathbf{r}},\tilde{\mathbf{r}}] = -\widetilde{[\mathbf{r},\mathbf{r}]}$. Therefore, (10.27) becomes

$$[w,w](g) = \overline{[\mathbf{r},\mathbf{r}]}(g) - \widetilde{[\mathbf{r},\mathbf{r}]}(g) = R_{g*}\left((\mathrm{Ad}g)[\mathbf{r},\mathbf{r}] - [\mathbf{r},\mathbf{r}]\right). \tag{10.28}$$

Hence $[w,w] = 0$ iff, $\forall g \in G$, $[\mathbf{r},\mathbf{r}] = (\mathrm{Ad}g)[\mathbf{r},\mathbf{r}]$. Q.e.d.

Because of a similarity with a famous equation of statistical mechanics, one defines

10.14. Definition. The equation

$$[\mathbf{r},\mathbf{r}] = 0 \qquad (\mathbf{r} \in \mathcal{G} \wedge \mathcal{G}), \tag{10.29}$$

where the operation is the algebraic Schouten bracket, is called the *classical Yang-Baxter* (YB) *equation* [GD]. The equation

$$[\mathbf{r},\mathbf{r}] = (\mathrm{Ad}G)[\mathbf{r},\mathbf{r}] \tag{10.30}$$

is the *group-generalized classical Yang-Baxter* (GYB) *equation.*

Accordingly, Proposition (10.13) offers a Poisson-Lie group structure for every solution of the generalized and, in particular, of the classical Yang-Baxter equation. We shall also refer to G with the ad-exact structure above as a *PL-group of the* GYB-*type* (*Poisson-Sklyanin groups* in [KS2]) or of the YB-type, respectively. The Lie bialgebra of a *PL*-group of the YB-type is also called *triangular*, and the corresponding solution $\mathbf{r}$ of the YB-equation is also called a *triangular classical* $\mathbf{r}$*-matrix* [ST1], a YB-*bivector* or a *triangular Jacobian potential.*

10.15. Remarks. 1) The Yang-Baxter equation (10.29) is obviously a condition that is equivalent to the fact that the left and right invariant bivectors $\bar{\mathbf{r}}, \tilde{\mathbf{r}}$ are Poisson structures on G.

2) The GYB-equation (G connected) can also be seen as a necessary and sufficient condition for the Jacobi identity of the bracket $[\ ,\ ]^*$ on $\mathcal{G}^*$, dual of the 1-cocycle (10.23), to be satisfied (Theorem 10.9). The expression of this dual bracket for $\alpha, \beta \in \mathcal{G}^*$, $X \in \mathcal{G}$ is

$$[\alpha, \beta]^*(X) = -\mathbf{r}((\text{coad } X)\alpha, \beta) - \mathbf{r}(\alpha, (\text{coad } X)\beta). \tag{10.31}$$

3) In other papers (e.g., [KS2]) the name *generalized Yang-Baxter equation* is used for

$$(\text{ad } X)[\mathbf{r}, \mathbf{r}] = 0 \qquad (\forall X \in \mathcal{G}). \tag{10.32}$$

We shall call (10.32) the *$\mathcal{G}YB$-equation.* Of course, (10.30) implies (10.32), and, if G is connected, the converse is also true.

4) If $\mathbf{r} \in \Lambda^2\mathcal{G}$ is a solution of the YB-equation, then, by arguments used in the proof of Proposition 10.13, it is clear that $\bar{r}$ and $\tilde{r}$ are compatible Poisson bivectors (i.e., $[\bar{\mathbf{r}}, \bar{\mathbf{r}}] = [\tilde{\mathbf{r}}, \tilde{\mathbf{r}}] = [\bar{\mathbf{r}}, \tilde{\mathbf{r}}] = 0$. See Definition 1.12). Also, for two solutions $\mathbf{r}_1, \mathbf{r}_2$ of the YB-equation $[\bar{\mathbf{r}}_1, \tilde{\mathbf{r}}_2] = 0$, and $(\bar{\mathbf{r}}_1, \tilde{\mathbf{r}}_2), (\tilde{\mathbf{r}}_1, \bar{\mathbf{r}}_2)$ are pairs of compatible Poisson structures.

5) Since a solution of the YB-equation is just a left invariant Poisson bivector on G, we can write a covariant version of the YB-equation. Namely, let g be a left invariant Riemannian metric on G. Then, by Proposition 1.9, $\mathbf{r}$ is a YB-bivector iff $\rho = \#_g^{-1}\bar{\mathbf{r}}$ is a left invariant 2-form such that

$$\delta_g(\rho \wedge \rho) = 2\rho \wedge \delta_g \rho, \tag{10.33}$$

and this can be transformed into an algebraic equation at $e \in G$.

In order to obtain another important equation related to those of Definition 10.14, let us show first

10.16. Lemma. *The algebraic Schouten bracket $[\mathbf{r}, \mathbf{r}]$, $\mathbf{r} \in \Lambda^2\mathcal{G}$, has also the following expression*

$$[\mathbf{r}, \mathbf{r}](\xi, \eta, \zeta) = -2 \sum_{\text{Cycl.}(\xi,\eta,\zeta)} \xi\left([\eta^{\#_r}, \zeta^{\#_r}]\right), \tag{10.34}$$

where $\xi, \eta, \zeta \in \mathcal{G}^*$, and $\#_r$ is defined as in Chapter 1.

Proof. Let $\{X_i\}$ be a basis of $\mathcal{G}$, and $\mathbf{r} = \frac{1}{2}r^{ij}X_i \wedge X_j$. Then the left invariant bivector $\bar{\mathbf{r}}$ is

$$\bar{\mathbf{r}} = \frac{1}{2}r^{ij}\bar{X}_i \wedge \bar{X}_j \qquad (r^{ij} = \text{const.}),$$

and formula (1.12) yields

$$[\bar{\mathbf{r}},\bar{\mathbf{r}}] = -\frac{1}{4}r^{ij}r^{hk}\left\{[\bar{X}_i,\bar{X}_h]\wedge\bar{X}_j\wedge\bar{X}_k - \right. \tag{10.35}$$

$$\left. -[\bar{X}_i,\bar{X}_k]\wedge\bar{X}_j\wedge\bar{X}_h - [\bar{X}_j,\bar{X}_h]\wedge\bar{X}_i\wedge\bar{X}_k + [\bar{X}_j,\bar{X}_k]\wedge\bar{X}_i\wedge\bar{X}_h\right\}.$$

Now, if we introduce the constant coefficients $r^{\cdot\cdot}$ into the Lie brackets, and notice that $r^{ij}X_j = (\xi^i)^{\#_r}$, where ξ^i is the dual cobasis $(\xi^i(X_j) = \delta^i_j)$, (10.35) becomes

$$[\mathbf{r},\mathbf{r}] = -\left[\#_r\xi^i,\#_r\xi^j\right]\wedge X_i\wedge X_j. \tag{10.36}$$

This relation implies

$$[\mathbf{r},\mathbf{r}](\xi^a,\xi^b,\xi^c) = -2\sum_{\mathrm{Cycl}}\xi^a([\#_r\xi^b,\#_r\xi^c]),$$

which is (10.34) on the cobasis ξ^a. (The factor 2 appears in view of our convention of using the value of an exterior product as in [Lh1] and [Mt].) Q.e.d.

Now, let us assume that we are in the case of a Lie algebra $\mathcal{G}$ which is endowed with a nondegenerate, ad-invariant symmetric bilinear form b, and let $\mathbf{R}:\mathcal{G}\to\mathcal{G}$ be defined by $\mathbf{R} = \#_r\circ\#_b^{-1}$. Then, if we take $\xi = \#_b^{-1}X$, $\eta = \#_b^{-1}Y$, $\zeta = \#_b^{-1}Z$ in (10.34), we get

$$[\mathbf{r},\mathbf{r}](\xi,\eta,\zeta) = -2\sum_{\mathrm{Cycl.}(X,Y,Z)} b(X,[\mathbf{R}Y,\mathbf{R}Z]). \tag{10.37}$$

Before going on, remember that the ad-invariance of $b\in\mathcal{G}\otimes\mathcal{G}$ means

$$b([U,X],Y) + b(X,[U,Y]) = 0 \quad (U,X,Y\in\mathcal{G}). \tag{10.38}$$

Using this formula it is easy to get

$$\#_b^{-1}[U,X] = (\mathrm{coad}\ U)\#_b^{-1}\mathrm{X}. \tag{10.39}$$

On the other hand, the skew-symmetry of $\mathbf{r}$ and the symmetry of b imply

$$b(X,\mathbf{R}Y) = -b(\mathbf{R}X,Y). \tag{10.40}$$

Using the last four formulas we get for the previous ξ,η,ζ

$$((\mathrm{ad}\ U)[\mathbf{r},\mathbf{r}])(\xi,\eta,\zeta) = -[\mathbf{r},\mathbf{r}](\mathrm{coad}\ U(\xi),\eta,\zeta) - \tag{10.41}$$

$$-[\mathbf{r},\mathbf{r}](\xi,\ \mathrm{coad}\ U(\eta),\zeta) - [\mathbf{r},\mathbf{r}])(\xi,\eta,\mathrm{coad}\ U(\zeta)) =$$

$$= 2\{b([U,X],[\mathbf{R}Y,\mathbf{R}Z]) + b(Y,[\mathbf{R}Z,\mathbf{R}[U,X]]) +$$

$$+ b(Z,[\mathbf{R}[U,X],\mathbf{R}Y]) + b(X,[\mathbf{R}[U,Y],\mathbf{R}Z]) +$$

$$+ b([U,Y],[\mathbf{R}Z,\mathbf{R}X]) + b(Z,[\mathbf{R}X,\mathbf{R}[U,Y]]) +$$

$$+b(X,[\mathbf{R}Y,\mathbf{R}[U,Z]]) + b(Y,[\mathbf{R}[U,Z],\mathbf{R}X]) + b([U,Z],[\mathbf{R}X,\mathbf{R}Y])\} =$$

$$= 2\sum_{\mathrm{Cycl.}(X,Y,Z)} b([U,X],[\mathbf{R},\mathbf{R}](Y,Z)),$$

where we denoted

$$[\mathbf{R},\mathbf{R}](Y,Z)=[\mathbf{R}Y,\mathbf{R}Z]-\mathbf{R}[\mathbf{R}Y,Z]-\mathbf{R}[Y,\mathbf{R}Z]. \tag{10.42}$$

As a consequence of this computation we have

10.17. Proposition. *The $\mathcal{G}YB$-equation* (10.32) *is equivalent to*

$$\sum_{\mathrm{Cycl.}(X,Y,Z)} [X,[\mathbf{R},\mathbf{R}](Y,Z)]=0 \quad (\forall X,Y,Z\in\mathcal{G}). \tag{10.43}$$

Proof. Ask the result of (10.41) to vanish $\forall U\in\mathcal{G}$, and use (10.38) and the nondegeneracy of b. Q.e.d.

10.18. Corollary. *The equation*

$$[\mathbf{R},\mathbf{R}](Y,Y)=\alpha[X,Y] \qquad (\alpha=\text{ const.}) \tag{10.44}$$

implies the $\mathcal{G}YB$-equation (10.32), *and* $\mathbf{r}$ *corresponding to a solution* $\mathbf{R}$ *of* (10.40) *and* (10.44) *defines a structure of a Poisson-Lie group on any connected Lie group G of Lie algebra $\mathcal{G}$.*

Proof. Obvious.

10.19. Definition [ST1], [KS3]. The equation (10.44) is called the *modified Yang-Baxter equation* of *coefficient* α (MYB_α). Its solutions are called *classical* $\mathbf{R}$*-matrices.*

A typical situation where the MYB_α-equation makes sense is that of a semisimple Lie algebra $\mathcal{G}$, where we may use as b the Killing form β of $\mathcal{G}$ (e.g., [Hl]). On the other hand, notice that $\mathbf{R}$ can also be seen as a left invariant field of endomorphisms of TG, where G is a connected Lie group of Lie algebra $\mathcal{G}$. For instance, if b exists, then a left invariant complex structure of G is a solution of MYB with $\alpha=1$, and a left invariant product structure of G is a solution of MYB with $\alpha=-1$. Indeed, these are well known integrability conditions of an almost complex, respectively, an almost product, structure (e.g., [KN], [Ya2]). In order to use Corollary 10.18 in these situations we must still require (10.40).

Another interesting result is

10.20. Proposition [ST1]. *With the notation of Proposition* 10.13, *if* $\mathbf{r}_1,\mathbf{r}_2\in\Lambda^2\mathcal{G}$ *correspond to two solutions,* $\mathbf{R}_1,\mathbf{R}_2$ *of* MYB_α, *then* $w=\bar{\mathbf{r}}_1-\tilde{\mathbf{r}}_2$ *is a (not necessarily multiplicative) Poisson bivector on the (connected) group G.*

Proof. Like in Proposition 10.13, we get $[w,w]=\overline{[\mathbf{r}_1,\mathbf{r}_1]}-[\mathbf{r}_2\tilde{,}\mathbf{r}_2]$. But, since MYB implies $\mathcal{G}YB, (\mathrm{ad}X)[\mathbf{r}_a,\mathbf{r}_a]=0,\ \forall X\in\mathcal{G},\ a=1,2$, and $\mathcal{G}YB$ implies GYB since G is connected, we get $\overline{[\mathbf{r}_a,\mathbf{r}_a]}=\widetilde{[\mathbf{r}_a,\mathbf{r}_a]}$ hence,

$$[w,w]=\overline{[\mathbf{r}_1,\mathbf{r}_1]}-\overline{[\mathbf{r}_2,\mathbf{r}_2]}. \tag{10.45}$$

Using (10.37), and the MYB_α-equation (10.44), we shall prove that $[\mathbf{r},\mathbf{r}]$ is the same for all the solutions $\mathbf{r}$ of MYB_α ($\alpha=$ const.), whence $[w,w]=0$ as requested. Indeed, modulo (10.44), (10.37) becomes

$$[\mathbf{r},\mathbf{r}](\xi,\eta,\zeta)=-2\sum_{\mathrm{Cycl.}(X,Y,Z)} b(X,\mathbf{R}[\mathbf{R}Y,Z]+\mathbf{R}[Y,\mathbf{R}Z]+\alpha[Y,Z]),$$

and after repeated applications of (10.38), (10.40), we see that the first two terms have equal cyclic sums, i.e.,

$$[\mathbf{r},\mathbf{r}](\xi,\eta,\zeta)=-4\sum_{\mathrm{Cycl.}} b(X,\mathbf{R}[\mathbf{R}Y,Z])-2\sum_{\mathrm{Cycl.}} b(X,\alpha[Y,Z]).$$

Applying again (10.38), (10.40) and (10.37), we get

$$[\mathbf{r},\mathbf{r}](\xi,\eta,\zeta)=4\sum_{\mathrm{Cycl.}} b(Z,[\mathbf{R}X,\mathbf{R}Y])-2\alpha\sum_{\mathrm{Cycl.}} b(X,[Y,Z])=$$

$$=-2[\mathbf{r},\mathbf{r}](\xi,\eta,\zeta)-2\alpha\sum_{\mathrm{Cycl.}} b(X,[Y,Z]),$$

which implies

$$[\mathbf{r},\mathbf{r}](\xi,\eta,\zeta)=-\frac{2}{3}\alpha\sum_{\mathrm{Cycl.}} b(X,[Y,Z]), \tag{10.46}$$

i.e., $[\mathbf{r},\mathbf{r}]$ is independent of $\mathbf{r}$. Q.e.d.

10.21. Proposition. *If* $\mathbf{R}$ *satisfies the* MYB_o*-equation (i.e.,* $\alpha=0$*) then* $\mathbf{r}$ *satisfies the* YB*-equation.*

Proof. Apply formula (10.46). Q.e.d.

This is one more reason for the name of a modified Yang-Baxter equation.

Again, let us give some simple examples:

1) [Am], [KS4]. Take $\mathcal{G}=sl(2,\mathbb{C})$ with the same basis e_1,e_2,e_3 as in example 3) of a Poisson-Lie algebra structure in Section 10.3, and take $\mathbf{r}=e_1\wedge e_3$. Using (1.12), we get $[\mathbf{r},\mathbf{r}]=0$ i.e., the classical YB-equation is satisfied. Moreover, using (10.23) and the duality of $d_e w$ with $[\ ,\ ]^*$, one sees that the

Lie bialgebra structure associated with this $\mathbf{r}$ is exactly the one that we gave at the end of the previous section.

2) [Am], [KS4]. $\mathcal{G} = gl(2,\mathbb{C})$, with the basis (e_i) $(i = 1,2,3,4)$ like in example 2), Section 10.3, of a Poisson-Lie algebra structure. Then, acting like in the previous example we see that the Poisson-Lie group structure mentioned is associated with the $\mathbf{r}$-matrix $\mathbf{r} = e_1 \wedge e_2$, which is a solution of the classical YB-equation.

3) [CG-]. $\mathcal{G} = sl(2,\mathbb{C})$ seen as the real Lie algebra $so(1,3)$. (Remember that $Sl(2,\mathbb{C})$ covers the Lorentz group $\mathrm{SO}_o(1,3)$.) One has the real basis

$$X_j = E_{oj} + E_{jo} \quad (1 \le j \le 3),$$
$$Y_j = \epsilon_{\mathrm{jkl}} E_{kl} \qquad (1 \le j,k,l \le 3),$$

where E_{ij} $(i,j = 0,1,2,3)$ is the matrix that has 1 at the $(i+1, j+1)$-entry and 0 elsewhere, and ϵ_{jkl} is the sign of the respective permutation. The structure equations are

$$[X_j, X_k] = \epsilon_{jkl} Y_l, \quad [X_j, Y_k] = -\epsilon_{jkl} X_l, \quad [Y_j, Y_k] = -\epsilon_{jkl} Y_l.$$

An ad-invariant nondegenerate bilinear form is given by

$$b(X_i, X_j) = -\delta_{jk}, \quad b(X_j, Y_k) = 0, \quad b(Y_j, Y_k) = \delta_{jk}.$$

Take $\mathbf{r} = a(X_1 \wedge X_2 - Y_1 \wedge Y_2)$, and define as usual $\mathbf{R} = \#_r \circ \#_b^{-1}$. Then it can be checked that $\mathbf{R}$ satisfies the MYB-equation of coefficient $\alpha = a^2$.

As a matter of fact, in [CG-] a whole important class of solutions of the MYB-equation for $sl(2,\mathbb{C})$ is constructed.

10.5 Manin triples

Now, we come back to the general notion of a Lie bialgebra, and we shall look at it from a different angle. Namely, the examination of the pair $(\mathcal{G}, \mathcal{G}^*)$ of a Lie bialgebra leads to

10.22. Definition [Dr2]. A *Manin triple* is a triple of Lie algebras $(\mathcal{G}, \mathcal{G}_+, \mathcal{G}_-)$ endowed with a symmetric, nondegenerate, ad-invariant bilinear form $\langle\ ,\ \rangle$ on $\mathcal{G}$ such that: i) $\mathcal{G}_+$ and $\mathcal{G}_-$ are Lie subalgebras of $\mathcal{G}$, and $\mathcal{G} = \mathcal{G}_+ \oplus \mathcal{G}_-$ *as vector spaces*, and ii) $\langle\ ,\ \rangle$ vanishes on $\mathcal{G}_+$ and on $\mathcal{G}_-$ (i.e., the latter are *isotropic* subspaces of $\mathcal{G}$).

We refer to finite dimensional Lie algebras only. Then, we have

10.23. Theorem [Dr2]. *Up to isomorphisms, the Lie bialgebras are in a one-to-one correspondence with the Manin triples.*

Proof. Let $(\mathcal{G}, \mathcal{G}^*)$ be a Lie bialgebra, and take the vector space $\tilde{\mathcal{G}} = \mathcal{G} \oplus \mathcal{G}^*$, $\tilde{\mathcal{G}}_+ = \mathcal{G}$, $\tilde{\mathcal{G}}_- = \mathcal{G}^*$. Then, define on $\tilde{\mathcal{G}}$ the bracket

$$\begin{aligned}[X \oplus \xi, Y \oplus \eta] = \{[X,Y] + (\mathrm{coad}^*\xi)(Y) - (\mathrm{coad}^*\eta)(X)\} \oplus \\ \oplus \{[\xi,\eta]^* + (\mathrm{coad}X)(\eta) - (\mathrm{coad}Y)(\xi)\},\end{aligned} \tag{10.47}$$

where coad^* is the coadjoint representation of the Lie structure $(\mathcal{G}^*, [\ ,\ ]^*)$ on its dual, that is $\mathcal{G}$. Furthermore define $\langle\ ,\ \rangle$ on $\tilde{\mathcal{G}}$ by

$$\langle X \oplus \xi, Y \oplus \eta\rangle = \xi(Y) + \eta(X). \tag{10.48}$$

Now, we can check that $(\tilde{\mathcal{G}}, \tilde{\mathcal{G}}_+, \tilde{\mathcal{G}}_-)$ is a Manin triple. What is not obvious is the Jacobi identity for (10.47), and the ad-invariance of (10.48). It suffices to verify them by working out separately all the cases of elements that have only a $+$ or only a $-$ component.

The Jacobi identity is already known to hold for a triple (X, Y, Z) and for a triple (ξ, η, ζ). We check it for (X, Y, ξ), and we leave the case (X, ξ, η) to the reader. Following (10.47), we get

$$\begin{aligned}&\mathcal{E} \stackrel{\mathrm{def}}{=} [X,[Y,\xi]] + [Y,[\xi,X]] + [\xi,[X,Y]] = \\ &= [X, -(\mathrm{coad}^*\xi)(Y) + (\mathrm{coad}\ Y)(\xi)] + \\ &+ [Y, (\mathrm{coad}^*\xi)(X) - (\mathrm{coad}\ X)(\xi)] + (\mathrm{coad}^*\xi)([X,Y]) - (\mathrm{coad}[X,Y])(\xi) = \\ &= (\mathrm{coad}^*\xi)([X,Y]) - [X, (\mathrm{coad}^*\xi)(Y)] + [Y, (\mathrm{coad}^*\xi)(X)] - \\ &- \mathrm{coad}^*((\mathrm{coad}\ Y)(\xi))(X) + (\mathrm{coad}\ X)(\mathrm{coad}\ Y)(\xi) + \\ &+ \mathrm{coad}^*((\mathrm{coad}\ X)(\xi))(Y) - (\mathrm{coad}\ Y)(\mathrm{coad}X)(\xi) - (\mathrm{coad}\ [X,Y])(\xi) = \\ &= (\mathrm{coad}^*\xi)([X,Y]) - [X, (\mathrm{coad}^*\xi)(Y)] + [Y, (\mathrm{coad}^*\xi)(X)] - \\ &- \mathrm{coad}^*((\mathrm{coad}\ Y)(\xi))(X) + \mathrm{coad}^*((\mathrm{coad}\ X)(\xi))(Y).\end{aligned} \tag{10.49}$$

The other terms cancel because coad is a Lie algebra representation. (The reader will notice by himself the places where the sums of the previous calculation are direct sums.)

Now, in order to clarify the remaining expression, which is entirely in $\mathcal{G}$, we compute $\alpha(\mathcal{E})$ for $\alpha \in \mathcal{G}^*$. Taking into account the relation between the representations ad and coad, we get

$$\begin{aligned}\alpha(\mathcal{E}) = -[\xi,\alpha]^*([X,Y]) - [\xi, (\mathrm{coad}\ X)(\alpha)]^*(Y) + \\ + [\xi, (\mathrm{coad}\ Y)\alpha]^*(X) + [(\mathrm{coad}\ Y)(\xi), \alpha]^*(X) - [(\mathrm{coad}\ X(\xi), \alpha]^*(Y),\end{aligned} \tag{10.50}$$

and this vanishes because of (10.22).

Therefore, the bracket (10.47) satisfies the Jacobi identity.

Similarly, we check one of the invariance conditions of (10.48), and leave the others to the reader:

$$\begin{aligned}\langle [X,Y],\xi\rangle + \langle Y,[X,\xi]\rangle &= \xi([X,Y])+\\ &+\langle Y, -(\mathrm{coad}^*\xi)(X) + (\mathrm{coad}\ X)(\xi)\rangle =\\ &= \xi([X,Y]) + (\mathrm{coad}\ X)(\xi)(Y) = \xi([X,Y]) - \xi([X,Y]) = 0.\end{aligned} \tag{10.51}$$

Etc.

This ends the construction of the correspondence from Lie bialgebras to Manin triples. We shall denote by $\tilde{\mathcal{G}} = \mathcal{G} \bowtie \mathcal{G}^*$ the Lie algebra defined by the bracket (10.47), and call it the *Manin product* of $(\mathcal{G}, \mathcal{G}^*)$.

Conversely, let $(\mathcal{G}, \mathcal{G}_+, \mathcal{G}_-, \langle\ ,\ \rangle)$ be a Manin triple. Define $\phi : \mathcal{G}_- \to \mathcal{G}_+$ by

$$(\phi X_-)(Y_+) = \langle X_-, Y_+\rangle, \tag{10.52}$$

where, here and in what follows, the signs indicate the belonging of the vectors to $\mathcal{G}_\pm$, respectively. From the nondegeneracy of the form $\langle\ ,\ \rangle$ together with condition ii) of Definition 10.22, it is easy to check the injectivity and the surjectivity of ϕ hence, ϕ is an isomorphism.

Now, we may transfer the Lie bracket of $\mathcal{G}$ to $\mathcal{G}_+ \oplus \mathcal{G}_+^*$ by means of the isomorphism id $\oplus\phi$, and we get a Lie algebra structure on $\mathcal{G}_+ \oplus \mathcal{G}_+^*$. In particular, since $\mathcal{G}_-$ is a subalgebra of $\mathcal{G}$, we get a Lie bracket $[\ ,\ ]^*$ on $\mathcal{G}_+^*$. Clearly, if we shall check that the obtained structure of $\mathcal{G}_+ \oplus \mathcal{G}_+^*$ is exactly (10.47) and that $[\ ,\ ]^*$ satisfies (10.22), Theorem 10.23 will be completely proven. Indeed, (10.22) will show that $(\mathcal{G}_+, \mathcal{G}_+^*)$ is a Lie bialgebra, and the verification of (10.47) will show that the Manin triple that corresponds to $(\mathcal{G}_+, \mathcal{G}_+^*)$ by the construction in the first part of the present proof is isomorphic to the given Manin triple hence, the two constructions made above are inverse to each other.

That the restriction of the total structure to $\mathcal{G}_+, \mathcal{G}_+^*$ satisfies (10.47) is obvious. Furthermore, we have

$$[X_+, \phi Y_-] = [X_+, Y_-]_+ + \phi([X_+, Y_-]_-). \tag{10.53}$$

We compute $[X_+, Y_-]_+$ by evaluating

$$\begin{aligned}\phi(Z_-)([X_+, Y_-]_+) &= \langle [X_+, Y_-]_+, Z_-\rangle = \langle [X_+, Y_-], Z_-\rangle =\\ &= \langle X_+, [Y_-, Z_-]\rangle = (\phi[Y_-, Z_-])(X_+) =\\ &= [\phi Y_-, \phi Z_-]^*(X_+) = -((\mathrm{coad}^*(\phi Y_-))(X_+))(\phi Z_-).\end{aligned}$$

Hence

$$[X_+, Y_-]_+ = -(\mathrm{coad}^*(\phi Y_-))(X_+). \tag{10.54}$$

Similarly, we get

$$\phi([X_+,Y_-]_-)(Z_+) = \langle [X_+,Y_-],Z_+\rangle = -\langle Y_-,[X_+,Z_+]\rangle =$$
$$= -(\phi Y_-)([X_+,Z_+]) = ((\text{coad } X_+)(\phi Y_-))(Z_+),$$

whence

$$\phi[X_+,Y_-]_- = (\text{coad } X_+)(\phi Y_-). \tag{10.55}$$

Adding up all these results, we see that the complete formula (10.47) holds good.

Finally, let us notice that, in fact, the computations (10.49), (10.50) establish that the relation (10.22) is equivalent to the Jacobi identity for the bracket (10.47). Now, we know that the Jacobi identity holds since $\mathcal{G}$ is a Lie algebra. Therefore, (10.22) holds too. Q.e.d.

Notice that Theorem 10.23 shows the symmetric role played by $\mathcal{G}$ and $\mathcal{G}^*$ in the structure of a Lie bialgebra, i.e., if $(\mathcal{G},\mathcal{G}^*)$ is a Lie bialgebra so is $(\mathcal{G}^*,\mathcal{G})$ also.

The following is an important example of a Manin triple

10.24. Example [LW2]. If the classical Gram orthonormalization procedure is applied to the columns of a matrix $A \in Gl(n,\mathbb{C})$ one obtains a matrix $U \in U(n)$. The Gram procedure means multiplication by an upper triangular matrix Θ with a real positive diagonal. Hence we have $A\Theta = U$, and if $T = \Theta^{-1}$, we get a unique decomposition

$$A = UT. \tag{10.56}$$

In particular, if $A \in Sl(n,\mathbb{C})$, it follows easily that $U \in SU(n)$, and $T \in S\Delta(n,\mathbb{C}) \overset{\text{def}}{=}$ the subgroup of the upper triangular matrices of $Sl(n,\mathbb{C})$, that have a positive real diagonal. Furthermore, if we apply (10.56) to a path $A(t)$ through the unit of $Sl(n,\mathbb{C})$, and take $(d/dt)\,|_{t=0}$, we get a vector space decomposition involving the Lie algebras corresponding to the previous Lie groups

$$sl(n,\mathbb{C}) = su(n) \oplus s\Delta(n,\mathbb{C}) \tag{10.57}$$

($s\Delta(n,\mathbb{C})$ consists of the trace-vanishing upper triangular complex matrices with a real diagonal).

Now, let us define

$$\langle X,Y\rangle = \text{Im}(\text{trace}(XY)) \qquad (X,Y \in sl(n,\mathbb{C})), \tag{10.58}$$

where Im denotes the imaginary part of a complex number. Since the matrices of $sl(n,\mathbb{C})$ have a vanishing trace, the form (10.58) is symmetric. It is also ad-invariant, since the trace is Ad-invariant. Moreover, if $X = (x_{ik})$ is such that

$\forall Y \in sl(n,\mathbb{C})$ $\langle X,Y\rangle = 0$, then looking at matrices Y which have either 1 or $\sqrt{-1}$ at a fixed entry (i,k) with $i \neq k$, and 0 elsewhere, we get $x_{ik} = 0$ for $i \neq k$. Then, one remains with $x_{ik} = \lambda_i\delta_{ik}$, $\Sigma\lambda_i = 0$, and using $Y = \sqrt{-1}\bar{X}$ (bar = complex conjugation) it follows that all $\lambda_i = 0$. Therefore, the form (10.58) is nondegenerate. Its restrictions to the terms of (10.57) obviously vanish.

Hence (10.57), (10.58) satisfy all the conditions of Definition 10.22, and $(sl(n,\mathbb{C}), su(n), s\Delta(n,\mathbb{C}))$ is a Manin triple.

10.25. Corollary. *The groups $SU(n)$ and $S\Delta(n,\mathbb{C})$ have Poisson-Lie group structures.*

The previous example and corollary were extended by Lu and Weinstein [LW2] (see also [Mj]) to the general framework of semisimple Lie algebras and groups by using the so-called *Iwasawa decomposition* (e.g., [Hl]). Using this technique, these authors prove that *every connected compact semisimple Lie group G has a nontrivial Poisson-Lie group structure.* The study of the Poisson-Lie group structures on semisimple groups is continued in [LR] and [GW].

10.6 Actions and dressing transformations

Now, let us develop briefly a few notions that have a natural place in group theory [LW2], [Lu1,2], [MW].

10.26. Definition. A left action $\varphi : G \times M \to M$ of a Poisson-Lie group (G,π) on a Poisson manifold (M,w) is called a *Poisson action* if φ is a Poisson mapping with respect to the product structure of $G \times M$. *Right Poisson actions* are defined similarly.

10.27. Proposition [LW2]. *$\varphi : G \times M \to M$ is a Poisson action iff one of the following equivalent conditions is satisfied:*

1) *$\forall g \in G$, $\forall x \in M$, one has*

$$w(g(x)) = \varphi_{g*}(w(x)) + \varphi^x_*(\pi(g)); \tag{10.59}$$

2) *if G is connected, $\forall X \in \mathcal{G}$ one has*

$$L_{X_M} w = -[(d_e\pi)(X)]_M; \tag{10.60}$$

3) *if G is connected, for any 1-forms α, β on M, and $x \in M$, one has*

$$(L_{X_M} w)_x(\alpha,\beta) = [\varphi^{x*}\alpha, \varphi^{x*}\beta]^*(X). \tag{10.61}$$

Proof. First, we must explain the notation used in the above formulation. φ_g and φ^x are as in (7.12), X_M is defined by (7.13), $d_e\pi$ and $[\ ,\ ]^*$ are as in (10.13) and (10.15), and the right-hand side of (10.60) is obtained by extending (7.13) to tensors.

Now, the proof of (10.59) is the same as that of (10.5), i.e., using (10.2) for φ instead of μ, φ_g instead of L_g, and φ^x instead of R_g.

Then, $\forall x \in M$, we have

$$(L_{X_M}w)(x) = \left.\frac{d}{ds}\right|_{s=0} \varphi_{\exp(sX)*}w(\exp(-sX)(x)) =$$

$$\overset{(10.59)}{=} \left.\frac{d}{ds}\right|_{s=0} \varphi_{\exp(sX)*}\left(\varphi_{\exp(-sX)*}w(x) + \varphi^x_*\pi(\exp(-sX))\right).$$

Here, the first term is constant $w(x)$, and its derivative is zero, and, by (10.13), the second term is exactly the right hand side of (10.60). Conversely, if we do this computation backwards, we see that (10.60) implies (10.59) for $g = \exp X$, and since G is connected we are done.

Finally, (10.61) is just the value of (10.60) on the two arguments α, β. The right-hand side takes the written form since (7.13) can be read as $X_M(x) = -\varphi^x_*(X)$, and this extends to the bivector $(d_e\pi)(X)$. Q.e.d.

10.28. Definition. Let H be a Lie subgroup of the Poisson-Lie group G. Then H is a *Poisson-Lie subgroup* if it is also a Poisson submanifold of G i.e., H has also a Poisson-Lie group structure, and $i : M \subseteq G$ is a Poisson map.

10.29. Proposition [ST2]. *The connected subgroup H of G is a Poisson-Lie subgroup of G iff the annihilator* $\mathrm{Ann}\mathcal{H}$ *of the Lie subalgebra $\mathcal{H}$ of H is an ideal in $(\mathcal{G}^*, [\ ,\ ]^*)$.*

Proof. Let w_G be the Poisson-Lie bivector of G, and let H be a Poisson-Lie subgroup of G with the bivector w_H. Take $\alpha \in \mathcal{G}^*, \xi \in \mathrm{Ann}\mathcal{H}$, and $Y \in \mathcal{H}$. Then, using (10.15) we have

$$[\alpha, \xi]^*(Y) = Y(w_G(\bar{\alpha}, \bar{\xi})) = Y(w_H(\bar{\alpha}/_H, \bar{\xi}/_H)) = 0$$

where $\bar{\alpha}, \bar{\xi}$ are obtained from α, ξ, say, by left translations, and, therefore, $\bar{\xi}/_H = 0$. This shows that $\mathrm{Ann}\mathcal{H}$ is an ideal.

Conversely, if $\mathcal{M}$ is any complementary subspace of $\mathcal{H}$ in $\mathcal{G}$, and if $h \in \mathcal{H}$, we have

$$T_hG = L_{h*}\mathcal{H} \oplus L_{h*}\mathcal{M} = T_hH \oplus L_{h*}\mathcal{M}.$$

This decomposition shows that $(L_{h*}\mathcal{M})^* = \mathrm{Ann}(T_hH)$, and it follows that w_G restricts to a structure w_H iff $w_G(\bar{\alpha}, \bar{\xi}) = 0$ along H and for every $\xi \in \mathrm{Ann}\,\mathcal{H}$, and where $\bar{\alpha}$, $\bar{\xi}$ are as above.

Now, since Ann $\mathcal{H}$ is an ideal of $\mathcal{G}, \forall Y \in \mathcal{H}$ we have

$$Y(w_G(\bar{\alpha},\bar{\xi})) = (L_Y w_G)_e(\alpha,\xi) = [\alpha,\xi]^*(Y) = 0.$$

But, for $\bar{Y} =$ the left invariant field of Y, $L_{\bar{Y}}w_G$ is a left invariant tensor field (Proposition 10.5, 3)). Hence, $\forall h \in H$,

$$(L_{\bar{Y}}w_G)_h(\bar{\alpha}_h,\bar{\xi}_h) = [L_{h*}(L_{\bar{Y}}w_G)_e](L^*_{(h^{-1})}\alpha_h, L^*_{(h^{-1})}\xi_h) = (L_{\bar{Y}}w_G)_e(\alpha,\xi) = 0,$$

and we see that $w_G(\bar{\alpha},\bar{\xi}) =$ constant along H. Since $w_e = 0$, this constant value must be zero, and we are done. Q.e.d.

10.30. Proposition [ST2]. *If H is a Poisson subgroup of G, then the natural projection $p : G \to G/H = \{gH/g \in G\}$ coinduces a Poisson structure on G/H, and the natural left action of G on G/H is a Poisson action.*

Proof. For the first assertion, we shall use Proposition 7.3 which tells us that we have the coinduced structure iff $\forall \varphi,\psi \in C^\infty(G/H), \{\varphi \circ p, \psi \circ p\}_G$ is constant along the fibers of p. For $g \in G, h \in H$, we have

$$\begin{aligned}\{\varphi\circ p,\psi\circ p\}_G(gh) &= w_{gh}(d_{gh}(\varphi\circ p), d_{gh}(\psi\circ p)) = \\ &= (L_{g*}w_h + R_{h*}w_g)(d_{gh}(\varphi\circ p), d_{gh}(\psi\circ p)) = \\ &= w_h(L^*_g d_{gh}(\varphi\circ p), L^*_g d_{gh}(\psi\circ p)) + \\ &+ w_g(R^*_h d_{gh}(\varphi\circ p), R^*_h d_{gh}(\psi\circ p)) = \\ &= w_h(d_h(\varphi\circ p\circ L_g), d_h(\psi\circ p\circ L_g)) + \\ &+ w_g(d_g(\varphi\circ p\circ R_h), d_g(\psi\circ p\circ R_h)).\end{aligned}$$

Since $\varphi\circ p\circ R_h = \varphi\circ p$, the second term above is $\{\varphi\circ p,\psi\circ p\}_G(g)$, while the first term is $\{\varphi\circ p\circ L_g,\psi\circ p\circ L_g\}_G(h)$. But, because H is a Poisson submanifold of G we have

$$\{\varphi\circ p\circ L_g,\psi\circ p\circ L_g\}_G(h) = \{\varphi\circ p\circ L_g,\psi\circ p\circ L_g\}_H(h) = 0,$$

since the two functions are constant on H. This proves the desired result, and provides us with the Poisson structure of G/H.

Finally, it is clear that (10.59) will be induced by the multiplicative property of w on G. Q.e.d.

We end these considerations by indicating an important case of a Poisson action that comes from the so-called *dressing transformations* [ST2], [We7].

If G is a Poisson-Lie group, as we know, there is a well defined associated Lie algebra structure $[\ ,\ ]^*$ on $\mathcal{G}^*$. Therefore, there exists a connected and simply connected Lie group G^* whose Lie algebra is just $(\mathcal{G}^*, [\ ,\ ]^*)$, and G^* is called the *dual group* of G. Obviously, G^* is also a Poisson-Lie group, and the dual group of G^* is the universal covering group of the connected component of the unit of G.

On the other hand, $\alpha \in \mathcal{G}^*$ defines the left invariant 1-form $\bar{\alpha}$ and the right invariant 1-form $\tilde{\alpha}$ on G, and we obtain mappings $\lambda, \rho : \mathcal{G}^* \to \chi(G) = \mathcal{V}^1(G)$ by putting

$$\lambda(\alpha) = \#_w\bar{\alpha}, \rho(\alpha) = -\#_w\tilde{\alpha}, \tag{10.62}$$

where w is the Poisson bivector of G. More precisely, it follows from formula (10.15′), Proposition 10.7, and formula (4.4) that λ is a homomorphism, and ρ is an anti-homomorphism (changes sign) of Lie algebras, $\chi(G)$ being endowed with the usual Lie bracket of vector fields. Accordingly, λ can be seen as an infinitesimal left action of G^* on G (remember that we use formula (7.13) for such an action), and ρ can be seen as an infinitesimal right action. The vector fields $\lambda(\alpha)$, $\rho(\alpha)$ are called the *left* and *right dressing vector fields associated with* α, respectively. Furthermore, as usually (e.g., [Pl]), these infinitesimal actions produce either local or global (if the dressing vector fields are complete) actions of the dual group G^* on G, and one calls them the left (right) action of G^* on G by *dressing transformations.* We denote these dressing actions by

$$\tilde{\lambda} : G^* \times G \to G, \tilde{\rho} : G \times G^* \to G. \tag{10.63}$$

The dressing transformations, which generalize the coadjoint action (the latter is obtained if G is the coadjoint space $\mathcal{G}^*$ of a Lie algebra $\mathcal{G}$, with the additive structure, and with its Lie Poisson structure) have been studied in detail in [ST2,3], [We7], [LW2], [Lu1,2], etc. In particular, they were used to compute the Poisson cohomology of G^* if G is compact, the result being $H^*_{LP}(G^*) = H^*(\mathcal{G}) \otimes \{\text{Casimirs}\,(G^*)\}$ [GW]. Here, we give only one result namely.

10.31. Theorem. [LW2]. *The dressing actions of G^* on G are Poisson actions.*

Proof. We discuss only the left action, since the right action can be discussed similarly. The idea is to check that (10.61) holds in our case, which amounts at proving that

$$(L_{\lambda(\alpha)}w)_g(\bar{\xi},\bar{\eta}) = \alpha\left(\left[\tilde{\lambda}^{g*}\bar{\xi}_g, \tilde{\lambda}^{g*}\bar{\eta}_g\right]\right), \tag{10.64}$$

where $\alpha, \xi, \eta \in \mathcal{G}^*, g \in G$, and $\bar{\alpha}$, $\bar{\xi}$, $\bar{\eta}$ are the left invariant 1-forms of α, ξ, η. We also have, $\forall \zeta \in \mathcal{G}^*$,

$$\zeta(\tilde{\lambda}^{g*}\bar{\xi}_g) \overset{(7.13)}{=} -\bar{\xi}_g(\lambda(\zeta)) = -\bar{\xi}_g(\bar{\zeta}^{\#}_g) = \bar{\zeta}_g(\bar{\xi}^{\#}_g) =$$
$$= (L^*_{g^{-1}}\zeta)(\bar{\xi}^{\#}_g) = \zeta(L_{g^{-1}*}\bar{\xi}^{\#}_g),$$

whence

$$\tilde{\lambda}^{g*}\bar{\xi}_g = L_{g^{-1}*}(\lambda(\xi))_g, \quad \tilde{\lambda}^{g*}\bar{\eta}_g = L_{g^{-1}*}(\lambda(\eta))_g. \tag{10.65}$$

If we look at the *false Lie derivative* $L_{\bar{\alpha}}$ of (4.15), we have

$$L_{\bar{\alpha}}w = i_{\bar{\alpha}}(\sigma w) + \sigma(i_{\bar{\alpha}}w) = 0 + \sigma(\bar{\alpha}^{\#}) = -L_{\bar{\alpha}^{\#}}w = -L_{\lambda(\alpha)}w,$$

where the last L is a true Lie derivative. Now, (4.14) yields

$$\begin{aligned}(L_{\lambda(\alpha)}w)(\bar{\xi},\bar{\eta}) &= -\lambda(\alpha)(w(\bar{\xi},\bar{\eta})) + w(\{\bar{\alpha},\bar{\xi}\},\bar{\eta}) + (\bar{\xi},\{\bar{\alpha},\bar{\eta}\}) = \\ &= -\lambda(\alpha)(w(\bar{\xi},\bar{\eta})) + \bar{\eta}(\{\bar{\alpha},\bar{\xi}\}^{\#}) - \bar{\xi}(\{\bar{\alpha},\bar{\eta}\}^{\#}) = \\ &\overset{(4.4)}{=} -\lambda(\alpha)(w(\bar{\xi},\bar{\eta})) + \bar{\eta}([\lambda(\alpha),\lambda(\xi)]) - \bar{\xi}([\lambda(\alpha),\lambda(\eta)]).\end{aligned} \tag{10.66}$$

At $g \in G$, the first term of the previous result can be evaluated as follows (using, among others, Proposition 10.5, 1°)

$$\begin{aligned}&\lambda(\alpha)(w(\bar{\xi},\bar{\eta}))(g) = [d_g(w(\bar{\xi},\bar{\eta}))](\lambda(\alpha)_g) = \\ &= [L_g^* d_g(w(\bar{\xi},\bar{\eta}))](L_{g^{-1}*}\lambda(\alpha)_g) \overset{(10.65)}{=} d_e(w(\bar{\xi},\bar{\eta})\circ L_g)(\tilde{\lambda}^{g*}\bar{\alpha}_g) = \\ &= (\tilde{\lambda}^{g*}\bar{\alpha}_g)(w(\bar{\xi},\bar{\eta})\circ L_g) = \left.\frac{d}{ds}\right|_{s=0}[w(\bar{\xi},\bar{\eta})(g\exp(s(\tilde{\lambda}^{g*}\bar{\alpha}_g)))] = \\ &= \left.\frac{d}{ds}\right|_{s=0}[w_l(g\exp(s(\tilde{\lambda}^{g*}\bar{\alpha}_g)))(\xi,\eta)] = \left.\frac{d}{ds}\right|_{s=0}[w_l(\exp(s(\tilde{\lambda}^{g*}\bar{\alpha}_g)))(\xi,\eta) + \\ &+ (\mathrm{Ad}(\exp(-s(\tilde{\lambda}^{g*}\bar{\alpha}_g))w_l(g))(\xi,\eta)] = \\ &= [\xi,\eta]^*(\tilde{\lambda}^{g*}\bar{\alpha}_g) - w_l(g)(\mathrm{coad}(\tilde{\lambda}^{g*}\bar{\alpha}_g)\xi,\eta) - w_l(g)(\xi,\mathrm{coad}(\tilde{\lambda}^{g*}\bar{\alpha}_g)\eta) = \\ &\overset{(10.65)}{=} -\bar{\alpha}(\lambda([\xi,\eta]^*))_g - w(g)(L^*_{g^{-1}}\,\mathrm{coad}(\tilde{\lambda}^{g*}\bar{\alpha}_g)\xi, L^*_{g^{-1}}\eta) - \\ &- w(g)(L^*_{g^{-1}}\xi, L^*_{g^{-1}}\mathrm{coad}(\tilde{\lambda}^{g*}\bar{\alpha}_g)\eta) = \\ &= -\bar{\alpha}([\lambda(\xi),\lambda(\eta)])_g - \xi([\tilde{\lambda}^{g*}\bar{\alpha}_g,\tilde{\lambda}^{g*}\bar{\eta}_g]) + \eta([\tilde{\lambda}^{g*}\bar{\alpha},\tilde{\lambda}^{g*}\bar{\xi}_g]).\end{aligned} \tag{10.67}$$

Now, (10.66) becomes

$$\begin{aligned}(L_{\lambda(\alpha)}w)_g(\bar{\xi},\bar{\eta}) &= \xi([\tilde{\lambda}^{g*}\bar{\alpha}_g,\tilde{\lambda}^{g*}\bar{\eta}_g]) - \eta([\tilde{\lambda}^{g*}\bar{\alpha}_g,\tilde{\lambda}^{g*}\bar{\xi}_g]) + \\ &+ \sum_{\mathrm{Cycl}(\alpha,\xi,\eta)} \bar{\alpha}([\lambda(\xi),\lambda(\eta)])_g.\end{aligned} \tag{10.68}$$

But

$$\begin{aligned}&\sum_{\mathrm{Cycl}} \bar{\alpha}([\lambda(\xi),\lambda(\eta)]) \overset{(4.4)}{=} \sum_{\mathrm{Cycl}} \bar{\alpha}(\{\bar{\xi},\bar{\eta}\}^{\#}) = \\ &= \sum_{\mathrm{Cycl}} w(\{\bar{\xi},\bar{\eta}\},\bar{\alpha}) = \sum_{\mathrm{Cycl}} \bar{\alpha}^{\#}(w(\bar{\xi},\bar{\eta})),\end{aligned} \tag{10.69}$$

where the last equality follows from $\sigma w = 0$, and from the expression (4.8) of the operator σ.

Finally, if (10.69) is replaced in (10.68), and if we use (10.67) in order to express the terms of the last cyclic sum of (10.69), we get

$$(L_{\lambda(\alpha)}w)_g(\bar{\xi},\bar{\eta}) = 2\alpha\left(\left[\tilde{\lambda}^{g*}\bar{\xi}_g, \tilde{\lambda}^{g*}\bar{\eta}_g\right]\right) - (L_{\lambda(\alpha)}w)_g(\bar{\xi},\bar{\eta}), \tag{10.70}$$

which is equivalent to the required result. Q.e.d.

10.32. Remark. It is clear from the definition of the dressing vector fields that, if G is a Poisson-Lie group, then the symplectic leaves of G are exactly the orbits of the dressing actions $\tilde{\lambda}$ and $\tilde{\rho}$.

We end this chapter by indicating some generalizations of the developed theory. First, we mention that it is possible to make a much more extensive use of the duality between G and G^*, by unifying these two groups in the framework of a *double Lie group* [LW2]. The latter is defined as a triple of Lie groups (G, G_+, G_-) such that $G_\pm$ are closed subgroups of G, and the multiplication sends $G_+ \times G_-$ diffeomorphically onto G. In the case of a Poisson-Lie group, one should consider $G = G_+$, $G^* = G_-$. The corresponding infinitesimal notion will be that of a *double Lie algebra* $(\mathcal{G}, \mathcal{G}_+, \mathcal{G}_-)$, where $\mathcal{G}_\pm$ are Lie subalgebras of the Lie algebra $\mathcal{G}$, and $\mathcal{G} = \mathcal{G}_+ \oplus \mathcal{G}_-$ as vector spaces. The *Manin product* $\tilde{\mathcal{G}} = \mathcal{G} \bowtie \mathcal{G}^*$ with the bracket (10.47) is such a double Lie algebra, and, under certain conditions [LW2], the connected and simply connected group $\tilde{G}$ defined by $\tilde{\mathcal{G}}$ is a double Lie group $\tilde{G} = G \bowtie G^*$. The dressing actions of G^* on G can be found again by using the double Lie group G [LW2].

The notion of a double Lie algebra was preceded by [KSM1], where a straightforward generalization of formula (10.47) is introduced. Namely let $\mathcal{G}$ and $\mathcal{H}$ be two Lie algebras, and suppose that A is a linear representation of $\mathcal{G}$ on $\mathcal{H}$ i.e., a $\mathcal{G}$-module structure on $\mathcal{H}$, and B is an $\mathcal{H}$-module structure on $\mathcal{G}$. Define on the vector space $\mathcal{G} \oplus \mathcal{H}$ a bracket by (10.47) where $[\ ,\ ]$ is in $\mathcal{G}$, $[\ ,\ ]^*$ is in $\mathcal{H}$, and coad, coad* are replaced by A and B respectively. Then, if this is a Lie bracket (i.e., it satisfies the Jacobi identity), the obtained Lie algebra is called the *twilled extension* of $(\mathcal{G}, \mathcal{H})$ or a *twilled Lie algebra*. We might denote it by $\mathcal{G} \bowtie_{AB} \mathcal{H}$. In [KSM1], it is shown that $\mathcal{G} \bowtie_{AB} \mathcal{H}$ has infinitesimal actions on the connected and simply connected corresponding Lie groups G, H, that generalize the dressing actions.

In a different direction, we would like to quote a generalization of the method of looking for Lie bialgebras via Yang-Baxter equations. Namely [KSM1], instead of looking at a *triangular potential* $\mathbf{r} \in \mathbf{\Lambda}^2\mathcal{G}$, we may look at an $\mathbf{r} \in \mathcal{G} \otimes \mathcal{G}$, and take the corresponding cocycle $\partial\mathbf{r}(X) = (\text{ad}\ X)\mathbf{r}$. If we decompose $\mathbf{r} = \mathbf{s} + \mathbf{a}$, where $\mathbf{s}$ is symmetric and $\mathbf{a}$ is skew-symmetric, and if $\mathbf{s}$ is ad-invariant (i.e., $(\text{ad}\ X)\mathbf{s} = 0$), then $\partial\mathbf{r} = \partial\mathbf{a}$ is a $\Lambda^2\mathcal{G}$-valued cocycle, and we may try to use it via Theorem 10.9, to get a Lie bialgebra structure. The corresponding Jacobi identity is a condition on $\mathbf{r}$ that replaces the $\mathcal{G}$YB-equation, and is of the form $(\text{ad}\ X)(S(\mathbf{r})) = 0$, where $S(\mathbf{r})$ is an expression called the *Schouten curvature* in [KSM1]. In particular, if $S(\mathbf{r}) = 0$, and the symmetric part $\mathbf{s}$ of $\mathbf{r}$ is nonsingular, $\mathbf{r}$ is called a *quasitriangular Jacobian potential* [KSM1], [AKS] (the non-singularity condition is dropped by other authors [WX2]), and $\mathbf{R} = \#_a \circ \#_s^{-1}$ satisfies the MYB-equation [KSM1].

Finally, we mention that, motivated by a new notion of *quasi-quantum groups* [Dr3], a generalization of the Poisson-Lie groups and Poisson-Lie algebras has been defined and studied by Y. Kosmann-Schwarzbach [KS6], [KS7], under the name of *quasi-Poisson-Lie groups* and *algebras*. The basic idea of this generalization is to replace the condition $[w, w] = 0$, by the demand that $[w, w]$ is a coboundary, in a certain sense. If we look at the formula (10.24) we see that it is natural to give the following definition: a triple (G, w, φ) where G is a Lie group, w is a multiplicative bivector on G, and $\varphi \in \Lambda^3\mathcal{G}$ is a *quasi-Poisson-Lie group* if i) $[w, w](g) = L_{g*}\varphi - R_{g*}\varphi(g \in G)$, and ii) $[w, \bar{\varphi}] = 0$ $(\bar{\varphi}(g) = L_{g*}\varphi)$. The corresponding structure on $\mathcal{G}$ is that of a quasi-Poisson-Lie algebra.

References

[AM] R. Abraham and J. Marsden, Foundations of Mechanics, 2nd. ed. Benjamin Cummings, Reading, Massachusetts, 1978.

[AR] R. Abraham and J. Robin, Transversal mappings and flows. W.A. Benjamin, Inc., New York-Amsterdam, 1967.

[AD1] C. Albert and P. Dazord, Groupoïdes de Lie et groupoïdes symplectiques. In: Symplectic Geometry, Groupoids, and Integrable Systems. Séminaire Sud-Rhodanien de Géométrie à Berkeley (1989) (P. Dazord and A. Weinstein, eds.). MSRI Publ. 20, Springer-Verlag, Berlin-Heidelberg-New York, 1991, 1.–12.

[AD2] C. Albert and P. Dazord, Théorie des groupoïdes symplectiques, Chapitre I, Théorie générale des groupoïdes; Chapitre II, Groupoïdes symplectiques. Publ. Dept. Math. Lyon 1988, 53–105; 1990, 27–99.

[AC1] F. Alcalde Cuesta, Groupoïde d'homotopie d'un feuilletage riemannien et réalisation symplectique de certaines variétés de Poisson. Publicacions Matemàtiques 33 (1989), 395–410.

[AC2] F. Alcalde Cuesta, Intégration symplectique des variétés de Poisson régulières sans cycle évanouissant. Thèse. Univ. Lyon I, France, 1993.

[ADH] F. Alcalde, P. Dazord and G. Hector, Sur l'intégration symplectique de la structure de Poisson singulière $\Lambda = (x^2 + y^2)\frac{\partial}{\partial x} \wedge \frac{\partial}{\partial y}$ de $\mathbb{R}^2$. Publicacions Matemàtiques 33(1989), 411–415.

[Am] R. Aminou, Groupes de Lie-Poisson et bigèbres de Lie. Thèse, Univ. de Lille, 1988.

[AKS] R. Aminou and Y. Kosmann-Schwarzbach, Bigèbres de Lie, doubles et carrés.Ann. Inst. H.Poincaré, Série A (Physique théorique), 49(4) (1988), 461–478.

[ACG] J.M. Arms, R.H. Cushman and M.J. Gotay, A universal reduction procedure for Hamiltonian group actions. In: The Geometry of Hamiltonian Systems (T. Ratiu, ed.). MSRI Publ. 22, Springer-Verlag, Berlin-Heidelberg-New York, 1991, 33–51.

[BF-] F. Bayen, M. Flato, C. Fronsdal, A. Lichnerowicz and D. Sternheimer, Deformation theory and quantization. Annals of Physics, 111 (1978), 61–110 and 111–152.

[Bh] K.H. Bhaskara, Affine Poisson Structures. Proc. Indian Academy of Sci. 100 (1990), 189–202.

[BR] K.H. Bhaskara and K. Rama, Quadratic Poisson Structures. J. of Math. Physics, 32 (1991), 2319–2322.

[BV1] B.H. Bhaskara and K. Viswanath, Calculus on Poisson manifolds. Bull. London Math. Soc. 20 (1988), 68–72.

[BV2] K.H. Bhaskara and K. Viswanath, Poisson algebras and Poisson manifolds. Pitman Research Notes in Math. 174, Longman Sci., Harlow and New York, 1988.

[Bl] D.E. Blair, Contact manifolds in Riemannian geometry, Lect. Notes in Math. 509, Springer-Verlag, Berlin-Heidelberg-New York, 1976.

[Bt] R. Bott, Lectures on characteristic classes and foliations. In: Lectures on Algebraic and Differential Topology, Lect. Notes in Math. 279, Springer-Verlag, Berlin-Heidelberg-New York, 1972.

[BT] R. Bott and L.W. Tu, Differential forms in algebraic topology, Graduate Texts in Math., 82, Springer-Verlag, Berlin-Heidelberg-New York, 1982.

[Bk] N. Bourbaki, Variétés différentiables et analytiques, Paris, Hermann, 1971.

[Bz] R. Brouzet, La notion de Poisson-compatibilité: Exemples et propriétés géométriques. Sém. G. Darboux, Montpellier, 1987–88, 37–66.

[Br] J.-L. Brylinski, A differential complex for Poisson manifolds. J. Differential Geometry, 28 (1988), 93–114.

[CV] A. Cabras and A.M. Vinogradov, Prolongations of the Poisson bracket on differential forms and multi-vector fields. J. of Geometry and Physics, 9 (1992), 75–100.

[CG-] M. Cahen, S. Gutt, C. Ohn and M. Parker, Lie-Poisson Groups: Remarks and Examples. Lett. in Math. Physics 19 (1990), 343–353.

[CE] C. Chevalley and S. Eilenberg, Cohomology theory of Lie groups and Lie algebras, Transactions American Math. Soc. 63 (1948), 85–124.

[Co1] J. Conn, Normal forms for analytic Poisson structures. Annals of Math., 119 (1984), 577–601.

[Co2] J. Conn, Normal forms for smooth Poisson structures. Annals of Math., 121 (1985), 565–593.

[CDW] A. Coste, P. Dazord and A. Weinstein, Groupoïdes symplectiques, Publ. Dept. Math. Lyon 2/A (1987), 1–62.

[CS] A. Coste and D. Sondaz, Classification des submersions de Poisson isotropes. Séminaire Sud-Rhodanien, 1ere partie, Publ. Dépt. Math. Lyon 1/B (1988), 91–102.

[Cr] T.J. Courant, Dirac Manifolds. Thésis. Univ. of California, Berkeley, 1987.

[CW] T.J. Courant and A. Weinstein, Beyond Poisson structures. In: Actions hamiltoniennes de groupes. Troisième théorème de Lie. Séminaire Sud-Rhodanien de Géometrie VIII (P. Dazord, N. Desolneux-Moulis and J.M. Morvan, eds.), Travaux en Cours,27, Hermann, Paris 1988.

[Dz1] P. Dazord, Feuilletages à singularités. Indagationes Math. 47 (1985), 21–39.

[Dz2] P. Dazord, Stabilité et linéarisation dans les variétés de Poisson. In: Géometrie symplectique et mécanique (J.P. Dufour, ed.) Hermann, Paris, 1985, 59–75.

[Dz3] P. Dazord, Réalisations isotropes de Libermann. Travaux du Séminaire Sud-Rhodanien de Géométrie II. Publ. Dept. Math. Lyon 4/B (1988), 1–52.

[Dz4] P. Dazord, Groupoïdes symplectiques et troisième théorème de Lie "non linéare". In: Géométrie symplectique et Mécanique, Proceedings, 1988 (C. Albert, ed.), Lecture Notes in Math. 1416, Springer-Verlag, Berlin-Heidelberg-New York, 1990, 39–74.

[Dz5] P. Dazord, Autour du mouvement de Lagrange. Atti Accademia Scienze Torino, Supplemento al numero 124(1990), 145–158.

[DD] P. Dazord and T. Delzant, Le problème général des variables actions angles. J. Differential Geometry 26 (1987), 223–251.

[DH] P. Dazord and G. Hector, Intégration symplectique des variétés de Poisson totalement asphériques. In: Symplectic Geometry, Groupoids and Integrable Systems, Séminaire Sud Rhodanien de Géométrie à Berkeley (1989) (P. Dazord and A. Weinstein, eds.). MSRI Publ. 20, Springer-Verlag, Berlin - Heidelberg-New York, 1991, 37–72.

[DLM] P. Dazord, A. Lichnerowicz and Ch.-M. Marle, Structure locale des variétés de Jacobi. J. Math. pures et appl. 70 (1991), 101–152.

[DS1] P. Dazord and D. Sondaz, Variétés de Poisson - Algébroïdes de Lie. Publ. Math. Univ. Lyon 1, 1/B (1988), 1–68.

[DS2] P. Dazord and D. Sondaz, Groupes de Poisson affines. In: Symplectic Geometry, Groupoids, and Integrable Systems, Séminaire Sud-Rhodanien de Géométrie à Berkeley (1989) (P. Dazord and A. Weinstein, eds.) MSRI Publ. 20, Springer-Verlag, Berlin-Heidelberg-New York 1991, 99–128.

[DL] M. De Wilde and P.B.A. Lecomte, Existence of star-products and of formal deformations of the Poisson Lie algebra of arbitrary symplectic manifolds. Lett. in Math. Physics, 7 (1983), 487–496.

[Dr1] V.G.Drinfeld, Hamiltonian structures on Lie groups, Lie bialgebras, and the geometric meaning of the classical Yang-Baxter equation, Soviet Math. Doklady 27(1) (1983), 68–71.

[Dr2] V.G. Drinfeld, Quantum groups, Proc. Internat. Congress Math., Berkeley, 1986, vol. 1, 789–820.

[Dr3] V.G. Drinfeld, Quasi-Hopf algebras. Leningrad Math. J. 1(6) (1990), 1419–1457.

[Du] J.P. Dufour, Linéarisation de certaines structures de Poisson. J. Differential Geometry 32(1990) 415–428.

[DuH] J.P. Dufour and A. Haraki, Rotationnels et structures de Poisson quadratiques. Comptes Rendus Acad. Sc. Paris, Série I, 312 (1991), 137–140.

[Dt] J. J. Duistermaat, On global action-angle coordinates. Comm. Pure Appl. Math. 33 (1980), 687–706.

[Fd] B. Fedosov, A simple geometrical construction of deformation quantization. J. Differential Geometry (to appear).

[FLS] M. Flato, A. Lichnerowicz and D. Sternheimer, Déformations 1-différentiables des algèbres de Lie attachées à une variété symplectique ou de contact. Compositio Math. 31 (1975), 47–82.

[Gt] F.R. Gantmacher, The theory of matrices. Chelsea Publ., New York, 1964.

[GD] I.M. Gel'fand and I. Ya. Dorfman, Hamiltonian operators and the classical Yang-Baxter equation. Funct. Anal. Appl. 16 (1982), 241–248.

[Gh] M. Gerstenhaber, Deformation theory of algebraic structures. Ann. Math. 79 (1964), 59–90.

[GW] V.L. Ginzburg and A. Weinstein, Lie-Poisson structure on some Poisson-Lie groups. J. American Math. Soc., 5 (1992), 445–453.

[Gn] R. Godement, Topologie algèbrique et théorie des faisceaux, Paris, Hermann, 1973.

[Go] S.I. Goldberg, Curvature and Homology. Academic Press, New York, 1962.

[Gy] M.J. Gotay, Functional geometric quantization and van Hove's theorem. Int. J. of Theoretical Physics 19 (1980), 139–161.

[GL] F. Guédira and A. Lichnerowicz, Géométrie des algèbres de Lie de Kirillov. J. Math. pures et appl. 63 (1984), 407–484.

[HL] S. Halperin and D. Lehmann, Cohomologies et classes caractéristiques des choux de Bruxelles. In: Diff. topology and geometry, Dijon 1974 (G.P. Joubert, R.P. Moussu and R.H. Roussarie, eds.), Lecture Notes in Math. 484, Springer-Verlag, Berlin-Heidelberg-New York, 1975, 79–120.

[Hc] G. Hector, Une nouvelle obstruction à l'intégrabilité des variétés de Poisson régulières. Hokkaido Math. J. 21 (1992), 159–185.

[HMS] G. Hector, E. Macias and M. Saralegi, Lemme de Moser feuilleté et classification des variétés de Poisson régulières. Publicacions Matemàtiques 33(1989), 423–430.

[Hl] S. Helgason, Differential geometry, Lie groups, and symmetric spaces. Academic Press, New York, 1978.

[Hu1] J. Huebschmann, Poisson cohomology and quantization. J. für reine angew. Math. 408 (1990), 57–113.

[Hu2] J. Huebschmann, Graded Lie-Rinehart algebras, graded Poisson algebras, and BRST-quantization I. The finitely generated case. Univ. Heidelberg, Heft 74, 1990.

[Hu3] J. Huebschmann, Some remarks about Poisson homology. Univ. Heidelberg. Heft 75, 1990.

[Hu4] J. Huebschmann, On the quantization of Poisson algebras. In: Symplectic Geometry and Mathematical Physics, Actes du colloque en l'honneur de Jean-Marie Souriau (P. Donato, C. Duval, J. Elhadad, G.M. Tuynman, eds.), Progress in Math. 99, Birkhäuser, Boston, 1991, 204–233.

[Hr] N. E. Hurt, Geometric quantization in action. Mathematics and its Applications, 8. D. Reidel Publ. Comp., Dordrecht - Boston, 1983.

[Kr] M.V. Karasev, Analogues of objects of the Lie group theory for nonlinear Poisson brackets. Soviet Math. Izvestia 28 (1987), 497–527.

[KM1] M.V. Karasev and V.P. Maslov, Operators with general commutation relations and their applications. I. Unitary non linear operator equations. J. Soviet Math. 15(1981), 273–368.

[KM2] M.V. Karasev and V.P. Maslov, Asymptotic and geometric quantization, Russian Math. Surveys 39 (6) (1984), 133–205.

[Kv] A. Kirillov, Local Lie algebras. Russian Math. Surveys 31 (1976), 55–75.

[Kb] S. Kobayashi, Principal fibre bundles with the 1-dimensional toroidal group. Tôhoku Math. J., 8 (1956), 29–45.

[KN] S. Kobayashi and K. Nomizu, Foundations of Differential Geometry. I, II, Intersci. Publ., New York-London, 1963, 1969.

[KMS] I. Kolár, P.W. Michor and J. Slovák. Natural Operations in Differential Geometry. Springer-Verlag. Berlin-Heidelberg-New York. 1993.

[KS1] Y. Kosmann-Schwarzbach, Géométrie des systèmes bihamiltoniens. In: Systèmes dynamiques, intégrabilité et comportement qualitatif (P. Winternitz, ed.), Univ. de Montreal, 1986.

[KS2] Y. Kosmann-Schwarzbach, Poisson-Drinfeld groups. In: Topics in soliton theory and exactly solvable nonlinear equations (M. Ablowitz, B. Fuchssteiner and M. Kruskal, eds.), World Scientific, Singapore, 1987, 191–215.

[KS3] Y. Kosmann-Schwarzbach, Equations de Yang-Baxter et structures de Poisson. Journées Relativistes de Chambery, 1987.

[KS4] Y. Kosmann-Schwarzbach, Groupes de Lie-Poisson quasitriangulaires. In: Géométrie Symplectique et Mécanique, Proceedings 1988 (C. Albert, ed.), Lecture Notes in Math. 1416, Springer-Verlag, Berlin-Heidelberg-New York 1990, 161–177.

[KS5] Y. Kosmann-Schwarzbach, Grand crochet, crochets de Schouten et cohomologies d'algèbres de Lie. Comptes Rendus Acad. Sc. Paris, Série I, 312 (1991), 123–126.

[KS6] Y. Kosmann-Schwarzbach, Quasi bigèbres de Lie et groupes de Lie quasi-Poisson. Comptes Rendues Acad. Sc. Paris Série I, 312 (1991), 391–394.

[KS7] Y. Kosmann-Schwarzbach, Jacobian quasi-bialgebras and quasi-Poisson Lie groups. In: Mathematical aspects of classical field theory (M. Gotay, J.E. Marsden and V. Moncrief, eds.), Contemporary Mathematics 132 (1992), 459–489.

[KSM1] Y. Kosmann-Schwarzbach and F. Magri, Poisson-Lie groups and complete integrability. Part I. Drinfeld bigebras, dual extensions and their canonical representations. Ann. Inst. H. Poincaré, série A (Physique théorique), 49 (1988), 433–460.

[KSM2] Y. Kosmann-Schwarzbach and F. Magri, Poisson-Nijenhuis Structures, Ann. Inst. H. Poincaré, série A (Physique théorique), 53(1990), 35–81.

[Kt] B. Kostant, Quantization and unitary representations. In: Lectures in modern analysis and applications III (C.T. Taam, ed.). Lecture Notes in Math. 170, Springer-Verlag, Berlin-Heidelberg-New York, 1970, 87–207.

[Kz] J.L. Koszul, Crochet de Schouten-Nijenhuis et cohomologie. In: É. Cartan et les mathématiques d'aujourd'hui, Soc. Math. de France, Astérisque, hors série, 1985, 257–271.

[LMR] P.B.A. Lecomte, D. Mélotte and C. Roger, Explicit form and convergence of 1-differential formal deformations of the Poisson Lie algebra. Lett. in Math. Physics 18 (1989), 275–285.

[Lf] J. Lefebvre, Propiétés du group des transformations conformes et du groupe des automorphismes d'une variété localement conformément symplectique. Comptes Rendus Acad. Sc. Paris, Série A, 268 (1969), 717–719.

[Ln] D. Lehmann, Théorie homotopique des formes différentielles. Astérisque 45, Soc. Math. de France, Paris, 1977.

[Lb1] P. Libermann, Sur le problème d'équivalence de certaines structures infinitésimales régulières. Ann. Mat. Pura Appl. 36 (1954), 27–120.

[Lb2] P. Libermann, Problèmes d'équivalence en géométrie symplectique. In: IIIe. Rencontre de Géom. du Schnepfenried, Vol. 1, Astérisque 107/108 (1983), 43–68.

[Lb3] P. Libermann, Sous-variétés et feuilletages symplectiquement réguliers. In: Symplectic geometry (A. Crumeyrolle and J. Grifone, eds.), Research Notes in Math. 80, Pitmann, Boston-London, 1983, 81–106.

[LM] P. Libermann and Ch.-M. Marle, Symplectic geometry and analytical mechanics. D. Reidel Publ. Comp., Dordrecht - Boston, 1987.

[Lh1] A. Lichnerowicz, Global theory of connections and holonomy groups. Noordhoof Intern. Publ., Leyden, 1976.

[Lh2] A. Lichnerowicz, Les variétés de Poisson et leurs algèbres de Lie associées. J. Differential Geometry 12 (1977), 253–300.

[Lh3] A. Lichnerowicz, Les variétés de Jacobi et leurs algèbres de Lie associées. J. Math. pures et appl. 57 (1978), 453–488.

[Lh4] A. Lichnerowicz, Quantum mechanics and deformations of geometrical dynamics. In: Quantum Theory, Groups, Fields and Particles (A.O. Barat, ed.), D. Reidel Publ. Comp. Dordrecht-Boston, 1983, 3–82.

[Lh5] A. Lichnerowicz, Variétés de Jacobi complexes et application coadjointe quotient, Comptes Rendus Acad. Sc. Paris, Série I, 302 (1986), 315–319.

[Lie] S. Lie, Theorie der Transformationsgruppen (Zweiter Abschnitt, unter Mitwirkung von Prof. Dr. Friederich Engel). Teubner, Leipzig, 1890.

[LX] Z.-J. Liu and P. Xu, On quadratic Poisson structures. Letters in Math. Physics 26 (1992), 33–42.

[Lu1] J.-H. Lu, Multiplicative and affine Poisson structures on Lie groups. Thesis, Univ. of California, Berkeley, 1990.

[Lu2] J.-H. Lu, Momentum mappings and reduction of Poisson actions. In: Symplectic Geometry, Groupoids, and Integrable Systems, Séminaire Sud-Rhodanien de Géometrie à Berkeley (1989) (P. Dazord and A. Weinstein, eds.) MSRI Publ. 20, Springer-Verlag, Berlin-Heidelberg-New York, 1991, 209–226.

[LR] J.-H. Lu and T. Ratiu, On the nonlinear convexity theorem of Kostant. J. American Math. Soc. 4 (1991), 349–363.

[LW1] J.-H. Lu and A. Weinstein, Groupoïdes symplectiques doubles des groupes de Lie-Poisson. Comptes Rendues Acad. Sc. Paris, Série I, t.309 (1989), 951–954.

[LW2] J.-H. Lu and A. Weinstein, Poisson Lie groups, dressing transformations and Bruhat decompositions. J. Differential Geometry 31 (1990), 501–526.

[Mz] K. Mackenzie, Lie groupoids and Lie algebroids in differential geometry. London Math. Soc. Lecture Notes Series 124, Cambridge Univ. Press, Cambridge, 1987.

[MX] K. Mackenzie and P. Xu, Li bialgebroids and Poisson groupoids. Preprint MSRI Berkeley, California, 022 (1993).

[MM] F. Magri and C. Morosi, A geometrical characterization of integrable Hamiltonian systems through the theory of Poisson-Nijenhuis manifolds, Quaderno S. 19 (1984) (University of Milan).

[Mj] S. Majid, Matched pairs of Lie groups associated to solutions of the Yang-Baxter equations. Pacific J. of Math. 141 (1990), 311–332.

[Mr] Ch-M. Marle, Lie group actions on a canonical manifold. In: Symplectic geometry (A Crumeyrolle and J. Grifone, eds.), Research Notes in Math. 80, Pitman, Boston-London, 1983, 144–166.

[MR] J.E. Marsden and T. Ratiu. Reduction of Poisson manifolds. Lett. Math. Phys. 11 (1986), 161–169.

[Mt] Y.Matsushima, Differentiable Manifolds, M. Dekker, Inc., New York, 1972.

[Me] A. Medina, Structures de Poisson affines. In: Symplectic Geometry and Mathematical Physics, Actes du colloque en l'honneur de Jean-Marie Souriau (P. Donato, C. Duval, J. Elhadad, G.M. Tuynman, eds.), Progress in Math., 99, Birkhäuser, Boston, 1991, 288–302.

[Ml] D. Mélotte, Cohomologie de Chevalley associée aux variétés de Poisson. Bull. Soc. Royale Sc. Liège, 58e année, 5 (1989), 319–413.

[Mh] P. Michor, A generalization of Hamiltonian mechanics. J. of Geometry and Physics 2 (1985), 67–82.

[Mk] K. Mikami, Local Lie algebra structure and momentum mapping. J. Math. Soc. Japan, 39 (1987), 233–246.

[MW] K. Mikami and A. Weinstein, Moments and reduction for symplectic groupoid actions. Publ. RIMS, Kyoto Univ. 24 (1988), 121–140.

[MT] M. Mimura and H. Toda, Topology of Lie Groups I and II. Transl. of Math. Monographs 91, American Math. Soc. Providence R.I., 1991.

[Mn1] P. Molino, Riemannian Foliations. Progress in Math. 73, Birkhäuser, Boston, 1988.

[Mn2] P. Molino, Dualité symplectique, feuilletages et géométrie du moment. Publicacions Matemàtiques 33 (1989), 533–541.

[MV] C. Moreno and L. Valero, Star product and quantization for Poisson-Lie groups. Preprint, 1992.

[My] J.E. Moyal, Quantum Mechanics as a statistical theory. Proc. Cambridge Philos. Soc. 45(1949), 99–124.

[Na] R. Narasimhan, Analysis on real and complex manifolds. Masson & Cie., Paris and North-Holland Publ. Comp. Amsterdam, 1968.

[Nj] A. Nijenhuis, Jacobi-type identities for bilinear differential concomitants of certain tensor fields I. Indagationes Math. 17 (1955), 390–403.

[OMY1] H. Omori, Y. Maeda and A. Yoshioka, Weyl manifolds and Deformation Quantization. Advances in Math. 85 (1991), 224–255.

[OMY2] H. Omori, Y. Maeda and A. Yoshioka, Global calculus on Weyl manifolds. Japanese J. of Math. 17 (1991), 57–82.

[Oz1] R. Ouzilou, Hamiltonian actions on Poisson manifolds. In: Symplectic Geometry (A. Crumeyrolle and J. Grifone, eds.) Res. Notes in Math. 80, Pitmann Boston-London, 1983, 172–183.

[Oz2] R. Ouzilou, Quelques propriétés des algèbres de Poisson appliqués à la géométrie. Comptes Rendues Acad. Sc. Paris, Série I, 299 (1984), 1021–1024.

[Pl] R.S. Palais, A global formulation of the Lie theory of transformation groups. Memoirs American Math. Soc. 22, American Math. Soc., Providence R.I., 1957.

[Pt] V. Patrangenaru, private communication.

[Pr1] J. Pradines, Théorie de Lie pour les groupoïdes différentiels. Calcul différentiel dans la categorie des groupoïdes infinitésimaux. Comptes Rendues Acad. Sc. Paris, Série A, 264 (1967), 245–248.

[Pr2] J. Pradines, Troisième théorème de Lie sur les groupoïdes différentiables. Comptes Rendues Acad. Sc. Paris, Série A, 267 (1968), 21–23.

[Pu] M. Puta, On the geometric prequantization of Poisson manifolds. Lett. Math. Phys. 15 (1988), 187–192.

[Rw] J. Rawnsley, Deformation quantization of Kähler manifolds. In: Symplectic Geometry and Mathematical Physics, Actes du colloque en l'honneur de Jean-Marie Souriau (P. Donato, C. Duval, J. Elhadad, G.M. Tuynman, eds.), Progress in Math., 99, Birkhäuser, Boston, 1991, 366–373.

[Ri] M.A. Rieffel, Deformation quantization of Heisenberg manifolds. Commun. Math. Phys. 122(1989), 531–562.

[Rg] C. Roger, Algèbres de Lie gradués et quantification. In: Symplectic Geometry and Mathematical Physics, Actes du colloque en l'honneur de Jean-Marie Souriau (P. Donato, C. Duval, J. Elhadad, G.M. Tuynman, eds.), Progress in Math., 99, Birkhäuser, Boston, 1991, 374–421.

[ST1] M.A. Semenov-Tian-Shansky, What is a classical r-matrix? Funct. Anal. Appl. 17(4) (1983), 259–272.

[ST2] M.A.Semenov-Tian-Shansky, Dressing transformations and Poisson group actions. Publ. RIMS, Kyoto Univ. 21 (1985), 1237–1260.

[ST3] M.A. Semenov-Tian-Shansky, Classical r-matrices, Lax equations, Poisson-Lie groups and dressing transformations. In: Field Theory, Quantum Gravity and Strings (H.J. de Vega, N. Sanchez, eds.), Lecture Notes in Physics, Springer-Verlag, Berlin-Heidelberg-New York, 1987, p.174–214.

[Sn1] J. Sniatycki, Dirac brackets in geometric dynamics. Ann. Inst. H. Poincaré, série A (Physique théorique), 20 (1974), 365–372.

[Sn2] J. Sniatycki, Geometric quantization and quantum mechanics. Springer-Verlag, Berlin-Heidelberg-New York, 1980.

[So] J.M. Souriau, Structures des systèmes dynamiques. Dunod, Paris, 1969.

[Sp] E.H. Spanier, Algebraic Topology. Mc Graw-Hill, New York, 1966.

[St] P. Stefan, Accessible sets, orbits and foliations with singularities. Proc. London Math. Soc. 29 (1974), 699–713.

[Sg] S. Sternberg, Lectures on differential geometry. Prentice-Hall, Englewood Cliffs, N.J. 1964.

[Su] H.J. Sussmann, Orbits of families of vector fields and integrability of distributions. Transactions American Math. Soc. 170 (1973), 171–188.

[Tl1] W.M. Tulczyjew, Poisson brackets and canonical manifolds. Bull. Polon. Sci. Série Sci. Math. Astronom. 22 (1974), 931–934.

[Tl2] W.M. Tulczyjew, Poisson reductions. In: Singularités, feuilletages et mécanique hamiltonienne (J.-P.Dufour, ed.), Travaux en Cours 13, Hermann, Paris, 1985, 113–120.

[TW] G. M. Tuynman and W.A.J.J. Wiegerinck, Central extensions and physics. J. of Geometry and Physics, 4 (1987), 207–258.

[Ur] R.W. Urwin, The prequantization representations of the Poisson Lie algebra. Advances in Math. 50 (1983), 126–154.

[Va1] I. Vaisman, Remarques sur la théorie des formes jets, Comptes Rendues Acad. Sc. Paris, Série A, 264 (1967), 351–354.

[Va2] I. Vaisman, Cohomology and Differential Forms, M. Dekker Inc., New York, 1973.

[Va3] I. Vaisman, Basic ideas of geometric quantization. Rend. Sem. Mat. Torino, 37 (1979), 31–41.

[Va4] I. Vaisman, A coordinatewise formulation of geometric quantization. Ann. Inst. H. Poincaré, série A (Physique théorique) 31 (1979), 5–24.

[Va5] I. Vaisman, Riemannian manifolds with a complex field of parallel planes. J. of Geometry 17 (1981), 174–192.

[Va6] I. Vaisman, Symplectic curvature tensors. Monatshefte für Math., 100 (1985), 299–327.

[Va7] I. Vaisman, Locally conformal symplectic manifolds. Internat. J. Math. & Math. Sci., 8 (1985), 521–536.

[Va8] I. Vaisman, Symplectic geometry and secondary characteristic classes. Progress in Math. 72, Birkhäuser, Boston, 1987.

[Va9] I. Vaisman, Remarks on the Lichnerowicz-Poisson cohomology. Ann. Inst. Fourier Grenoble, 40 (1990), 951–963.

[Va10] I. Vaisman, On the geometric quantization of Poisson manifolds. J. of Math. Physics 32 (1991), 3339–3345.

[vE] W.T. van Est, Rapport sur les S-atlas. In: Structure transverse des feuilletages, Toulouse 1982, Astérisque 116 (1984), 235–292.

[Vd] J.-L. Verdier, Groupes quantiques. Séminaire Bourbaki 1986–87, Astérisque 152–153 (1987), 305–319.

[Ve] J. Vey, Déformation du crochet de Poisson sur une variété symplectique. Comment. Math. Helvetici 50 (1975), 421–454.

[Vf] V.P. Viflyantsev, Frobenius theorem for differential systems with singularities. Vestnik Moskow. Univ. (1980) (3), 11–14.

[VK] Yu. M. Vorob'ev and M.V. Karasev, Poisson manifolds and their Schouten bracket. Funct. Anal. Appl. 22 (1988), 1–9.

[Wl] A. Weil, Introduction à l'étude des variétés Kählériennes. Hermann, Paris, 1971.

[We1] A. Weinstein, Lectures on symplectic manifolds. CBMS, Conf. Series in Math. 29, American Math. Soc. Providence R.I., 1977.

[We2] A. Weinstein, Symplectic geometry. Bull. American Math. Soc. 5 (1981), 1–13.

[We3] A. Weinstein, The local structure of Poisson manifolds. J. Differential Geometry 18 (1983), 523–557.

[We4] A. Weinstein, Poisson structures and Lie algebras. In: É. Cartan et les mathématiques d'aujourd'hui, Soc. Math. de France, Astérisque, hors série, 1985, 421–434.

[We5] A. Weinstein, Symplectic groupoids and Poisson manifolds. Bull. American Math. Soc. 16 (1987), 101–103.

[We6] A. Weinstein, Coisotropic calculus and Poisson groupoids. J. Math. Soc. Japan 40 (1988), 705–727.

[We7] A. Weinstein, Some remarks on dressing transformations. J. Fac. Sci. Univ. Tokyo, Sect. 1A, Math. 36 (1988), 163–167.

[We8] A. Weinstein, Blowing up realizations of Heisenberg-Poisson manifolds. Bull. Sc. Math. 113 (1989), 381–406.

[We9] A. Weinstein, Noncommutative geometry and geometric quantization. In: Symplectic Geometry and Mathematical Physics, Actes du colloque en l'honeur de Jean-Marie Souriau (P. Donato, C. Duval, J. Elhadad, G.M. Tuynman, eds.), Progress in Math., 99, Birkhäuser, Boston, 1991, 446–461.

[We10] A. Weinstein, Symplectic groupoids, geometric quantization, and irrational rotation algebras. In: Symplectic Geometry, Groupoids and Integrable Systems, Séminaire Sud-Rhodanien de Géométrie à Berkeley (1989) (P. Dazord and A. Weinstein, eds.). MSRI Publ. 20, Springer-Verlag, Berlin-Heidelberg-New York, 1991, 281–290.

[WX1] A. Weinstein and P. Xu, Extensions of symplectic groupoids and quantization. J. Reine angew. Math. 417 (1991), 159–189.

[WX2] A. Weinstein and P. Xu, Classical solutions of the quantum Yang-Baxter equation. Comm. Math. Physics, 148 (1992), 309–343.

[WA] D.C. Wilbour and J.M. Arms, Reduction procedures for Poisson manifolds. In: Symplectic Geometry and Mathematical Physics, Actes du colloque en l'honneur de Jean-Marie Souriau (P. Donato, C. Duval, J. Elhadad, G.M. Tuynman, eds.), Progress in Math. 99, Birkhäuser, Boston, 1991, 462–476.

[Wd] N. Woodhouse, Geometric quantization. Clarendon Press, Oxford, 1980.

[Xu1] P. Xu, Poisson cohomology of regular Poisson manifolds. Ann. Inst. Fourier Grenoble, 42 (1992), 967–988.

[Xu2] P. Xu, Morita equivalent symplectic groupoids. In: Symplectic Geometry, Groupoids, and Integrable Systems, Séminaire Sud-Rhodanien de Géométrie à Berkeley (1989) (P. Dazord and A. Weinstein, eds.) MSRI Publ. 20, Springer-Verlag, Berlin-Heidelberg-New York, 1991, 291–311.

[Xu3] P. Xu, Morita equivalence and symplectic realizations of Poisson manifolds. Ann. Éc. Norm. Sup. 25 (1992), 307–333.

[Xu4] P. Xu, Poisson manifolds associated with group actions, and classical triangular r-matrices. J. Funct. Analysis, 112 (1993), 218–240.

[Ya1] K. Yano, The theory of Lie derivatives and its applications. North Holland Publ., Amsterdam, 1957.

[Ya2] K. Yano, Differential geometry of complex and almost complex spaces. Pergamon Press, New York, 1965.

[Zk] S. Zakrzewski, Quantum and classical pseudogroups I, II. Comm. Math. Phys. 134 (1990), 347–370, 371–395.

Index

action, Hamiltonian, 107
action, infinitesimal, 107
action, of a Lie group, 107
action, Poisson, 107
affine Poisson groups, 166
almost-symplectic manifold, 36
anchor map, 148
annihilator, 37

bi-cross section, 146
bi-cross section, Lagrangian, 146
Bohr-Sommerfeld condition, 88
bracket, Jacobi, 3
bracket, Poisson, 1

canonical coordinates, 29
Casimir functions, 30
central extension, 35
characteristic (Chern) class of the isotropic realization, 130
characteristic form-class, 131
Chevalley-Eilenberg cohomology, 90
clean intersection, 99
cofoliation, 54
coisotropic submanifold, 99
contact manifold, 36
contravariant derivative, 55

deformation, 93
differential, contravariant exterior, 43
Dirac bracket, 37
distinguished cross section, 126
distribution, characteristic, 19
distribution, completely integrable, 20
distribution, differentiable, 19
distribution, general, 19
distribution, invariant, 20
distribution, involutive, 21
distribution, leaf of completely integrable, 20
distribution, rank of, 19
distribution, regular, 19
distribution, subcharacteristic, 104
divergence, generalized, 12
double Lie algebra, 187
double Lie group, 187
dressing transformations, 185
dressing vector field, 185
dual group, 185

foliation, general, 20
foliation, leafwise symplectic, 36
foliation, Libermann, 115
foliation, regular, 20
foliation, symplectically complete, 115
function group, 121
function group, polar, 121

gauge groupoid, 140
groupoid, 138
groupoid, banal, 138
groupoid, double symplectic, 144
groupoid, homomorphism, 144
groupoid, local symplectic, 151
groupoid, Poisson, 143
groupoid, symplectic, 143
groupoid, transitive, 139
groupoid, zero, 139

Hamiltonian vector field, 5
Heisenberg-Poisson (HP) manifold, 153
Hochschild cohomology, 94

i-vector, 6
intrinsic derivative, 166
isotropic realizations, 123
isotropic realizations, connected complete (c.c.i.), 124
isotropic realizations, Libermann, 124

Jacobian cocycle, 169

Lagrangian submanifold, 100
Legendrian submanifold, 155
Libermann foliation, 54
Lichnerowicz-Poisson cohomology, 63
Lie algebra cohomology, 90
Lie algebroid, 43, 148
Lie bialgebra, 170
Lie bigebra, 170
Lie groupoid, local, 150
Lie groupoid, 140
Lie-Drinfeld algebra, 169
Lie-Poisson structure, 32
linear approximation, 34
local Lie algebra, 3
LP-Poincaré lemma, 75
LP-simple neighbourhood, 75

manifold, contact, 3
manifold, Dirac, 3
manifold, Jacobi, 3, 17
manifold, locally conformal symplectic, 3
manifold, Poisson, 2
manifold, symplectic, 6
Manin triple, 178
Mayer-Victoris exact sequence, 65
momentum map, 109
momentum map, equivariant, 109
multiplicative tensor field, 162

net, 128, 129

observables, 83, 93
orbit, 139

phase space, 83
point, regular, 19
point, singular, 19
Poisson action, 182
Poisson algebra, 1
Poisson automorphism, 16, 97
Poisson automorphism, infinitesimal, 26
Poisson bivector, 5
Poisson characteristic classes, 85
Poisson codifferential, 45
Poisson connection, 11, 29
Poisson equivalence, 16, 97
Poisson homology, 77
Poisson isomorphism, 97
Poisson manifold, 2
Poisson manifold, exact, 63
Poisson manifold, integrable, 150
Poisson manifold, quantizable, 86
Poisson manifold, reduced, 102
Poisson manifold, regular, 27
Poisson manifolds, dual, 121
Poisson mapping, 16
Poisson morphism, 16, 97
Poisson product, 17
Poisson relation, 100
Poisson space, 2
Poisson structure, affine, 34
Poisson structure, coinduced, 98
Poisson structure, constant, 31
Poisson structure, linear, 31
Poisson structure, linearizable, 34
Poisson structure, nondegenerate, 26
Poisson structure, quadratic, 35
Poisson structure, rank of, 26
Poisson structure, reduced, 106
Poisson structure, transverse, 29
Poisson structures, compatible, 15
Poisson submanifold, 16
Poisson-Chern classes, 85
Poisson-Godbillon-Vey class, 64
Poisson-Lie algebra, 169
Poisson-Lie group, 161
Poisson-Lie subgroup, 183
Poisson-Nijenhuis structure, 62
polarization, 87
prequantization, 83
prequantization bundle, 84
prequantization formula, 84
prequantization representation, 89
product, interior, 13
Product, Poisson, 17

quantization, 83, 87
quantization, deformation, 83, 92
quantization, geometric, 83
quasi-Poisson-Lie groups, 188

realizations, equivalent, 119
reducibility condition, 103
reducible triple, 102
reduction, 101
reduction, leafwise, 104
reductive structure, 102
representation, adjoint, 32
representation, coadjoint, 33

Schouten bracket, algebraic, 172
Schouten-Nijenhuis bracket, 9
Schouten-Nijenhuis bracket, covariant, 59
set of units, 139
splitting theorem, 27
symbol, 89
symplectic connection, 16
symplectic connections, 29
symplectic foliation, 26
symplectic leaves, 26
symplectic realization, 115

Theorem, generalized Frobenius, 25
Theorem, Noether's, 114
Theorem, Sussmann-Stefan-Frobenius, 20
Theorem, Viflyantsev-Frobenius, 21
Triangular Poisson-Lie algebra, 173
twilled Lie algebra, 187

unit submanifold, 140

vanishing cycle, 157

Yang-Baxter (YB) equation, classical, 173
Yang-Baxter equation, generalized ($\mathcal{G}$YB), 174
Yang-Baxter equation, group-generalized classical (GYB), 173
Yang-Baxter equation, modified (MYB), 176

Progress in Mathematics

Edited by:

J. Oesterlé
Départment de Mathématiques
Université de Paris VI
4, Place Jussieu
75230 Paris Cedex 05, France

A. Weinstein
Department of Mathematics
University of California
Berkeley, CA 94720
U.S.A.

Progress in Mathematics is a series of books intended for professional mathematicians and scientists, encompassing all areas of pure mathematics. This distinguished series, which began in 1979, includes authored monographs, and edited collections of papers on important research developments as well as expositions of particular subject areas.

We encourage preparation of manuscripts in such form of TeX for delivery in camera-ready copy which leads to rapid publication, or in electronic form for interfacing with laser printers or typesetters.

Proposals should be sent directly to the editors or to: Birkhäuser Boston, 675 Massachusetts Avenue, Cambridge, MA 02139, U.S.A.

53 LAURENT. Théorie de la 2ième Microlocalisation dans le Domaine Complexe
54 VERDIER. Module des Fibres Stables sur les Courbes Algébriques
55 EICHLER/ZAGIER. The Theory of Jacobi Forms
56 SHIFFMAN/SOMMESE. Vanishing Theorems on Complex Manifolds
57 RIESEL. Prime Numbers and Computer Methods for Factorization
58 HELFFER/NOURRIGAT. Hypoellipticité Maximale pour des Operateurs Polynomes de Champs de Vecteurs
59 GOLDSTEIN. Séminaire de Théorie de Nombres, Paris 83-84
60 ARBARELLO. Geometry Today
62 GUILLOU. A la Recherche de la Topologie Perdue
63 GOLDSTEIN. Séminaire de Théorie des Nombres, Paris 84-85
64 MYUNG. Malcev-Admissible Algebras
65 GRUBB. Functional Calculus of Pseudo-Differential Boundary Problems
66 CASSOU-NOGUÈS/TAYLOR. Elliptic Functions and Rings of Integers
67 HOWE. Discrete Groups in Geometry and Analysis
68 ROBERT. Autour de l'Approximation Semi-Classique
69 FARAUT/HARZALLAH. Analyse Harmonique: Fonctions Speciales et Distributions Invariantes
70 YAGER. Analytic Number Theory and Diophantine Problems
71 GOLDSTEIN. Séminaire de Théorie de Nombres, Paris 85-86
72 VAISMAN. Symplectic Geometry and Secondary Characteristic Classes
73 MOLINO. Riemannian Foliations
74 HENKIN/LEITERER. Andreotti-Grauert Theory by Integral Formulas
75 GOLDSTEIN. Séminaire de Théorie de Nombres, Paris 86-87
76 COSSEC/DOLGACHEV. Enriques Surfaces I

77 REYSSAT. Quelques Aspects des Surfaces de Riemann
78 BORHO/BRYLINSKI/MCPHERSON. Nilpotent Orbits, Primitive Ideals, and Characteristic Classes
79 MCKENZIE/VALERIOTE. The Structure of Decidable Locally Finite Varieties
80 KRAFT/ SCHWARZ/PETRIE (eds.) Topological Methods in Algebraic Transformation Groups
81 GOLDSTEIN. Séminaire de Théorie des Nombres, Paris 87–88
82 DUFLO/PEDERSEN/VERGNE (eds.) The Orbit Method in Representation Theory
83 GHYS/DE LA HARPE (eds.) Sur les Groupes Hyperboliques d'après M. Gromov
84 ARAKI/KADISON (eds.) Mappings of Operator Algebras
85 BERNDT/DIAMOND/HALBERSTAM/ HILDEBRAND (eds.) Analytic Number Theory
89 VAN DER GEER/OORT/STEENBRINK (eds.) Arithmetic Algebraic Geometry
90 SRINIVAS. Algebraic K-Theory
91 GOLDSTEIN. Séminaire de Théorie des Nombres, Paris 1988-89
92 CONNES/DUFLO/JOSEPH/RENTSCHLER. Operator Algebras, Unitary Representations, Enveloping Algebras, and Invariant Theory. A Collection of Articles in Honor of the 65th Birthday of Jacques Dixmier
93 AUDIN. The Topology of Torus Actions on Symplectic Manifolds
94 MORA/TRAVERSO (eds.) Effective Methods in Algebraic Geometry
95 MICHLER/RINGEL (eds.) Representation Theory of Finite Groups and Finite–Dimensional Algebras
96 MALGRANGE. Equations Différentielles à Coefficients Polynomiaux
97 MUMFORD/NORMAN/NORI. Tata Lectures on Theta III
98 GODBILLON. Feuilletages, Etudes géométriques
99 DONATO/DUVAL/ELHADAD/ TUYNMAN. Symplectic Geometry and Mathematical Physics. A Collection of Articles in Honor of J.-M. Souriau
100 TAYLOR. Pseudodifferential Operators and Nonlinear PDE
101 BARKER/SALLY. Harmonic Analysis on Reductive Groups
102 DAVID. Séminaire de Théorie des Nombres, Paris 1989-90
103 ANGER/PORTENIER. Radon Integrals
104 ADAMS/BARBASCH/VOGAN. The Langlands Classification and Irreducible Characters for Real Reductive Groups
105 TIRAO/WALLACH. New Developments in Lie Theory and Their Applications
106 BUSER. Geometry and Spectra of Compact Riemann Surfaces
107 BRYLINSKI. Loop Spaces,Characteristic Classes and Geometric Quantization
108 DAVID. Séminaire de Théorie des Nombres, Paris 1990-91
109 EYSSETTE/GALLIGO. Computational Algebraic Geometry
110 LUSZTIG. Introduction to Quantum Groups
111 SCHWARZ. Morse Homology
112 DONG/LEPOWSKY. Generalized Vertex-Algebras and Relative Vertex Operators
113 MOEGLIN/WALDSPURGER. Décomposition Spectrale et Series d'Eisenstein
114 BERENSTEIN/GAY/VIDRAS/YGER. Residue Currents and Bezout Identities
115 BABELON/CARTIER/KOSMANN-SCHWARZBACH (eds.) Integrable Systems. The Verdier Memorial Conference
116 DAVID (ed.) Séminaire de Théorie des Nombres, Paris, 1991-1992
117 AUDIN/LAFONTAINE (eds.) Holomorphic Curves in Symplectic Geometry
118 VAISMAN. Lectures on the Geometry of Poisson Manifolds